INTERNATIONAL
PLUMBING
CODE®

A Member of the International Code Family®

2012 IPC®

Become a **Building Safety Professional Member** and Learn More about the Code Council

GO TO WWW.ICCSAFE.ORG for All Your Technical and Professional Needs Including:

❯ Codes, Standards and Guidelines

❯ Membership Benefits

❯ Education and Certification

❯ Communications on Industry News

2012 International Plumbing Code®

First Printing: April 2011

ISBN: 978-1-60983-053-3 (soft-cover edition)
ISBN: 978-1-60983-052-6 (loose-leaf edition)

COPYRIGHT © 2011
by
INTERNATIONAL CODE COUNCIL, INC.

PRINTED IN THE U.S.A.

PREFACE

Introduction

Internationally, code officials recognize the need for a modern, up-to-date plumbing code addressing the design and installation of plumbing systems through requirements emphasizing performance. The *International Plumbing Code*®, in this 2012 edition, is designed to meet these needs through model code regulations that safeguard the public health and safety in all communities, large and small.

This comprehensive plumbing code establishes minimum regulations for plumbing systems using prescriptive and performance-related provisions. It is founded on broad-based principles that make possible the use of new materials and new plumbing designs. This 2012 edition is fully compatible with all of the *International Codes*® (I-Codes®) published by the International Code Council (ICC)®, including the *International Building Code*®, *International Energy Conservation Code*®, *International Existing Building Code*®, *International Fire Code*®, *International Fuel Gas Code*®, *International Green Construction Code*™ (to be available March 2012), *International Mechanical Code*®, ICC *Performance Code*®, *International Private Sewage Disposal Code*®, *International Property Maintenance Code*®, *International Residential Code*®, *International Swimming Pool and Spa Code*™ (to be available March 2012), *International Wildland-Urban Interface Code*® and *International Zoning Code*®.

The *International Plumbing Code* provisions provide many benefits, among which is the model code development process that offers an international forum for plumbing professionals to discuss performance and prescriptive code requirements. This forum provides an excellent arena to debate proposed revisions. This model code also encourages international consistency in the application of provisions.

Development

The first edition of the *International Plumbing Code* (1995) was the culmination of an effort initiated in 1994 by a development committee appointed by the ICC and consisting of representatives of the three statutory members of the International Code Council at that time, including: Building Officials and Code Administrators International, Inc. (BOCA), International Conference of Building Officials (ICBO) and Southern Building Code Congress International (SBCCI). The intent was to draft a comprehensive set of regulations for plumbing systems consistent with and inclusive of the scope of the existing model codes. Technical content of the latest model codes promulgated by BOCA, ICBO and SBCCI was utilized as the basis for the development. This 2012 edition presents the code as originally issued, with changes as reflected in the subsequent editions through 2009 and with changes approved through the ICC Code Development Process through 2010. A new edition such as this is promulgated every three years.

This code is founded on principles intended to establish provisions consistent with the scope of a plumbing code that adequately protects public health, safety and welfare; provisions that do not unnecessarily increase construction costs; provisions that do not restrict the use of new materials, products or methods of construction; and provisions that do not give preferential treatment to particular types or classes of materials, products or methods of construction.

Adoption

The *International Plumbing Code* is available for adoption and use by jurisdictions internationally. Its use within a governmental jurisdiction is intended to be accomplished through adoption by reference in accordance with proceedings establishing the jurisdiction's laws. At the time of adoption, jurisdictions should insert the appropriate information in provisions requiring specific local information, such as the name of the adopting jurisdiction. These locations are shown in bracketed words in small capital letters in the code and in the sample ordinance. The sample adoption ordinance on page xi addresses several key elements of a code adoption ordinance, including the information required for insertion into the code text.

Maintenance

The *International Plumbing Code* is kept up to date through the review of proposed changes submitted by code enforcing officials, industry representatives, design professionals and other interested parties. Proposed changes are carefully considered through an open code development process in which all interested and affected parties may participate.

The contents of this work are subject to change both through the Code Development Cycles and the governmental body that enacts the code into law. For more information regarding the code development process, contact the Codes and Standards Development Department of the International Code Council.

While the development procedure of the *International Plumbing Code* ensures the highest degree of care, ICC and ICC's members and those participating in the development of this code do not accept any liability resulting from compliance or noncompliance with the provisions, since ICC and its members do not have the power or authority to police or enforce compliance with the contents of this code. Only the governmental body that enacts the code into law has such authority.

Code Development Committee Responsibilities
(Letter Designations in Front of Section Numbers)

In each code development cycle, proposed changes to the code are considered at the Code Development Hearings by the International Plumbing Code Development Committee, whose action constitutes a recommendation to the voting membership for final action on the proposed change. Proposed changes to a code section that has a number beginning with a letter in brackets are considered by a different code development committee. For example, proposed changes to code sections that have [B] in front of them (e.g. [B] 309.2) are considered by the appropriate International Building Code Development Committee (IBC-General) at the code development hearings.

The content of sections in this code that begin with a letter designation is maintained by another code development committee in accordance with the following:

[A] = Administrative Code Development Committee;

[B] = International Building Code Development Committee (IBC—Fire Safety, General, Means of Egress or Structural);

[E] = International Energy Conservation Còde Development Committee;

[F] = International Fire Code Development Committee; and

[M] = International Mechanical Code Development Committee.

Note that, for the development of the 2015 edition of the I-Codes, there will be two groups of code development committees and they will meet in separate years. The groupings are as follows:

Group A Codes (Heard in 2012, Code Change Proposals Deadline: January 3, 2012)	Group B Codes (Heard in 2013, Code Change Proposals Deadline: January 3, 2013)
International Building Code	Administrative Provisions (Chapter 1 all codes except IRC and ICC PC, administrative updates to currently referenced standards, and designated definitions)
International Fuel Gas Code	International Energy Conservation Code
International Mechanical Code	International Existing Building Code
International Plumbing Code	International Fire Code
International Private Sewage Disposal Code	International Green Construction Code
	ICC Performance Code
	International Property Maintenance Code
	International Residential Code
	International Swimming Pool and Spa Code
	International Wildland-Urban Interface Code
	International Zoning Code

Code change proposals submitted for code sections that have a letter designation in front of them will be heard by the respective committee responsible for such code sections. Because different committees will meet in different years, it is possible that some proposals for this code will be heard by a committee in a different year than the year in which the primary committee for this code meets.

For example, every section of Chapter 1 of this code is designated as the responsibility of the Administrative Code Development Committee, and that committee is part of the Group B portion of the hearings. This committee will conduct its code development hearings in 2013 to consider all code change proposals for Chapter 1 of this code and proposals for Chapter 1 of all I-Codes except the *International Residential Code* and the ICC *Performance Code*. Therefore, any proposals received for Chapter 1 of this code will be deferred for consideration in 2013 by the Administrative Code Development Committee.

Another example is Section 607.5 of this code which is designated as the responsibility of the International Energy Conservation Code Development Committee. This committee will conduct its code development hearings in 2013 to consider code change proposals in its purview, which includes any proposals to Section 607.5.

In some cases, another committee in Group A will be responsible for a section of this code. For example, Section 314.2 has a [M] in front of the numbered section, indicating that this section of the code is the responsibility of the International Mechanical Code Development Committee. The International Mechanical Code is in Group A; therefore, any code change proposals to this section will be due before the Group A deadline of January 3, 2012, and these code change proposals will be assigned to the International Mechanical Code Development Committee for consideration.

It is very important that anyone submitting code change proposals understand which code development committee is responsible for the section of the code that is the subject of the code change proposal. For further information on the code development committee responsibilities, please visit the ICC web site at www.iccsafe.org/scoping.

Marginal Markings

Solid vertical lines in the margins within the body of the code indicate a technical change from the requirements of the 2009 edition. Deletion indicators in the form of an arrow (➡) are provided in the margin where an entire section, paragraph, exception or table has been deleted or an item in a list of items or a table has been deleted.

A single asterisk [*] placed in the margin indicates that text or a table has been relocated within the code. A double asterisk [**] placed in the margin indicates that the text or table immediately following it has been relocated there from elsewhere in the code. The following table indicates such relocations in the 2012 *International Plumbing Code*.

2012 LOCATION	2009 LOCATION
316.1 through 316.1.6	105.4 through 105.4.6
315.1	305.4
405.3.3, 405.3.4, 405.3.5	310.2, 310.4, 310.5
904	903
903	904
909	906
910	907
911	908
912	909
913	910
914	911
915	912
916	913
908	914
907	915
906	916

Italicized Terms

Selected terms set forth in Chapter 2, Definitions, are italicized where they appear in code text. Such terms are not italicized where the definition set forth in Chapter 2 does not impart the intended meaning in the use of the term. The terms selected have definitions which the user should read carefully to facilitate better understanding of the code.

Effective Use of the International Plumbing Code

The *International Plumbing Code* (IPC) is a model code that regulates the design and installation of plumbing systems including the plumbing fixtures in all types of buildings except for detached one- and two-family dwellings and townhouses that are not more than three stories above grade in height. The regulations for plumbing systems in one- and two-family dwellings and townhouses are covered by Chapters 25 through 33 of the *International Residential Code* (IRC). The IPC addresses general plumbing regulations, fixture requirements, water heater installations and systems for water distribution, sanitary drainage, special wastes, venting, storm drainage and medical gases. The IPC does not address fuel gas piping systems as those systems are covered by the *International Fuel Gas Code* (IFGC). The IPC also does not regulate swimming pool piping systems, process piping systems, or utility-owned piping and systems. The purpose of the IPC is to the establish the minimum acceptable level of safety to protect life and property from the potential dangers associated with supplying potable water to plumbing fixtures and outlets and the conveyance of bacteria-laden waste water from fixtures.

The IPC is primarily a specification-oriented (prescriptive) code with some performance-oriented text. For example, Section 405.1 is a performance statement but Chapter 6 contains the prescriptive requirements that will cause Section 405.1 to be satisfied.

Where a building contains plumbing fixtures, those fixtures requiring water must be provided with an adequate supply of water for proper operation. The number of required plumbing fixtures for a building is specified by this code and is based upon the anticipated maximum number of occupants for the building and the type of building occupancy. This code provides prescriptive criteria for sizing piping systems connected to those fixtures. Through the use of code-approved materials and the installation requirements specified in this code, plumbing systems will perform their intended function over the life of the building. In summary, the IPC sets forth the minimum requirements for providing safe water to a building as well as a safe manner in which liquid-borne wastes are carried away from a building.

Arrangement and Format of the 2012 IPC

The format of the IPC allows each chapter to be devoted to a particular subject with the exception of Chapter 3 which contains general subject matters that are not extensive enough to warrant their own independent chapter. The IPC is divided into thirteen different parts:

Chapters	Subjects
1-2	Administration and Definitions
3	General Regulations
4	Fixtures, Faucets and Fixture Fittings
5	Water Heaters
6	Water Supply and Distribution
7	Sanitary Drainage
8	Indirect/Special Wastes
9	Vents
10	Traps, Interceptors and Separators
11	Storm Drainage
12	Special Piping (Medical Gas)
13	Gray Water Recycling Systems
14	Referenced Standards
Appendices A-G	Appendices

The following is a chapter-by-chapter synopsis of the scope and intent of the provisions of the *International Plumbing Code*:

Chapter 1 Scope and Administration. This chapter contains provisions for the application, enforcement and administration of subsequent requirements of the code. In addition to establishing the scope of the code, Chapter 1 identifies which buildings and structures come under its purview. Chapter 1 is largely concerned with maintaining "due process of law" in enforcing the requirements contained in the body of this code. Only through careful observation of the administrative provisions can the building official reasonably expect to demonstrate that "equal protection under the law" has been provided.

Chapter 2 Definitions. Chapter 2 is the repository of the definitions of terms used in the body of the code. Codes are technical documents and every word, term and punctuation mark can impact the meaning of the code text and the intended results. The code often uses terms that have a unique meaning in the code and the code meaning can differ substantially from the ordinarily understood meaning of the term as used outside of the code.

The terms defined in Chapter 2 are deemed to be of prime importance in establishing the meaning and intent of the code text that uses the terms. The user of the code should be familiar with and consult this chapter because the definitions are essential to the correct interpretation of the code and because the user may not be aware that a term is defined.

Where understanding of a term's definition is especially key to or necessary for understanding of a particular code provision, the term is shown in *italics* wherever it appears in the code. This is true only for those terms that have a meaning that is unique to the code. In other words, the generally understood meaning of a term or phrase might not be sufficient or consistent with the meaning prescribed by the code; therefore, it is essential that the code-defined meaning be known.

Guidance regarding tense, gender and plurality of defined terms as well as guidance regarding terms not defined in this code is provided.

Chapter 3 General Regulations. The content of Chapter 3 is often referred to as "miscellaneous," rather than general regulations. This is the only chapter in the code whose requirements do not interrelate. If a requirement cannot be located in another chapter, it should be located in this chapter. Chapter 3 contains safety requirements for the installation of plumbing and nonplumbing requirements for all types of fixtures. This chapter also has requirements for the identification of pipe, pipe fittings, traps, fixtures, materials and devices used in plumbing systems.

The safety requirements of this chapter provide protection for the building's structural members, as well as prevent undue stress and strain on pipes. The building's structural stability is protected by the regulations for cutting and notching of structural members. Additional protection for the building occupants includes requirements to maintain the plumbing in a safe and sanitary condition, as well as privacy for those occupants.

Chapter 4 Fixtures, Faucets and Fixture Fittings. This chapter regulates the minimum number of plumbing fixtures that must be provided for every type of building. This chapter also regulates the quality of fixtures and faucets by requiring those items to comply with nationally recognized standards. Because fixtures must be properly installed so that they are usable by the occupants of the building, this chapter contains the requirements for the installation of fixtures. Because the requirements for the number of plumbing fixtures affects the design of a building, Chapter 29 of the *International Building Code* (IBC) includes, verbatim, many of the requirements listed in Chapter 4 of this code.

Chapter 5 Water Heaters. Chapter 5 regulates the design, approval and installation of water heaters and related safety devices. The intent is to minimize the hazards associated with the installation and operation of water heaters. Although this code does not regulate the size of a water heater, it does regulate all other aspects of the water heater installation such as temperature and pressure relief valves, safety drip pans, installation and connections. Where a water heater also supplies water for space heating, this chapter regulates the maximum water temperature supplied to the water distribution system.

Chapter 6 Water Supply and Distribution. This chapter regulates the supply of potable water from both public and individual sources to every fixture and outlet so that it remains potable and uncontaminated. Chapter 6 also regulates the design of the water distribution system, which will allow fixtures to function properly and also help prevent backflow conditions. The unique requirements of the water supply for health care facilities are addressed separately. It is critical that the potable water supply system remain free of actual or potential sanitary hazards by providing protection against backflow.

Chapter 7 Sanitary Drainage. The purpose of Chapter 7 is to regulate the materials, design and installation of sanitary drainage piping systems as well as the connections made to the system. The intent is to design and install sanitary drainage systems that will function reliably, that are neither undersized nor oversized and that are constructed from materials, fittings and connections as prescribed herein. This chapter addresses the proper use of fittings for directing the flow into and within the sanitary drain piping system. Materials and provisions necessary for servicing the drainage system are also included in this chapter.

Chapter 8 Indirect/Special Waste. This chapter regulates drainage installations that require an indirect connection to the sanitary drainage system. Fixtures and plumbing appliances, such as those associated with food preparation or handling, health care facilities and potable liquids, must be protected from contamination that can result from connection to the drainage system. An indirect connection prevents sewage from backing up into a fixture or appliance, thus providing protection against potential health hazards. The chapter also regulates special wastes containing hazardous chemicals. Special waste must be treated to prevent any damage to the sanitary drainage piping and to protect the sewage treatment processes.

Chapter 9 Vents. Chapter 9 covers the requirements for vents and venting. Knowing why venting is required makes it easier to understand the intent of this chapter. Venting protects every trap against the loss of its seal. Provisions set forth in this chapter are geared toward limiting the pressure differentials in the drainage system to a maximum of 1 inch of water column (249 Pa) above or below atmospheric pressure (i.e., positive or negative pressures).

Chapter 10 Traps, Interceptors and Separators. This chapter contains design requirements and installation limitations for traps. Prohibited types of traps are specifically identified. Where fixtures do not frequently replenish the water in traps, a method is provided to ensure that the water seal of the trap will be maintained. Requirements for the design and location of various types of interceptors and separators are provided. Specific venting requirements are given for separators and interceptors as those requirements are not addressed in Chapter 9.

Chapter 11 Storm Drainage. Chapter 11 regulates the removal of storm water typically associated with rainfall. The proper installation of a storm drainage system reduces the possibility of structural collapse of a flat roof, prevents the leakage of water through the roof, prevents damage to the footings and foundation of the building and prevents flooding of the lower levels of the building.

Chapter 12 Special Piping and Storage Systems. This chapter contains the requirements for the design, installation, storage, handling and use of nonflammable medical gas systems, including inhalation anesthetic and vacuum piping systems, bulk oxygen storage systems and oxygen-fuel gas systems used for welding and cutting operations. The intent of these requirements is to minimize the potential fire and explosion hazards associated with the gases used in these systems.

Chapter 13 Gray Water Recycling Systems. This chapter regulates the design and installation of gray water collection and disposal systems. The reduction of the use of potable water in buildings has led to the use of gray water for flushing of water closets and urinals and subsurface irrigation. As such, this chapter provides the overall requirements for these systems.

Chapter 14 Referenced Standards. The code contains numerous references to standards that are used to regulate materials and methods of construction. Chapter 14 contains a comprehensive list of all standards that are referenced in the code. The standards are part of the code to the extent of the reference to the standard. Compliance with the referenced standard is necessary for compliance with this code. By providing specifically adopted standards, the construction and installation requirements necessary for compliance with the code can be readily determined. The basis for code compliance is, therefore, established and available on an equal basis to the code official, contractor, designer and owner.

Chapter 14 is organized in a manner that makes it easy to locate specific standards. It lists all of the referenced standards, alphabetically, by acronym of the promulgating agency of the standard. Each agency's standards are then listed in either alphabetical or numeric order based upon the standard identification. The list also contains the title of the standard; the edition (date) of the standard referenced; any addenda included as part of the ICC adoption; and the section or sections of this code that reference the standard.

Appendix A Plumbing Permit Fee Schedule. Appendix A provides a format for a fee schedule.

Appendix B Rates of Rainfall for Various Cities. Appendix B provides specific rainfall rates for major cities in the United States.

Appendix C Vacuum Drainage System. Appendix C offers basic information on how a vacuum drainage system relates to the code, should a vacuum drainage system be used for a building.

Appendix D Degree Day and Design Temperatures. This appendix provides valuable temperature information for designers and installers of plumbing systems in areas where freezing temperatures might exist.

Appendix E Sizing of Water Piping System. Appendix E provides two recognized methods for sizing the water service and water distribution piping for any structure. The method under Section E103 provides friction loss diagrams which require the user to "plot" points and read values from the diagrams in order to perform the required calculations and necessary checks. This method is the most accurate of the two presented in this appendix. The method under Section E201 is known to be conservative; however, very few calculations are necessary in order to determine a pipe size that satisfies the flow requirements of any application.

Appendix F Structural Safety. Appendix F is provided so that the user does not have to refer to another code book for limitations for cutting, notching and boring of sawn lumber and cold-formed steel framing.

LEGISLATION

The *International Codes* are designed and promulgated to be adopted by reference by legislative action. Jurisdictions wishing to adopt the 2012 *International Plumbing Code* as an enforceable regulation governing plumbing systems should ensure that certain factual information is included in the adopting legislation at the time adoption is being considered by the appropriate governmental body. The following sample adoption legislation addresses several key elements, including the information required for insertion into the code text.

SAMPLE LEGISLATION FOR ADOPTION OF THE *INTERNATIONAL PLUMBING CODE* ORDINANCE NO._____

A[N] [ORDINANCE/STATUTE/REGULATION] of the [JURISDICTION] adopting the 2012 edition of the *International Plumbing Code*, regulating and governing the design, construction, quality of materials, erection, installation, alteration, repair, location, relocation, replacement, addition to, use or maintenance of plumbing systems in the [JURISDICTION]; providing for the issuance of permits and collection of fees therefor; repealing [ORDINANCE/STATUTE/REGULATION] No. _____ of the [JURISDICTION] and all other ordinances or parts of laws in conflict therewith.

The [GOVERNING BODY] of the [JURISDICTION] does ordain as follows:

Section 1. That a certain document, three (3) copies of which are on file in the office of the [TITLE OF JURISDICTION'S KEEPER OF RECORDS] of [NAME OF JURISDICTION], being marked and designated as the *International Plumbing Code*, 2012 edition, including Appendix Chapters [FILL IN THE APPENDIX CHAPTERS BEING ADOPTED], as published by the International Code Council, be and is hereby adopted as the Plumbing Code of the [JURISDICTION], in the State of [STATE NAME] regulating and governing the design, construction, quality of materials, erection, installation, alteration, repair, location, relocation, replacement, addition to, use or maintenance of plumbing systems as herein provided; providing for the issuance of permits and collection of fees therefor; and each and all of the regulations, provisions, penalties, conditions and terms of said Plumbing Code on file in the office of the [JURISDICTION] are hereby referred to, adopted, and made a part hereof, as if fully set out in this legislation, with the additions, insertions, deletions and changes, if any, prescribed in Section 2 of this ordinance.

Section 2. The following sections are hereby revised:

Section 101.1. Insert: [NAME OF JURISDICTION]

Section 106.6.2. Insert: [APPROPRIATE SCHEDULE]

Section 106.6.3. Insert: [PERCENTAGES IN TWO LOCATIONS]

Section 108.4. Insert: [OFFENSE, DOLLAR AMOUNT, NUMBER OF DAYS]

Section 108.5. Insert: [DOLLAR AMOUNT IN TWO LOCATIONS]

Section 305.6.1. Insert: [NUMBER OF INCHES IN TWO LOCATIONS]

Section 904.1. Insert: [NUMBER OF INCHES]

Section 3. That [ORDINANCE/STATUTE/REGULATION] No. _____ of [JURISDICTION] entitled [FILL IN HERE THE COMPLETE TITLE OF THE LEGISLATION OR LAWS IN EFFECT AT THE PRESENT TIME SO THAT THEY WILL BE REPEALED BY DEFINITE MENTION] and all other ordinances or parts of laws in conflict herewith are hereby repealed.

Section 4. That if any section, subsection, sentence, clause or phrase of this legislation is, for any reason, held to be unconstitutional, such decision shall not affect the validity of the remaining portions of this ordinance. The [GOVERNING BODY] hereby declares that it would have passed this law, and each section, subsection, clause or phrase thereof, irrespective of the fact that any one or more sections, subsections, sentences, clauses and phrases be declared unconstitutional.

Section 5. That nothing in this legislation or in the Plumbing Code hereby adopted shall be construed to affect any suit or proceeding impending in any court, or any rights acquired, or liability incurred, or any cause or causes of action acquired or existing, under any act or ordinance hereby repealed as cited in Section 3 of this law; nor shall any just or legal right or remedy of any character be lost, impaired or affected by this legislation.

Section 6. That the [JURISDICTION'S KEEPER OF RECORDS] is hereby ordered and directed to cause this legislation to be published. (An additional provision may be required to direct the number of times the legislation is to be published and to specify that it is to be in a newspaper in general circulation. Posting may also be required.)

Section 7. That this law and the rules, regulations, provisions, requirements, orders and matters established and adopted hereby shall take effect and be in full force and effect [TIME PERIOD] from and after the date of its final passage and adoption.

TABLE OF CONTENTS

CHAPTER 1

SCOPE AND ADMINISTRATION

PART 1—SCOPE AND APPLICATION

SECTION 101
GENERAL

[A] 101.1 Title. These regulations shall be known as the *International Plumbing Code* of [NAME OF JURISDICTION] hereinafter referred to as "this code."

[A] 101.2 Scope. The provisions of this code shall apply to the erection, installation, alteration, repairs, relocation, replacement, addition to, use or maintenance of plumbing systems within this jurisdiction. This code shall also regulate nonflammable medical gas, inhalation anesthetic, vacuum piping, nonmedical oxygen systems and sanitary and condensate vacuum collection systems. The installation of fuel gas distribution piping and equipment, fuel-gas-fired water heaters and water heater venting systems shall be regulated by the *International Fuel Gas Code*. Provisions in the appendices shall not apply unless specifically adopted.

> **Exception:** Detached one- and two-family dwellings and multiple single-family dwellings (townhouses) not more than three stories high with separate means of egress and their accessory structures shall comply with the *International Residential Code*.

[A] 101.3 Intent. The purpose of this code is to provide minimum standards to safeguard life or limb, health, property and public welfare by regulating and controlling the design, construction, installation, quality of materials, location, operation and maintenance or use of plumbing equipment and systems.

[A] 101.4 Severability. If any section, subsection, sentence, clause or phrase of this code is for any reason held to be unconstitutional, such decision shall not affect the validity of the remaining portions of this code.

SECTION 102
APPLICABILITY

[A] 102.1 General. Where there is a conflict between a general requirement and a specific requirement, the specific requirement shall govern. Where, in any specific case, different sections of this code specify different materials, methods of construction or other requirements, the most restrictive shall govern.

[A] 102.2 Existing installations. Plumbing systems lawfully in existence at the time of the adoption of this code shall be permitted to have their use and maintenance continued if the use, maintenance or repair is in accordance with the original design and no hazard to life, health or property is created by such plumbing system.

[A] 102.3 Maintenance. All plumbing systems, materials and appurtenances, both existing and new, and all parts thereof, shall be maintained in proper operating condition in accordance with the original design in a safe and sanitary condition. All devices or safeguards required by this code shall be maintained in compliance with the code edition under which they were installed.

The owner or the owner's designated agent shall be responsible for maintenance of plumbing systems. To determine compliance with this provision, the code official shall have the authority to require any plumbing system to be reinspected.

[A] 102.4 Additions, alterations or repairs. Additions, alterations, renovations or repairs to any plumbing system shall conform to that required for a new plumbing system without requiring the existing plumbing system to comply with all the requirements of this code. Additions, alterations or repairs shall not cause an existing system to become unsafe, insanitary or overloaded.

Minor additions, alterations, renovations and repairs to existing plumbing systems shall meet the provisions for new construction, unless such work is done in the same manner and arrangement as was in the existing system, is not hazardous and is *approved*.

[A] 102.5 Change in occupancy. It shall be unlawful to make any change in the *occupancy* of any structure that will subject the structure to any special provision of this code applicable to the new *occupancy* without approval of the code official. The code official shall certify that such structure meets the intent of the provisions of law governing building construction for the proposed new *occupancy* and that such change of *occupancy* does not result in any hazard to the public health, safety or welfare.

[A] 102.6 Historic buildings. The provisions of this code relating to the construction, alteration, repair, enlargement, restoration, relocation or moving of buildings or structures shall not be mandatory for existing buildings or structures identified and classified by the state or local jurisdiction as historic buildings when such buildings or structures are judged by the code official to be safe and in the public interest of health, safety and welfare regarding any proposed construction, alteration, repair, enlargement, restoration, relocation or moving of buildings.

[A] 102.7 Moved buildings. Except as determined by Section 102.2, plumbing systems that are a part of buildings or structures moved into or within the jurisdiction shall comply with the provisions of this code for new installations.

[A] 102.8 Referenced codes and standards. The codes and standards referenced in this code shall be those that are listed in Chapter 14 and such codes and standards shall be considered as part of the requirements of this code to the prescribed extent of each such reference and as further regulated in Sections 102.8.1 and 102.8.2.

[A] 102.8.1 Conflicts. Where conflicts occur between provisions of this code and the referenced standards, the provisions of this code shall apply.

[A] 102.8.2 Provisions in referenced codes and standards. Where the extent of the reference to a referenced code or standard includes subject matter that is within the scope of this code, the provisions of this code, as applicable, shall take precedence over the provisions in the referenced code or standard.

[A] 102.9 Requirements not covered by code. Any requirements necessary for the strength, stability or proper operation of an existing or proposed plumbing system, or for the public safety, health and general welfare, not specifically covered by this code shall be determined by the code official.

[A] 102.10 Other laws. The provisions of this code shall not be deemed to nullify any provisions of local, state or federal law.

[A] 102.11 Application of references. Reference to chapter section numbers, or to provisions not specifically identified by number, shall be construed to refer to such chapter, section or provision of this code.

PART 2—ADMINISTRATION AND ENFORCEMENT

SECTION 103
DEPARTMENT OF PLUMBING INSPECTION

[A] 103.1 General. The department of plumbing inspection is hereby created and the executive official in charge thereof shall be known as the code official.

[A] 103.2 Appointment. The code official shall be appointed by the chief appointing authority of the jurisdiction.

[A] 103.3 Deputies. In accordance with the prescribed procedures of this jurisdiction and with the concurrence of the appointing authority, the code official shall have the authority to appoint a deputy code official, other related technical officers, inspectors and other employees. Such employees shall have powers as delegated by the code official.

[A] 103.4 Liability. The code official, member of the board of appeals or employee charged with the enforcement of this code, while acting for the jurisdiction in good faith and without malice in the discharge of the duties required by this code or other pertinent law or ordinance, shall not thereby be rendered liable personally, and is hereby relieved from all personal liability for any damage accruing to persons or property as a result of any act or by reason of an act or omission in the discharge of official duties.

Any suit instituted against any officer or employee because of an act performed by that officer or employee in the lawful discharge of duties and under the provisions of this code shall be defended by the legal representative of the jurisdiction until the final termination of the proceedings. The code official or any subordinate shall not be liable for costs in any action, suit or proceeding that is instituted in pursuance of the provisions of this code.

SECTION 104
DUTIES AND POWERS OF THE CODE OFFICIAL

[A] 104.1 General. The code official is hereby authorized and directed to enforce the provisions of this code. The code official shall have the authority to render interpretations of this code and to adopt policies and procedures in order to clarify the application of its provisions. Such interpretations, policies and procedures shall be in compliance with the intent and purpose of this code. Such policies and procedures shall not have the effect of waiving requirements specifically provided for in this code.

[A] 104.2 Applications and permits. The code official shall receive applications, review construction documents and issue permits for the installation and alteration of plumbing systems, inspect the premises for which such permits have been issued, and enforce compliance with the provisions of this code.

[A] 104.3 Inspections. The code official shall make all the required inspections, or shall accept reports of inspection by *approved* agencies or individuals. All reports of such inspections shall be in writing and be certified by a responsible officer of such *approved* agency or by the responsible individual. The code official is authorized to engage such expert opinion as deemed necessary to report on unusual technical issues that arise, subject to the approval of the appointing authority.

[A] 104.4 Right of entry. Whenever it is necessary to make an inspection to enforce the provisions of this code, or whenever the code official has reasonable cause to believe that there exists in any building or upon any premises any conditions or violations of this code that make the building or premises unsafe, insanitary, dangerous or hazardous, the code official shall have the authority to enter the building or premises at all reasonable times to inspect or to perform the duties imposed upon the code official by this code. If such building or premises is occupied, the code official shall present credentials to the occupant and request entry. If such building or premises is unoccupied, the code official shall first make a reasonable effort to locate the owner or other person having charge or control of the building or premises and request entry. If entry is refused, the code official shall have recourse to every remedy provided by law to secure entry.

When the code official shall have first obtained a proper inspection warrant or other remedy provided by law to secure entry, no owner or occupant or person having charge, care or control of any building or premises shall fail or neglect, after proper request is made as herein provided, to promptly permit entry therein by the code official for the purpose of inspection and examination pursuant to this code.

[A] 104.5 Identification. The code official shall carry proper identification when inspecting structures or premises in the performance of duties under this code.

[A] 104.6 Notices and orders. The code official shall issue all necessary notices or orders to ensure compliance with this code.

[A] 104.7 Department records. The code official shall keep official records of applications received, permits and certificates issued, fees collected, reports of inspections, and notices and orders issued. Such records shall be retained in the official records for the period required for the retention of public records.

SECTION 105
APPROVAL

[A] 105.1 Modifications. Whenever there are practical difficulties involved in carrying out the provisions of this code, the code official shall have the authority to grant modifications for individual cases, upon application of the owner or owner's representative, provided the code official shall first find that special individual reason makes the strict letter of this code impractical and the modification conforms to the intent and purpose of this code and that such modification does not lessen health, life and fire safety requirements. The details of action granting modifications shall be recorded and entered in the files of the plumbing inspection department.

[A] 105.2 Alternative materials, methods and equipment. The provisions of this code are not intended to prevent the installation of any material or to prohibit any method of construction not specifically prescribed by this code, provided that any such alternative has been *approved.* An alternative material or method of construction shall be *approved* where the code official finds that the proposed alternative material, method or equipment complies with the intent of the provisions of this code and is at least the equivalent of that prescribed in this code.

> **[A] 105.2.1 Research reports.** Supporting data, where necessary to assist in the approval of materials or assemblies not specifically provided for in this code, shall consist of valid research reports from approved sources.

[A] 105.3 Required testing. Whenever there is insufficient evidence of compliance with the provisions of this code, or evidence that a material or method does not conform to the requirements of this code, or in order to substantiate claims for alternate materials or methods, the code official shall have the authority to require tests as evidence of compliance to be made at no expense to the jurisdiction.

> **[A] 105.3.1 Test methods.** Test methods shall be as specified in this code or by other recognized test standards. In the absence of recognized and accepted test methods, the code official shall approve the testing procedures.

> **[A] 105.3.2 Testing agency.** All tests shall be performed by an *approved* agency.

> **[A] 105.3.3 Test reports.** Reports of tests shall be retained by the code official for the period required for retention of public records.

[A] 105.4 Approved materials and equipment. Materials, equipment and devices *approved* by the code official shall be constructed and installed in accordance with such approval.

> **[A] 105.4.1 Material and equipment reuse.** Materials, equipment and devices shall not be reused unless such elements have been reconditioned, tested, placed in good and proper working condition and *approved.*

SECTION 106
PERMITS

[A] 106.1 When required. Any owner, authorized agent or contractor who desires to construct, enlarge, alter, repair, move, demolish or change the *occupancy* of a building or structure, or to erect, install, enlarge, alter, repair, remove, convert or replace any plumbing system, the installation of which is regulated by this code, or to cause any such work to be done, shall first make application to the code official and obtain the required permit for the work.

[A] 106.2 Exempt work. The following work shall be exempt from the requirement for a permit:

1. The stopping of leaks in drains, water, soil, waste or vent pipe provided, however, that if any concealed trap, drainpipe, water, soil, waste or vent pipe becomes defective and it becomes necessary to remove and replace the same with new material, such work shall be considered as new work and a permit shall be obtained and inspection made as provided in this code.

2. The clearing of stoppages or the repairing of leaks in pipes, valves or fixtures, and the removal and reinstallation of water closets, provided such repairs do not involve or require the replacement or rearrangement of valves, pipes or fixtures.

Exemption from the permit requirements of this code shall not be deemed to grant authorization for any work to be done in violation of the provisions of this code or any other laws or ordinances of this jurisdiction.

[A] 106.3 Application for permit. Each application for a permit, with the required fee, shall be filed with the code official on a form furnished for that purpose and shall contain a general description of the proposed work and its location. The application shall be signed by the owner or an authorized agent. The permit application shall indicate the proposed *occupancy* of all parts of the building and of that portion of the site or lot, if any, not covered by the building or structure and shall contain such other information required by the code official.

> **[A] 106.3.1 Construction documents.** Construction documents, engineering calculations, diagrams and other such data shall be submitted in two or more sets with each application for a permit. The code official shall require construction documents, computations and specifications to be prepared and designed by a registered design professional when required by state law. Construction documents shall be drawn to scale and shall be of sufficient clarity to indicate the location, nature and extent of the work proposed and show in detail that the work conforms to the provisions of this code. Construction documents for buildings more than two stories in height shall indicate where penetrations will be made for pipes, fittings and components and shall indicate the materials and methods for maintaining required structural safety, fire-resistance rating and fireblocking.

> > **Exception:** The code official shall have the authority to waive the submission of construction documents, calculations or other data if the nature of the work applied

for is such that reviewing of construction documents is not necessary to determine compliance with this code.

[A] 106.3.2 Preliminary inspection. Before a permit is issued, the code official is authorized to inspect and evaluate the systems, equipment, buildings, devices, premises and spaces or areas to be used.

[A] 106.3.3 Time limitation of application. An application for a permit for any proposed work shall be deemed to have been abandoned 180 days after the date of filing, unless such application has been pursued in good faith or a permit has been issued; except that the code official shall have the authority to grant one or more extensions of time for additional periods not exceeding 180 days each. The extension shall be requested in writing and justifiable cause demonstrated.

[A] 106.4 By whom application is made. Application for a permit shall be made by the person or agent to install all or part of any plumbing system. The applicant shall meet all qualifications established by statute, or by rules promulgated by this code, or by ordinance or by resolution. The full name and address of the applicant shall be stated in the application.

[A] 106.5 Permit issuance. The application, construction documents and other data filed by an applicant for permit shall be reviewed by the code official. If the code official finds that the proposed work conforms to the requirements of this code and all laws and ordinances applicable thereto, and that the fees specified in Section 106.6 have been paid, a permit shall be issued to the applicant.

[A] 106.5.1 Approved construction documents. When the code official issues the permit where construction documents are required, the construction documents shall be endorsed in writing and stamped "APPROVED." Such *approved* construction documents shall not be changed, modified or altered without authorization from the code official. All work shall be done in accordance with the *approved* construction documents.

The code official shall have the authority to issue a permit for the construction of a part of a plumbing system before the entire construction documents for the whole system have been submitted or *approved*, provided adequate information and detailed statements have been filed complying with all pertinent requirements of this code. The holders of such permit shall proceed at their own risk without assurance that the permit for the entire plumbing system will be granted.

[A] 106.5.2 Validity. The issuance of a permit or approval of construction documents shall not be construed to be a permit for, or an approval of, any violation of any of the provisions of this code or any other ordinance of the jurisdiction. No permit presuming to give authority to violate or cancel the provisions of this code shall be valid.

The issuance of a permit based upon construction documents and other data shall not prevent the code official from thereafter requiring the correction of errors in said construction documents and other data or from preventing building operations being carried on thereunder when in

violation of this code or of other ordinances of this jurisdiction.

[A] 106.5.3 Expiration. Every permit issued by the code official under the provisions of this code shall expire by limitation and become null and void if the work authorized by such permit is not commenced within 180 days from the date of such permit, or if the work authorized by such permit is suspended or abandoned at any time after the work is commenced for a period of 180 days. Before such work can be recommenced, a new permit shall be first obtained and the fee therefor shall be one-half the amount required for a new permit for such work, provided no changes have been made or will be made in the original construction documents for such work, and provided further that such suspension or abandonment has not exceeded 1 year.

[A] 106.5.4 Extensions. Any permittee holding an unexpired permit shall have the right to apply for an extension of the time within which the permittee will commence work under that permit when work is unable to be commenced within the time required by this section for good and satisfactory reasons. The code official shall extend the time for action by the permittee for a period not exceeding 180 days if there is reasonable cause. No permit shall be extended more than once. The fee for an extension shall be one-half the amount required for a new permit for such work.

[A] 106.5.5 Suspension or revocation of permit. The code official shall have the authority to suspend or revoke a permit issued under the provisions of this code wherever the permit is issued in error or on the basis of incorrect, inaccurate or incomplete information, or in violation of any ordinance or regulation or any of the provisions of this code.

[A] 106.5.6 Retention of construction documents. One set of *approved* construction documents shall be retained by the code official for a period of not less than 180 days from date of completion of the permitted work, or as required by state or local laws.

One set of *approved* construction documents shall be returned to the applicant, and said set shall be kept on the site of the building or work at all times during which the work authorized thereby is in progress.

[A] 106.5.7 Previous approvals. This code shall not require changes in the construction documents, construction or designated occupancy of a structure for which a lawful permit has been heretofore issued or otherwise lawfully authorized, and the construction of which has been pursued in good faith within 180 days after the effective date of this code and has not been abandoned.

[A] 106.5.8 Posting of permit. The permit or a copy shall be kept on the site of the work until the completion of the project.

[A] 106.6 Fees. A permit shall not be issued until the fees prescribed in Section 106.6.2 have been paid, and an amendment to a permit shall not be released until the additional fee, if any, due to an increase of the plumbing systems, has been paid.

[A] 106.6.1 Work commencing before permit issuance. Any person who commences any work on a plumbing system before obtaining the necessary permits shall be subject to 100 percent of the usual permit fee in addition to the required permit fees.

[A] 106.6.2 Fee schedule. The fees for all plumbing work shall be as indicated in the following schedule:

[JURISDICTION TO INSERT APPROPRIATE SCHEDULE]

[A] 106.6.3 Fee refunds. The code official shall authorize the refunding of fees as follows:

1. The full amount of any fee paid hereunder that was erroneously paid or collected.

2. Not more than [SPECIFY PERCENTAGE] percent of the permit fee paid when no work has been done under a permit issued in accordance with this code.

3. Not more than [SPECIFY PERCENTAGE] percent of the plan review fee paid when an application for a permit for which a plan review fee has been paid is withdrawn or canceled before any plan review effort has been expended.

The code official shall not authorize the refunding of any fee paid except upon written application filed by the original permittee not later than 180 days after the date of fee payment.

SECTION 107
INSPECTIONS AND TESTING

[A] 107.1 General. The code official is authorized to conduct such inspections as are deemed necessary to determine compliance with the provisions of this code. Construction or work for which a permit is required shall be subject to inspection by the code official, and such construction or work shall remain accessible and exposed for inspection purposes until *approved*. Approval as a result of an inspection shall not be construed to be an approval of a violation of the provisions of this code or of other ordinances of the jurisdiction. Inspections presuming to give authority to violate or cancel the provisions of this code or of other ordinances of the jurisdiction shall not be valid. It shall be the duty of the permit applicant to cause the work to remain accessible and exposed for inspection purposes. Neither the code official nor the jurisdiction shall be liable for expense entailed in the removal or replacement of any material required to allow inspection.

[A] 107.2 Required inspections and testing. The code official, upon notification from the permit holder or the permit holder's agent, shall make the following inspections and such other inspections as necessary, and shall either release that portion of the construction or shall notify the permit holder or an agent of any violations that must be corrected. The holder of the permit shall be responsible for the scheduling of such inspections.

1. Underground inspection shall be made after trenches or ditches are excavated and bedded, piping installed, and before any backfill is put in place.

2. Rough-in inspection shall be made after the roof, framing, fireblocking, firestopping, draftstopping and bracing is in place and all sanitary, storm and water distribution piping is roughed-in, and prior to the installation of wall or ceiling membranes.

3. Final inspection shall be made after the building is complete, all plumbing fixtures are in place and properly connected, and the structure is ready for occupancy.

[A] 107.2.1 Other inspections. In addition to the inspections specified above, the code official is authorized to make or require other inspections of any construction work to ascertain compliance with the provisions of this code and other laws that are enforced.

[A] 107.2.2 Inspection requests. It shall be the duty of the holder of the permit or their duly authorized agent to notify the code official when work is ready for inspection. It shall be the duty of the permit holder to provide *access* to and means for inspections of such work that are required by this code.

[A] 107.2.3 Approval required. Work shall not be done beyond the point indicated in each successive inspection without first obtaining the approval of the code official. The code official, upon notification, shall make the requested inspections and shall either indicate the portion of the construction that is satisfactory as completed, or notify the permit holder or his or her agent wherein the same fails to comply with this code. Any portions that do not comply shall be corrected and such portion shall not be covered or concealed until authorized by the code official.

[A] 107.2.4 Approved agencies. The code official is authorized to accept reports of *approved* inspection agencies, provided that such agencies satisfy the requirements as to qualifications and reliability.

[A] 107.2.5 Evaluation and follow-up inspection services. Prior to the approval of a closed, prefabricated plumbing system and the issuance of a plumbing permit, the code official shall require the submittal of an evaluation report on each prefabricated plumbing system indicating the complete details of the plumbing system, including a description of the system and its components, the basis upon which the plumbing system is being evaluated, test results and similar information, and other data as necessary for the code official to determine conformance to this code.

[A] 107.2.5.1 Evaluation service. The code official shall designate the evaluation service of an *approved* agency as the evaluation agency, and review such agency's evaluation report for adequacy and conformance to this code.

[A] 107.2.5.2 Follow-up inspection. Except where ready *access* is provided to all plumbing systems, service equipment and accessories for complete inspection at the site without disassembly or dismantling, the code official shall conduct the frequency of in-plant inspections necessary to ensure conformance to the *approved* evaluation report or shall designate an independent, *approved* inspection agency to conduct such inspections. The inspection agency shall furnish the code official with the follow-up inspection manual and a report of inspections upon request, and the plumbing system shall have an identifying label permanently affixed to the system indicating that factory inspections have been performed.

[A] 107.2.5.3 Test and inspection records. All required test and inspection records shall be available to the code official at all times during the fabrication of the plumbing system and the erection of the building, or such records as the code official designates shall be filed.

[A] 107.3 Special inspections. Special inspections of *alternative engineered design* plumbing systems shall be conducted in accordance with Sections 107.3.1 and 107.3.2.

[A] 107.3.1 Periodic inspection. The registered design professional or designated inspector shall periodically inspect and observe the *alternative engineered design* to determine that the installation is in accordance with the *approved* construction documents. All discrepancies shall be brought to the immediate attention of the plumbing contractor for correction. Records shall be kept of all inspections.

[A] 107.3.2 Written report. The registered design professional shall submit a final report in writing to the code official upon completion of the installation, certifying that the *alternative engineered design* conforms to the *approved* construction documents. A notice of approval for the plumbing system shall not be issued until a written certification has been submitted.

[A] 107.4 Testing. Plumbing work and systems shall be tested as required in Section 312 and in accordance with Sections 107.4.1 through 107.4.3. Tests shall be made by the permit holder and observed by the code official.

[A] 107.4.1 New, altered, extended or repaired systems. New plumbing systems and parts of existing systems that have been altered, extended or repaired shall be tested as prescribed herein to disclose leaks and defects, except that testing is not required in the following cases:

1 In any case that does not include addition to, replacement, alteration or relocation of any water supply, drainage or vent piping.

2. In any case where plumbing equipment is set up temporarily for exhibition purposes.

[A] 107.4.2 Equipment, material and labor for tests. All equipment, material and labor required for testing a plumbing system or part thereof shall be furnished by the permit holder.

[A] 107.4.3 Reinspection and testing. Where any work or installation does not pass any initial test or inspection, the necessary corrections shall be made to comply with this code. The work or installation shall then be resubmitted to the code official for inspection and testing.

[A] 107.5 Approval. After the prescribed tests and inspections indicate that the work complies in all respects with this code, a notice of approval shall be issued by the code official.

[A] 107.5.1 Revocation. The code official is authorized to, in writing, suspend or revoke a notice of approval issued under the provisions of this code wherever the notice is issued in error, or on the basis of incorrect information supplied, or where it is determined that the building or structure, premise or portion thereof is in violation of any ordinance or regulation or any of the provisions of this code.

[A] 107.6 Temporary connection. The code official shall have the authority to authorize the temporary connection of the building or system to the utility source for the purpose of testing plumbing systems or for use under a temporary certificate of occupancy.

[A] 107.7 Connection of service utilities. A person shall not make connections from a utility, source of energy, fuel, power, water system or *sewer* system to any building or system that is regulated by this code for which a permit is required until authorized by the code official.

SECTION 108
VIOLATIONS

[A] 108.1 Unlawful acts. It shall be unlawful for any person, firm or corporation to erect, construct, alter, repair, remove, demolish or utilize any plumbing system, or cause same to be done, in conflict with or in violation of any of the provisions of this code.

[A] 108.2 Notice of violation. The code official shall serve a notice of violation or order to the person responsible for the erection, installation, alteration, extension, repair, removal or demolition of plumbing work in violation of the provisions of this code, or in violation of a detail statement or the *approved* construction documents thereunder, or in violation of a permit or certificate issued under the provisions of this code. Such order shall direct the discontinuance of the illegal action or condition and the abatement of the violation.

[A] 108.3 Prosecution of violation. If the notice of violation is not complied with promptly, the code official shall request the legal counsel of the jurisdiction to institute the appropriate proceeding at law or in equity to restrain, correct or abate such violation, or to require the removal or termination of the unlawful occupancy of the structure in violation of the provisions of this code or of the order or direction made pursuant thereto.

[A] 108.4 Violation penalties. Any person who shall violate a provision of this code or shall fail to comply with any of the requirements thereof or who shall erect, install, alter or repair plumbing work in violation of the *approved* construction documents or directive of the code official, or of a permit or certificate issued under the provisions of this code, shall be guilty of a [SPECIFY OFFENSE], punishable by a fine of not more than [AMOUNT] dollars or by imprisonment not exceeding [NUMBER OF DAYS], or both such fine and imprisonment. Each day that a violation continues after due notice has been served shall be deemed a separate offense.

[A] 108.5 Stop work orders. Upon notice from the code official, work on any plumbing system that is being done contrary to the provisions of this code or in a dangerous or unsafe manner shall immediately cease. Such notice shall be in writing and shall be given to the owner of the property, or to the owner's agent, or to the person doing the work. The notice shall state the conditions under which work is authorized to resume. Where an emergency exists, the code official shall not be required to give a written notice prior to stopping the work. Any person who shall continue any work in or about the structure after having been served with a stop work order, except such work as that person is directed to perform to remove a violation or unsafe condition, shall be liable to a fine of not less than [AMOUNT] dollars or more than [AMOUNT] dollars.

[A] 108.6 Abatement of violation. The imposition of the penalties herein prescribed shall not preclude the legal officer of the jurisdiction from instituting appropriate action to prevent unlawful construction or to restrain, correct or abate a violation, or to prevent illegal occupancy of a building, structure or premises, or to stop an illegal act, conduct, business or utilization of the plumbing on or about any premises.

[A] 108.7 Unsafe plumbing. Any plumbing regulated by this code that is unsafe or that constitutes a fire or health hazard, insanitary condition, or is otherwise dangerous to human life is hereby declared unsafe. Any use of plumbing regulated by this code constituting a hazard to safety, health or public welfare by reason of inadequate maintenance, dilapidation, obsolescence, fire hazard, disaster, damage or abandonment is hereby declared an unsafe use. Any such unsafe equipment is hereby declared to be a public nuisance and shall be abated by repair, rehabilitation, demolition or removal.

[A] 108.7.1 Authority to condemn equipment. Whenever the code official determines that any plumbing, or portion thereof, regulated by this code has become hazardous to life, health or property or has become insanitary, the code official shall order in writing that such plumbing either be removed or restored to a safe or sanitary condition. A time limit for compliance with such order shall be specified in the written notice. No person shall use or maintain defective plumbing after receiving such notice.

When such plumbing is to be disconnected, written notice as prescribed in Section 108.2 shall be given. In cases of immediate danger to life or property, such disconnection shall be made immediately without such notice.

[A] 108.7.2 Authority to disconnect service utilities. The code official shall have the authority to authorize disconnection of utility service to the building, structure or system regulated by the technical codes in case of an emergency, where necessary, to eliminate an immediate danger to life or property. Where possible, the owner and occupant of the building, structure or service system shall be notified of the decision to disconnect utility service prior to taking such action. If not notified prior to disconnecting, the owner or occupant of the building, structure or service systems shall be notified in writing, as soon as practical thereafter.

[A] 108.7.3 Connection after order to disconnect. No person shall make connections from any energy, fuel, power supply or water distribution system or supply energy, fuel or water to any equipment regulated by this code that has been disconnected or ordered to be disconnected by the code official or the use of which has been ordered to be discontinued by the code official until the code official authorizes the reconnection and use of such equipment.

When any plumbing is maintained in violation of this code, and in violation of any notice issued pursuant to the provisions of this section, the code official shall institute any appropriate action to prevent, restrain, correct or abate the violation.

SECTION 109
MEANS OF APPEAL

[A] 109.1 Application for appeal. Any person shall have the right to appeal a decision of the code official to the board of appeals. An application for appeal shall be based on a claim that the true intent of this code or the rules legally adopted thereunder have been incorrectly interpreted, the provisions of this code do not fully apply, or an equally good or better form of construction is proposed. The application shall be filed on a form obtained from the code official within 20 days after the notice was served.

[A] 109.2 Membership of board. The board of appeals shall consist of five members appointed by the chief appointing authority as follows: one for 5 years, one for 4 years, one for 3 years, one for 2 years and one for 1 year. Thereafter, each new member shall serve for 5 years or until a successor has been appointed.

[A] 109.2.1 Qualifications. The board of appeals shall consist of five individuals, one from each of the following professions or disciplines:

1. Registered design professional who is a registered architect; or a builder or superintendent of building construction with at least 10 years' experience, 5 years of which shall have been in responsible charge of work.

2. Registered design professional with structural engineering or architectural experience.

3. Registered design professional with mechanical and plumbing engineering experience; or a mechanical and plumbing contractor with at least 10 years' experience, 5 years of which shall have been in responsible charge of work.

4. Registered design professional with electrical engineering experience; or an electrical contractor with at least 10 years' experience, 5 years of which shall have been in responsible charge of work.

5. Registered design professional with fire protection engineering experience; or a fire protection contractor with at least 10 years' experience, 5 years of which shall have been in responsible charge of work.

[A] 109.2.2 Alternate members. The chief appointing authority shall appoint two alternate members who shall be called by the board chairman to hear appeals during the absence or disqualification of a member. Alternate members shall possess the qualifications required for board membership, and shall be appointed for 5 years or until a successor has been appointed.

[A] 109.2.3 Chairman. The board shall annually select one of its members to serve as chairman.

[A] 109.2.4 Disqualification of member. A member shall not hear an appeal in which that member has any personal, professional or financial interest.

[A] 109.2.5 Secretary. The chief administrative officer shall designate a qualified clerk to serve as secretary to the board. The secretary shall file a detailed record of all proceedings in the office of the chief administrative officer.

[A] 109.2.6 Compensation of members. Compensation of members shall be determined by law.

[A] 109.3 Notice of meeting. The board shall meet upon notice from the chairman, within 10 days of the filing of an appeal or at stated periodic meetings.

[A] 109.4 Open hearing. All hearings before the board shall be open to the public. The appellant, the appellant's representative, the code official and any person whose interests are affected shall be given an opportunity to be heard.

[A] 109.4.1 Procedure. The board shall adopt and make available to the public through the secretary procedures under which a hearing will be conducted. The procedures shall not require compliance with strict rules of evidence, but shall mandate that only relevant information be received.

[A] 109.5 Postponed hearing. When five members are not present to hear an appeal, either the appellant or the appellant's representative shall have the right to request a postponement of the hearing.

[A] 109.6 Board decision. The board shall modify or reverse the decision of the code official by a concurring vote of three members.

[A] 109.6.1 Resolution. The decision of the board shall be by resolution. Certified copies shall be furnished to the appellant and to the code official.

[A] 109.6.2 Administration. The code official shall take immediate action in accordance with the decision of the board.

[A] 109.7 Court review. Any person, whether or not a previous party of the appeal, shall have the right to apply to the appropriate court for a writ of certiorari to correct errors of law. Application for review shall be made in the manner and time required by law following the filing of the decision in the office of the chief administrative officer.

SECTION 110
TEMPORARY EQUIPMENT, SYSTEMS AND USES

[A] 110.1 General. The code official is authorized to issue a permit for temporary equipment, systems and uses. Such permits shall be limited as to time of service, but shall not be permitted for more than 180 days. The code official is authorized to grant extensions for demonstrated cause.

[A] 110.2 Conformance. Temporary equipment, systems and uses shall conform to the structural strength, fire safety, means of egress, accessibility, light, ventilation and sanitary requirements of this code as necessary to ensure the public health, safety and general welfare.

[A] 110.3 Temporary utilities. The code official is authorized to give permission to temporarily supply utilities before an installation has been fully completed and the final certificate of completion has been issued. The part covered by the temporary certificate shall comply with the requirements specified for temporary lighting, heat or power in the code.

[A] 110.4 Termination of approval. The code official is authorized to terminate such permit for temporary equipment, systems or uses and to order the temporary equipment, systems or uses to be discontinued.

CHAPTER 2

DEFINITIONS

201.1 Scope. Unless otherwise expressly stated, the following words and terms shall, for the purposes of this code, have the meanings shown in this chapter.

201.2 Interchangeability. Words stated in the present tense include the future; words stated in the masculine gender include the feminine and neuter; the singular number includes the plural and the plural the singular.

201.3 Terms defined in other codes. Where terms are not defined in this code and are defined in the *International Building Code, International Fire Code, International Fuel Gas Code* or the *International Mechanical Code,* such terms shall have the meanings ascribed to them as in those codes.

201.4 Terms not defined. Where terms are not defined through the methods authorized by this section, such terms shall have ordinarily accepted meanings such as the context implies.

SECTION 202
GENERAL DEFINITIONS

ACCEPTED ENGINEERING PRACTICE. That which conforms to accepted principles, tests or standards of nationally recognized technical or scientific authorities.

[M] ACCESS (TO). That which enables a fixture, appliance or equipment to be reached by ready *access* or by a means that first requires the removal or movement of a panel, door or similar obstruction (see "Ready *access*").

ACCESS COVER. A removable plate, usually secured by bolts or screws, to permit *access* to a pipe or pipe fitting for the purposes of inspection, repair or cleaning.

ADAPTER FITTING. An *approved* connecting device that suitably and properly joins or adjusts pipes and fittings which do not otherwise fit together.

AIR ADMITTANCE VALVE. One-way valve designed to allow air to enter the plumbing drainage system when negative pressures develop in the piping system. The device shall close by gravity and seal the vent terminal at zero differential pressure (no flow conditions) and under positive internal pressures. The purpose of an air admittance valve is to provide a method of allowing air to enter the plumbing drainage system without the use of a vent extended to open air and to prevent *sewer* gases from escaping into a building.

AIR BREAK (Drainage System). A piping arrangement in which a drain from a fixture, appliance or device discharges indirectly into another fixture, receptacle or interceptor at a point below the *flood level rim* and above the trap seal.

AIR GAP (Drainage System). The unobstructed vertical distance through the free atmosphere between the outlet of the waste pipe and the *flood level rim* of the receptacle into which the waste pipe is discharging.

AIR GAP (Water Distribution System). The unobstructed vertical distance through the free atmosphere between the lowest opening from any pipe or faucet supplying water to a tank, plumbing fixture or other device and the *flood level rim* of the receptacle.

ALTERNATIVE ENGINEERED DESIGN. A plumbing system that performs in accordance with the intent of Chapters 3 through 12 and provides an equivalent level of performance for the protection of public health, safety and welfare. The system design is not specifically regulated by Chapters 3 through 12.

ANCHORS. See "Supports."

ANTISIPHON. A term applied to valves or mechanical devices that eliminate siphonage.

APPROVED. Acceptable to the code official or other authority having jurisdiction.

[A] APPROVED AGENCY. An established and recognized agency approved by the code official and that is regularly engaged in conducting tests or furnishing inspection services.

AREA DRAIN. A receptacle designed to collect surface or storm water from an open area.

ASPIRATOR. A fitting or device supplied with water or other fluid under positive pressure that passes through an integral orifice or constriction, causing a vacuum. Aspirators are also referred to as suction apparatus, and are similar in operation to an ejector.

BACKFLOW. Pressure created by any means in the water distribution system, which by being in excess of the pressure in the water supply mains causes a potential backflow condition.

> **Backpressure, low head.** A pressure less than or equal to 4.33 psi (29.88 kPa) or the pressure exerted by a 10-foot (3048 mm) column of water.
>
> **Backsiphonage.** The backflow of potentially contaminated water into the potable water system as a result of the pressure in the potable water system falling below atmospheric pressure of the plumbing fixtures, pools, tanks or vats connected to the potable water distribution piping.
>
> **Drainage.** A reversal of flow in the drainage system.
>
> **Water supply system.** The flow of water or other liquids, mixtures or substances into the distribution pipes of a potable water supply from any source except the intended source.

BACKFLOW CONNECTION. Any arrangement whereby backflow is possible.

BACKFLOW PREVENTER. A device or means to prevent backflow.

BACKWATER VALVE. A device or valve installed in the *building drain* or *sewer* pipe where a *sewer* is subject to back-flow, and which prevents drainage or waste from backing up into a lower level or fixtures and causing a flooding condition.

BASE FLOOD ELEVATION. A reference point, determined in accordance with the building code, based on the depth or peak elevation of flooding, including wave height, which has a 1 percent (100-year flood) or greater chance of occurring in any given year.

BATHROOM GROUP. A group of fixtures consisting of a water closet, lavatory, bathtub or shower, including or excluding a bidet, an *emergency floor drain* or both. Such fixtures are located together on the same floor level.

BEDPAN STEAMER OR BOILER. A fixture utilized for scalding bedpans or urinals by direct application of steam or boiling water.

BEDPAN WASHER AND STERILIZER. A fixture designed to wash bedpans and to flush the contents into the sanitary drainage system. Included are fixtures of this type that provide for disinfecting utensils by scalding with steam or *hot water*.

BEDPAN WASHER HOSE. A device supplied with hot and cold water and located adjacent to a water closet or clinical sink to be utilized for cleansing bedpans.

BRANCH. Any part of the piping system except a riser, main or *stack*.

BRANCH INTERVAL. A vertical measurement of distance, 8 feet (2438 mm) or more in *developed length*, between the connections of horizontal branches to a drainage *stack*. Measurements are taken down the *stack* from the highest horizontal *branch* connection.

BRANCH VENT. A vent connecting one or more individual vents with a vent *stack* or *stack* vent.

[A] BUILDING. Any structure occupied or intended for supporting or sheltering any *occupancy*.

BUILDING DRAIN. That part of the lowest piping of a drainage system that receives the discharge from soil, waste and other drainage pipes inside and that extends 30 inches (762 mm) in *developed length* of pipe beyond the exterior walls of the building and conveys the drainage to the *building sewer*.

> **Combined.** A *building drain* that conveys both sewage and storm water or other drainage.

> **Sanitary.** A *building drain* that conveys sewage only.

> **Storm.** A *building drain* that conveys storm water or other drainage, but not sewage.

BUILDING SEWER. That part of the drainage system that extends from the end of the *building drain* and conveys the discharge to a *public sewer*, *private sewer*, individual sewage disposal system or other point of disposal.

> **Combined.** A *building sewer* that conveys both sewage and storm water or other drainage.

> **Sanitary.** A *building sewer* that conveys sewage only.

> **Storm.** A *building sewer* that conveys storm water or other drainage, but not sewage.

BUILDING SUBDRAIN. That portion of a drainage system that does not drain by gravity into the *building sewer*.

BUILDING TRAP. A device, fitting or assembly of fittings installed in the *building drain* to prevent circulation of air between the drainage system of the building and the *building sewer*.

CIRCUIT VENT. A vent that connects to a horizontal drainage *branch* and vents two traps to a maximum of eight traps or trapped fixtures connected into a battery.

CISTERN. A small covered tank for storing water for a home or farm. Generally, this tank stores rainwater to be utilized for purposes other than in the potable water supply, and such tank is placed underground in most cases.

CLEANOUT. An *access* opening in the drainage system utilized for the removal of obstructions. Types of cleanouts include a removable plug or cap, and a removable fixture or fixture trap.

[A] CODE. These regulations, subsequent amendments thereto, or any emergency rule or regulation that the administrative authority having jurisdiction has lawfully adopted.

[A] CODE OFFICIAL. The officer or other designated authority charged with the administration and enforcement of this code, or a duly authorized representative.

COMBINATION FIXTURE. A fixture combining one sink and laundry tray or a two- or three-compartment sink or laundry tray in one unit.

COMBINATION WASTE AND VENT SYSTEM. A specially designed system of waste piping embodying the horizontal wet venting of one or more sinks, lavatories, drinking fountains or floor drains by means of a common waste and vent pipe adequately sized to provide free movement of air above the flow line of the drain.

COMBINED BUILDING DRAIN. See "*Building drain*, combined."

COMBINED BUILDING SEWER. See "*Building sewer*, combined."

COMMON VENT. A vent connecting at the junction of two fixture drains or to a fixture *branch* and serving as a vent for both fixtures.

CONCEALED FOULING SURFACE. Any surface of a plumbing fixture which is not readily visible and is not scoured or cleansed with each fixture operation.

CONDUCTOR. A pipe inside the building that conveys storm water from the roof to a storm or combined *building drain*.

[A] CONSTRUCTION DOCUMENTS. All of the written, graphic and pictorial documents prepared or assembled for describing the design, location and physical characteristics of the elements of the project necessary for obtaining a building permit. The construction drawings shall be drawn to an appropriate scale.

CONTAMINATION. An impairment of the quality of the potable water that creates an actual hazard to the public health through poisoning or through the spread of disease by sewage, industrial fluids or waste.

CRITICAL LEVEL (C-L). An elevation (height) reference point that determines the minimum height at which a backflow preventer or vacuum breaker is installed above the *flood level rim* of the fixture or receptor served by the device. The critical level is the elevation level below which there is a potential for backflow to occur. If the critical level marking is not indicated on the device, the bottom of the device shall constitute the critical level.

CROSS CONNECTION. Any physical connection or arrangement between two otherwise separate piping systems, one of which contains potable water and the other either water of unknown or questionable safety or steam, gas or chemical, whereby there exists the possibility for flow from one system to the other, with the direction of flow depending on the pressure differential between the two systems (see "Backflow").

DEAD END. A *branch* leading from a soil, waste or vent pipe; a *building drain*; or a *building sewer*, and terminating at a *developed length* of 2 feet (610 mm) or more by means of a plug, cap or other closed fitting.

DEPTH OF TRAP SEAL. The depth of liquid that would have to be removed from a full trap before air could pass through the trap.

[B] DESIGN FLOOD ELEVATION. The elevation of the "design flood," including wave height, relative to the datum specified on the community's legally designated flood hazard map.

DEVELOPED LENGTH. The length of a pipeline measured along the centerline of the pipe and fittings.

DISCHARGE PIPE. A pipe that conveys the discharges from plumbing fixtures or appliances.

DRAIN. Any pipe that carries waste water or water-borne wastes in a building drainage system.

DRAINAGE FITTINGS. Type of fitting or fittings utilized in the drainage system. Drainage fittings are similar to cast-iron fittings, except that instead of having a bell and spigot, drainage fittings are recessed and tapped to eliminate ridges on the inside of the installed pipe.

DRAINAGE FIXTURE UNIT.

> **Drainage (dfu).** A measure of the probable discharge into the drainage system by various types of plumbing fixtures. The drainage fixture-unit value for a particular fixture depends on its volume rate of drainage discharge, on the time duration of a single drainage operation and on the average time between successive operations.

DRAINAGE SYSTEM. Piping within a *public* or *private* premise that conveys sewage, rainwater or other liquid wastes to a point of disposal. A drainage system does not include the mains of a *public sewer* system or a private or public sewage treatment or disposal plant.

> **Building gravity.** A drainage system that drains by gravity into the *building sewer*.

> **Sanitary.** A drainage system that carries sewage and excludes storm, surface and ground water.

> **Storm.** A drainage system that carries rainwater, surface water, subsurface water and similar liquid wastes.

EFFECTIVE OPENING. The minimum cross-sectional area at the point of water supply discharge, measured or expressed in terms of the diameter of a circle or, if the opening is not circular, the diameter of a circle of equivalent cross-sectional area. For faucets and similar fittings, the *effective opening* shall be measured at the smallest orifice in the fitting body or in the supply piping to the fitting.

EMERGENCY FLOOR DRAIN. A floor drain that does not receive the discharge of any drain or indirect waste pipe, and that protects against damage from accidental spills, fixture overflows and leakage.

ESSENTIALLY NONTOXIC TRANSFER FLUIDS. Fluids having a Gosselin rating of 1, including propylene glycol; mineral oil; polydimethylsiloxane; hydrochlorofluorocarbon, chlorofluorocarbon and carbon refrigerants; and FDA-approved boiler water additives for steam boilers.

ESSENTIALLY TOXIC TRANSFER FLUIDS. Soil, waste or gray water and fluids having a Gosselin rating of 2 or more including ethylene glycol, hydrocarbon oils, ammonia refrigerants and hydrazine.

EXISTING INSTALLATIONS. Any plumbing system regulated by this code that was legally installed prior to the effective date of this code, or for which a permit to install has been issued.

FAUCET. A valve end of a water pipe through which water is drawn from or held within the pipe.

FILL VALVE. A water supply valve, opened or closed by means of a float or similar device, utilized to supply water to a tank. An antisiphon fill valve contains an antisiphon device in the form of an *approved air gap* or vacuum breaker that is an integral part of the fill valve unit and that is positioned on the discharge side of the water supply control valve.

FIXTURE. See "Plumbing fixture."

FIXTURE BRANCH. A drain serving two or more fixtures that discharges to another drain or to a *stack*.

FIXTURE DRAIN. The drain from the trap of a fixture to a junction with any other drain pipe.

FIXTURE FITTING.

> **Supply fitting.** A fitting that controls the volume and/or directional flow of water and is either attached to or accessible from a fixture, or is used with an open or atmospheric discharge.

> **Waste fitting.** A combination of components that conveys the sanitary waste from the outlet of a fixture to the connection to the sanitary drainage system.

FIXTURE SUPPLY. The water supply pipe connecting a fixture to a *branch* water supply pipe or directly to a main water supply pipe.

[B] FLOOD HAZARD AREA. The greater of the following two areas:

1. The area within a flood plain subject to a 1-percent or greater chance of flooding in any given year.

2. The area designated as a *flood hazard area* on a community's flood hazard map or as otherwise legally designated.

FLOOD LEVEL RIM. The edge of the receptacle from which water overflows.

FLOW CONTROL (Vented). A device installed upstream from the interceptor having an orifice that controls the rate of flow through the interceptor and an air intake (vent) downstream from the orifice that allows air to be drawn into the flow stream.

FLOW PRESSURE. The pressure in the water supply pipe near the faucet or water outlet while the faucet or water outlet is wide open and flowing.

FLUSH TANK. A tank designed with a fill valve and flush valve to flush the contents of the bowl or usable portion of the fixture.

FLUSHOMETER TANK. A device integrated within an air accumulator vessel that is designed to discharge a predetermined quantity of water to fixtures for flushing purposes.

FLUSHOMETER VALVE. A valve attached to a pressurized water supply pipe and so designed that when activated it opens the line for direct flow into the fixture at a rate and quantity to operate the fixture properly, and then gradually closes to reseal fixture traps and avoid water hammer.

GRAY WATER. Waste discharged from lavatories, bathtubs, showers, clothes washers and laundry trays.

GREASE INTERCEPTOR.

> **Hydromechanical.** Plumbing appurtenances that are installed in the sanitary drainage system to intercept free-floating fats, oils and grease from waste water discharge. Continuous separation is accomplished by air entrainment, buoyancy and interior baffling.

> **Gravity.** Plumbing appurtenances of not less than 500 gallons (1893 L) capacity that are installed in the sanitary drainage system to intercept free-floating fats, oils and grease from waste water discharge. Separation is accomplished by gravity during a retention time of not less than 30 minutes.

GREASE-LADEN WASTE. Effluent discharge that is produced from food processing, food preparation or other sources where grease, fats and oils enter automatic dishwater prerinse stations, sinks or other appurtenances.

GREASE REMOVAL DEVICE, AUTOMATIC (GRD). A plumbing appurtenance that is installed in the sanitary drainage system to intercept free-floating fats, oils and grease from waste water discharge. Such a device operates on a time- or event-controlled basis and has the ability to remove free-floating fats, oils and grease automatically without intervention from the user except for maintenance.

GRIDDED WATER DISTRIBUTION SYSTEM. A water distribution system where every water distribution pipe is interconnected so as to provide two or more paths to each fixture supply pipe.

HANGERS. See "Supports."

HORIZONTAL BRANCH DRAIN. A drainage *branch* pipe extending laterally from a soil or waste *stack* or *building drain*, with or without vertical sections or branches, that receives the discharge from two or more fixture drains or branches and conducts the discharge to the soil or waste *stack* or to the *building drain*.

HORIZONTAL PIPE. Any pipe or fitting that makes an angle of less than 45 degrees (0.79 rad) with the horizontal.

HOT WATER. Water at a temperature greater than or equal to 110°F (43°C).

HOUSE TRAP. See "Building trap."

INDIRECT WASTE PIPE. A waste pipe that does not connect directly with the drainage system, but that discharges into the drainage system through an *air break* or *air gap* into a trap, fixture, receptor or interceptor.

INDIVIDUAL SEWAGE DISPOSAL SYSTEM. A system for disposal of domestic sewage by means of a septic tank, cesspool or mechanical treatment, designed for utilization apart from a public *sewer* to serve a single establishment or building.

INDIVIDUAL VENT. A pipe installed to vent a fixture trap and that connects with the vent system above the fixture served or terminates in the open air.

INDIVIDUAL WATER SUPPLY. A water supply that serves one or more families, and that is not an *approved* public water supply.

INTERCEPTOR. A device designed and installed to separate and retain for removal, by automatic or manual means, deleterious, hazardous or undesirable matter from normal wastes, while permitting normal sewage or wastes to discharge into the drainage system by gravity.

JOINT.

> **Expansion.** A loop, return bend or return offset that provides for the expansion and contraction in a piping system and is utilized in tall buildings or where there is a rapid change of temperature, as in power plants, steam rooms and similar occupancies.

> **Flexible.** Any joint between two pipes that permits one pipe to be deflected or moved without movement or deflection of the other pipe.

> **Mechanical.** See "Mechanical joint."

> **Slip.** A type of joint made by means of a washer or a special type of packing compound in which one pipe is slipped into the end of an adjacent pipe.

LEAD-FREE PIPE AND FITTINGS. Containing not more than 8.0-percent lead.

LEAD-FREE SOLDER AND FLUX. Containing not more than 0.2-percent lead.

LEADER. An exterior drainage pipe for conveying storm water from roof or gutter drains to an *approved* means of disposal.

LOCAL VENT STACK. A vertical pipe to which connections are made from the fixture side of traps and through which vapor or foul air is removed from the fixture or device utilized on bedpan washers.

MACERATING TOILET SYSTEMS. An assembly consisting of a water closet and sump with a macerating pump that is designed to collect, grind and pump wastes from the water closet and up to two other fixtures connected to the sump.

MAIN. The principal pipe artery to which branches are connected.

MANIFOLD. See "Plumbing appurtenance."

MECHANICAL JOINT. A connection between pipes, fittings, or pipes and fittings that is not screwed, caulked, threaded, soldered, solvent cemented, brazed or welded. A joint in which compression is applied along the centerline of the pieces being joined. In some applications, the joint is part of a coupling, fitting or adapter.

MEDICAL GAS SYSTEM. The complete system to convey medical gases for direct patient application from central supply systems (bulk tanks, manifolds and medical air compressors), with pressure and operating controls, alarm warning systems, related components and piping networks extending to station outlet valves at patient use points.

MEDICAL VACUUM SYSTEMS. A system consisting of central-vacuum-producing equipment with pressure and operating controls, shutoff valves, alarm-warning systems, gauges and a network of piping extending to and terminating with suitable station inlets at locations where patient suction may be required.

NONPOTABLE WATER. Water not safe for drinking, personal or culinary utilization.

NUISANCE. Public nuisance as known in common law or in equity jurisprudence; whatever is dangerous to human life or detrimental to health; whatever structure or premises is not sufficiently ventilated, sewered, drained, cleaned or lighted, with respect to its intended *occupancy*; and whatever renders the air, or human food, drink or water supply unwholesome.

[A] OCCUPANCY. The purpose for which a building or portion thereof is utilized or occupied.

OFFSET. A combination of *approved* bends that makes two changes in direction bringing one section of the pipe out of line but into a line parallel with the other section.

OPEN AIR. Outside the structure.

PLUMBING. The practice, materials and fixtures utilized in the installation, maintenance, extension and alteration of all piping, fixtures, plumbing appliances and plumbing appurtenances, within or adjacent to any structure, in connection with sanitary drainage or storm drainage facilities; venting systems; and public or private water supply systems.

PLUMBING APPLIANCE. Water-connected or drain-connected devices intended to perform a special function. These devices have their operation or control dependent on one or more energized components, such as motors, controls, or heating elements. Such devices are manually adjusted or controlled by the owner or operator, or are operated automatically through one or more of the following actions: a time cycle, a temperature range, a pressure range, a measured volume or weight.

PLUMBING APPURTENANCE. A manufactured device, prefabricated assembly or an on-the-job assembly of component parts that is an adjunct to the basic piping system and plumbing fixtures. An appurtenance demands no additional water supply and does not add any discharge load to a fixture or to the drainage system.

PLUMBING FIXTURE. A receptacle or device that is connected to a water supply system or discharges to a drainage system or both. Such receptacles or devices require a supply of water; or discharge liquid waste or liquid-borne solid waste; or require a supply of water and discharge waste to a drainage system.

PLUMBING SYSTEM. Includes the water supply and distribution pipes; plumbing fixtures and traps; water-treating or water-using equipment; soil, waste and vent pipes; and sanitary and storm sewers and building drains; in addition to their respective connections, devices and appurtenances within a structure or premises.

POLLUTION. An impairment of the quality of the potable water to a degree that does not create a hazard to the public health but that does adversely and unreasonably affect the aesthetic qualities of such potable water for domestic use.

POTABLE WATER. Water free from impurities present in amounts sufficient to cause disease or harmful physiological effects and conforming to the bacteriological and chemical quality requirements of the Public Health Service Drinking Water Standards or the regulations of the public health authority having jurisdiction.

PRIVATE. In the classification of plumbing fixtures, "*private*" applies to fixtures in residences and apartments, and to fixtures in nonpublic toilet rooms of hotels and motels and similar installations in buildings where the plumbing fixtures are intended for utilization by a family or an individual.

PUBLIC OR PUBLIC UTILIZATION. In the classification of plumbing fixtures, "*public*" applies to fixtures in general toilet rooms of schools, gymnasiums, hotels, airports, bus and railroad stations, public buildings, bars, public comfort stations, office buildings, stadiums, stores, restaurants and other installations where a number of fixtures are installed so that their utilization is similarly unrestricted.

PUBLIC WATER MAIN. A water supply pipe for public utilization controlled by public authority.

QUICK-CLOSING VALVE. A valve or faucet that closes automatically when released manually or that is controlled by a mechanical means for fast-action closing.

[M] READY ACCESS. That which enables a fixture, appliance or equipment to be directly reached without requiring the removal or movement of any panel, door or similar obstruction and without the use of a portable ladder, step stool or similar device.

REDUCED PRESSURE PRINCIPLE BACKFLOW PREVENTER. A backflow prevention device consisting of two independently acting check valves, internally force-loaded to a normally closed position and separated by an intermediate chamber (or zone) in which there is an automatic relief means of venting to the atmosphere, internally loaded to a normally open position between two tightly closing shutoff valves and with a means for testing for tightness of the checks and opening of the relief means.

[A] REGISTERED DESIGN PROFESSIONAL. An individual who is registered or licensed to practice professional architecture or engineering as defined by the statutory requirements of the professional registration laws of the state or jurisdiction in which the project is to be constructed.

RELIEF VALVE.

Pressure relief valve. A pressure-actuated valve held closed by a spring or other means and designed to relieve pressure automatically at the pressure at which such valve is set.

Temperature and pressure relief (T&P) valve. A combination relief valve designed to function as both a temperature relief and a pressure relief valve.

Temperature relief valve. A temperature-actuated valve designed to discharge automatically at the temperature at which such valve is set.

RELIEF VENT. A vent whose primary function is to provide circulation of air between drainage and vent systems.

RIM. An unobstructed open edge of a fixture.

RISER. See "Water pipe, riser."

ROOF DRAIN. A drain installed to receive water collecting on the surface of a roof and to discharge such water into a leader or a conductor.

ROUGH-IN. Parts of the plumbing system that are installed prior to the installation of fixtures. This includes drainage, water supply, vent piping and the necessary fixture supports and any fixtures that are built into the structure.

SELF-CLOSING FAUCET. A faucet containing a valve that automatically closes upon deactivation of the opening means.

SEPARATOR. See "Interceptor."

SEWAGE. Any liquid waste containing animal or vegetable matter in suspension or solution, including liquids containing chemicals in solution.

SEWAGE EJECTORS. A device for lifting sewage by entraining the sewage in a high-velocity jet of steam, air or water.

SEWER.

Building sewer. See "Building sewer."

Public sewer. A common *sewer* directly controlled by public authority.

Sanitary sewer. A *sewer* that carries sewage and excludes storm, surface and ground water.

Storm sewer. A *sewer* that conveys rainwater, surface water, subsurface water and similar liquid wastes.

SLOPE. The fall (pitch) of a line of pipe in reference to a horizontal plane. In drainage, the slope is expressed as the fall in units vertical per units horizontal (percent) for a length of pipe.

SOIL PIPE. A pipe that conveys sewage containing fecal matter to the *building drain* or *building sewer*.

SPILLPROOF VACUUM BREAKER. An assembly consisting of one check valve force-loaded closed and an air-inlet vent valve force-loaded open to atmosphere, positioned downstream of the check valve, and located between and including two tightly closing shutoff valves and a test cock.

STACK. A general term for any vertical line of soil, waste, vent or inside conductor piping that extends through at least one story with or without offsets.

STACK VENT. The extension of a soil or waste *stack* above the highest horizontal drain connected to the *stack*.

STACK VENTING. A method of venting a fixture or fixtures through the soil or waste *stack*.

STERILIZER.

Boiling type. A boiling-type sterilizer is a fixture of a nonpressure type utilized for boiling instruments, utensils or other equipment for disinfection. These devices are portable or are connected to the plumbing system.

Instrument. A device for the sterilization of various instruments.

Pressure (autoclave). A pressure vessel fixture designed to utilize steam under pressure for sterilizing.

Pressure instrument washer sterilizer. A pressure instrument washer sterilizer is a pressure vessel fixture designed to both wash and sterilize instruments during the operating cycle of the fixture.

Utensil. A device for the sterilization of utensils as utilized in health care services.

Water. A water sterilizer is a device for sterilizing water and storing sterile water.

STERILIZER VENT. A separate pipe or *stack*, indirectly connected to the building drainage system at the lower terminal, that receives the vapors from nonpressure sterilizers, or the exhaust vapors from pressure sterilizers, and conducts the vapors directly to the open air. Also called vapor, steam, atmospheric or exhaust vent.

STORM DRAIN. See "Drainage system, storm."

[A] STRUCTURE. That which is built or constructed or a portion thereof.

SUBSOIL DRAIN. A drain that collects subsurface water or seepage water and conveys such water to a place of disposal.

SUMP. A tank or pit that receives sewage or liquid waste, located below the normal grade of the gravity system and that must be emptied by mechanical means.

SUMP PUMP. An automatic water pump powered by an electric motor for the removal of drainage, except raw sewage, from a sump, pit or low point.

SUMP VENT. A vent from pneumatic sewage ejectors, or similar equipment, that terminates separately to the open air.

SUPPORTS. Devices for supporting and securing pipe, fixtures and equipment.

SWIMMING POOL. Any structure, basin, chamber or tank containing an artificial body of water for swimming, diving or recreational bathing having a depth of 2 feet (610 mm) or more at any point.

TEMPERED WATER. Water having a temperature range between 85°F (29°C) and 110°F (43°C).

THIRD-PARTY CERTIFICATION AGENCY. An *approved* agency operating a product or material certification sytem that incorporates initial product testing, assessment and surveillance of a manufacturer's quality control system.

THIRD-PARTY CERTIFIED. Certification obtained by the manufacturer indicating that the function and performance characteristics of a product or material have been determined by testing and ongoing surveillance by an *approved third-party certification agency*. Assertion of certification is in the form of identification in accordance with the requirements of the *third-party certification agency*.

THIRD-PARTY TESTED. Procedure by which an *approved* testing laboratory provides documentation that a product, material or system conforms to specified requirements.

TRAP. A fitting or device that provides a liquid seal to prevent the emission of *sewer* gases without materially affecting the flow of sewage or waste water through the trap.

TRAP SEAL. The vertical distance between the weir and the top of the dip of the trap.

UNSTABLE GROUND. Earth that does not provide a uniform bearing for the barrel of the *sewer* pipe between the joints at the bottom of the pipe trench.

VACUUM. Any pressure less than that exerted by the atmosphere.

VACUUM BREAKER. A type of backflow preventer installed on openings subject to normal atmospheric pressure that prevents backflow by admitting atmospheric pressure through ports to the discharge side of the device.

VENT PIPE. See "Vent system."

VENT STACK. A vertical vent pipe installed primarily for the purpose of providing circulation of air to and from any part of the drainage system.

VENT SYSTEM. A pipe or pipes installed to provide a flow of air to or from a drainage system, or to provide a circulation of air within such system to protect trap seals from siphonage and backpressure.

VERTICAL PIPE. Any pipe or fitting that makes an angle of 45 degrees (0.79 rad) or more with the horizontal.

WALL-HUNG WATER CLOSET. A wall-mounted water closet installed in such a way that the fixture does not touch the floor.

WASTE. The discharge from any fixture, appliance, area or appurtenance that does not contain fecal matter.

WASTE PIPE. A pipe that conveys only waste.

WATER-HAMMER ARRESTOR. A device utilized to absorb the pressure surge (water hammer) that occurs when water flow is suddenly stopped in a water supply system.

WATER HEATER. Any heating appliance or equipment that heats potable water and supplies such water to the potable *hot water* distribution system.

WATER MAIN. A water supply pipe or system of pipes, installed and maintained by a city, township, county, public utility company or other public entity, on public property, in the street or in an *approved* dedicated easement of public or community use.

WATER OUTLET. A discharge opening through which water is supplied to a fixture, into the atmosphere (except into an open tank that is part of the water supply system), to a boiler or heating system, or to any devices or equipment requiring water to operate but which are not part of the plumbing system.

WATER PIPE.

> **Riser.** A water supply pipe that extends one full story or more to convey water to branches or to a group of fixtures.

> **Water distribution pipe.** A pipe within the structure or on the premises that conveys water from the water service pipe, or from the meter when the meter is at the structure, to the points of utilization.

> **Water service pipe.** The pipe from the water main or other source of potable water supply, or from the meter when the meter is at the public right of way, to the water distribution system of the building served.

WATER SUPPLY SYSTEM. The water service pipe, water distribution pipes, and the necessary connecting pipes, fittings, control valves and all appurtenances in or adjacent to the structure or premises.

WELL.

> **Bored.** A well constructed by boring a hole in the ground with an auger and installing a casing.

> **Drilled.** A well constructed by making a hole in the ground with a drilling machine of any type and installing casing and screen.

> **Driven.** A well constructed by driving a pipe in the ground. The drive pipe is usually fitted with a well point and screen.

> **Dug.** A well constructed by excavating a large-diameter shaft and installing a casing.

WHIRLPOOL BATHTUB. A plumbing appliance consisting of a bathtub fixture that is equipped and fitted with a circulating piping system designed to accept, circulate and discharge bathtub water upon each use.

YOKE VENT. A pipe connecting upward from a soil or waste *stack* to a vent *stack* for the purpose of preventing pressure changes in the stacks.

CHAPTER 3
GENERAL REGULATIONS

SECTION 301
GENERAL

301.1 Scope. The provisions of this chapter shall govern the general regulations regarding the installation of plumbing not specific to other chapters.

301.2 System installation. Plumbing shall be installed with due regard to preservation of the strength of structural members and prevention of damage to walls and other surfaces through fixture usage.

301.3 Connections to drainage system. Plumbing fixtures, drains, appurtenances and appliances used to receive or discharge liquid wastes or sewage shall be directly connected to the sanitary drainage system of the building or premises, in accordance with the requirements of this code. This section shall not be construed to prevent indirect waste systems required by Chapter 8.

> **Exception:** Bathtubs, showers, lavatories, clothes washers and laundry trays shall not be required to discharge to the sanitary drainage system where such fixtures discharge to an approved gray water system for flushing of water closets and urinals or for subsurface landscape irrigation.

301.4 Connections to water supply. Every plumbing fixture, device or appliance requiring or using water for its proper operation shall be directly or indirectly connected to the water supply system in accordance with the provisions of this code.

301.5 Pipe, tube and fitting sizes. Unless otherwise specified, the pipe, tube and fitting sizes specified in this code are expressed in nominal or standard sizes as designated in the referenced material standards.

301.6 Prohibited locations. Plumbing systems shall not be located in an elevator shaft or in an elevator equipment room.

> **Exception:** Floor drains, sumps and sump pumps shall be permitted at the base of the shaft, provided that they are indirectly connected to the plumbing system and comply with Section 1003.4.

301.7 Conflicts. In instances where conflicts occur between this code and the manufacturer's installation instructions, the more restrictive provisions shall apply.

SECTION 302
EXCLUSION OF MATERIALS DETRIMENTAL TO THE SEWER SYSTEM

302.1 Detrimental or dangerous materials. Ashes, cinders or rags; flammable, poisonous or explosive liquids or gases; oil, grease or any other insoluble material capable of obstructing, damaging or overloading the building drainage or *sewer* system, or capable of interfering with the normal operation of the sewage treatment processes, shall not be deposited, by any means, into such systems.

302.2 Industrial wastes. Waste products from manufacturing or industrial operations shall not be introduced into the *public sewer* until it has been determined by the code official or other authority having jurisdiction that the introduction thereof will not damage the *public sewer* system or interfere with the functioning of the sewage treatment plant.

SECTION 303
MATERIALS

303.1 Identification. Each length of pipe and each pipe fitting, trap, fixture, material and device utilized in a plumbing system shall bear the identification of the manufacturer and any markings required by the applicable referenced standards.

303.2 Installation of materials. All materials used shall be installed in strict accordance with the standards under which the materials are accepted and *approved*. In the absence of such installation procedures, the manufacturer's instructions shall be followed. Where the requirements of referenced standards or manufacturer's installation instructions do not conform to minimum provisions of this code, the provisions of this code shall apply.

303.3 Plastic pipe, fittings and components. All plastic pipe, fittings and components shall be third-party certified as conforming to NSF 14.

303.4 Third-party certification. All plumbing products and materials shall be listed by a third-party certification agency as complying with the referenced standards. Products and materials shall be identified in accordance with Section 303.1.

SECTION 304
RODENTPROOFING

304.1 General. Plumbing systems shall be designed and installed in accordance with Sections 304.2 through 304.4 to prevent rodents from entering structures.

304.2 Strainer plates. All strainer plates on drain inlets shall be designed and installed so that all openings are not greater than $^1/_2$ inch (12.7 mm) in least dimension.

304.3 Meter boxes. Meter boxes shall be constructed in such a manner that rodents are prevented from entering a structure by way of the water service pipes connecting the meter box and the structure.

304.4 Openings for pipes. In or on structures where openings have been made in walls, floors or ceilings for the passage of pipes, the annular space between the pipe and the sides of the opening shall be sealed with caulking materials or closed with gasketing systems compatible with the piping materials and locations.

SECTION 305
PROTECTION OF PIPES AND
PLUMBING SYSTEM COMPONENTS

305.1 Corrosion. Pipes passing through concrete or cinder walls and floors or other corrosive material shall be protected against external corrosion by a protective sheathing or wrapping or other means that will withstand any reaction from the lime and acid of concrete, cinder or other corrosive material. Sheathing or wrapping shall allow for movement including expansion and contraction of piping. The wall thickness of the material shall be not less than 0.025 inch (0.64 mm).

305.2 Stress and strain. Piping in a plumbing system shall be installed so as to prevent strains and stresses that exceed the structural strength of the pipe. Where necessary, provisions shall be made to protect piping from damage resulting from expansion, contraction and structural settlement.

305.3 Pipes through foundation walls. Any pipe that passes through a foundation wall shall be provided with a relieving arch, or a pipe sleeve pipe shall be built into the foundation wall. The sleeve shall be two pipe sizes greater than the pipe passing through the wall.

305.4 Freezing. Water, soil and waste pipes shall not be installed outside of a building, in attics or crawl spaces, concealed in outside walls, or in any other place subjected to freezing temperatures unless adequate provision is made to protect such pipes from freezing by insulation or heat or both. Exterior water supply system piping shall be installed not less than 6 inches (152 mm) below the frost line and not less than 12 inches (305 mm) below grade.

305.4.1 Sewer depth. Building sewers that connect to private sewage disposal systems shall be installed not less than [NUMBER] inches (mm) below finished grade at the point of septic tank connection. Building sewers shall be installed not less than [NUMBER] inches (mm) below grade.

305.5 Waterproofing of openings. Joints at the roof and around vent pipes, shall be made water-tight by the use of lead, copper, galvanized steel, aluminum, plastic or other *approved* flashings or flashing material. Exterior wall openings shall be made water-tight.

305.6 Protection against physical damage. In concealed locations where piping, other than cast-iron or galvanized steel, is installed through holes or notches in studs, joists, rafters or similar members less than $1^1/_2$ inches (38 mm) from the nearest edge of the member, the pipe shall be protected by steel shield plates. Such shield plates shall have a thickness of not less than 0.0575 inch (1.463 mm) (No. 16 gage). Such plates shall cover the area of the pipe where the member is notched or bored, and shall extend not less than 2 inches (51 mm) above sole plates and below top plates.

305.7 Protection of components of plumbing system. Components of a plumbing system installed along alleyways, driveways, parking garages or other locations exposed to damage shall be recessed into the wall or otherwise protected in an *approved* manner.

SECTION 306
TRENCHING, EXCAVATION AND BACKFILL

306.1 Support of piping. Buried piping shall be supported throughout its entire length.

306.2 Trenching and bedding. Where trenches are excavated such that the bottom of the trench forms the bed for the pipe, solid and continuous load-bearing support shall be provided between joints. Bell holes, hub holes and coupling holes shall be provided at points where the pipe is joined. Such pipe shall not be supported on blocks to grade. In instances where the materials manufacturer's installation instructions are more restrictive than those prescribed by the code, the material shall be installed in accordance with the more restrictive requirement.

306.2.1 Overexcavation. Where trenches are excavated below the installation level of the pipe such that the bottom of the trench does not form the bed for the pipe, the trench shall be backfilled to the installation level of the bottom of the pipe with sand or fine gravel placed in layers not greater than 6 inches (152 mm) in depth and such backfill shall be compacted after each placement.

306.2.2 Rock removal. Where rock is encountered in trenching, the rock shall be removed to not less than 3 inches (76 mm) below the installation level of the bottom of the pipe, and the trench shall be backfilled to the installation level of the bottom of the pipe with sand tamped in place so as to provide uniform load-bearing support for the pipe between joints. The pipe, including the joints, shall not rest on rock at any point.

306.2.3 Soft load-bearing materials. If soft materials of poor load-bearing quality are found at the bottom of the trench, stabilization shall be achieved by overexcavating not less than two pipe diameters and backfilling to the installation level of the bottom of the pipe with fine gravel, crushed stone or a concrete foundation. The concrete foundation shall be bedded with sand tamped into place so as to provide uniform load-bearing support for the pipe between joints.

306.3 Backfilling. Backfill shall be free from discarded construction material and debris. Loose earth free from rocks, broken concrete and frozen chunks shall be placed in the trench in 6-inch (152 mm) layers and tamped in place until the crown of the pipe is covered by 12 inches (305 mm) of tamped earth. The backfill under and beside the pipe shall be compacted for pipe support. Backfill shall be brought up evenly on both sides of the pipe so that the pipe remains aligned. In instances where the manufacturer's instructions for materials are more restrictive than those prescribed by the code, the material shall be installed in accordance with the more restrictive requirement.

306.4 Tunneling. Where pipe is to be installed by tunneling, jacking or a combination of both, the pipe shall be protected from damage during installation and from subsequent uneven loading. Where earth tunnels are used, adequate supporting structures shall be provided to prevent future settling or caving.

SECTION 307
STRUCTURAL SAFETY

307.1 General. In the process of installing or repairing any part of a plumbing and drainage installation, the finished floors, walls, ceilings, tile work or any other part of the building or premises that must be changed or replaced shall be left in a safe structural condition in accordance with the requirements of the *International Building Code.*

307.2 Cutting, notching or bored holes. A framing member shall not be cut, notched or bored in excess of limitations specified in the *International Building Code.*

307.3 Penetrations of floor/ceiling assemblies and fire-resistance-rated assemblies. Penetrations of floor/ceiling assemblies and assemblies required to have a fire-resistance rating shall be protected in accordance with the *International Building Code.*

[B] 307.4 Alterations to trusses. Truss members and components shall not be cut, drilled, notched, spliced or otherwise altered in any way without written concurrence and approval of a registered design professional. Alterations resulting in the addition of loads to any member (e.g., HVAC equipment, water heater) shall not be permitted without verification that the truss is capable of supporting such additional loading.

307.5 Trench location. Trenches installed parallel to footings shall not extend below the 45-degree (0.79 rad) bearing plane of the footing or wall.

307.6 Piping materials exposed within plenums. All piping materials exposed within plenums shall comply with the provisions of the *International Mechanical Code.*

SECTION 308
PIPING SUPPORT

308.1 General. All plumbing piping shall be supported in accordance with this section.

308.2 Piping seismic supports. Where earthquake loads are applicable in accordance with the building code, plumbing piping supports shall be designed and installed for the seismic forces in accordance with the *International Building Code.*

308.3 Materials. Hangers, anchors and supports shall support the piping and the contents of the piping. Hangers and strapping material shall be of *approved* material that will not promote galvanic action.

308.4 Structural attachment. Hangers and anchors shall be attached to the building construction in an *approved* manner.

308.5 Interval of support. Pipe shall be supported in accordance with Table 308.5.

> **Exception:** The interval of support for piping systems designed to provide for expansion/contraction shall conform to the engineered design in accordance with Section 105.4.

308.6 Sway bracing. Rigid support sway bracing shall be provided at changes in direction greater than 45 degrees (0.79 rad) for pipe sizes 4 inches (102 mm) and larger.

TABLE 308.5
HANGER SPACING

PIPING MATERIAL	MAXIMUM HORIZONTAL SPACING (feet)	MAXIMUM VERTICAL SPACING (feet)
ABS pipe	4	10[b]
Aluminum tubing	10	15
Brass pipe	10	10
Cast-iron pipe	5[a]	15
Copper or copper-alloy pipe	12	10
Copper or copper-alloy tubing, 1¼-inch diameter and smaller	6	10
Copper or copper-alloy tubing, 1½-inch diameter and larger	10	10
Cross-linked polyethylene (PEX) pipe	2.67 (32 inches)	10[b]
Cross-linked polyethylene/aluminum/cross-linked polyethylene (PEX-AL-PEX) pipe	2.67 (32 inches)	4
CPVC pipe or tubing, 1 inch and smaller	3	10[b]
CPVC pipe or tubing, 1¼ inches and larger	4	10[b]
Steel pipe	12	15
Lead pipe	Continuous	4
Polyethylene/aluminum/polyethylene (PE-AL-PE) pipe	2.67 (32 inches)	4
Polyethylene of raised temperature (PE-RT) pipe	2.67 (32 inches)	10[b]
Polypropylene (PP) pipe or tubing 1 inch and smaller	2.67 (32 inches)	10[b]
Polypropylene (PP) pipe or tubing, 1¼ inches and larger	4	10[b]
PVC pipe	4	10[b]
Stainless steel drainage systems	10	10[b]

For SI: 1 inch = 25.4 mm, 1 foot = 304.8 mm.

a. The maximum horizontal spacing of cast-iron pipe hangers shall be increased to 10 feet where 10-foot lengths of pipe are installed.

b. Midstory guide for sizes 2 inches and smaller.

308.7 Anchorage. Anchorage shall be provided to restrain drainage piping from axial movement.

> **308.7.1 Location.** For pipe sizes greater than 4 inches (102 mm), restraints shall be provided for drain pipes at all changes in direction and at all changes in diameter greater than two pipe sizes. Braces, blocks, rodding and other suitable methods as specified by the coupling manufacturer shall be utilized.

308.8 Expansion joint fittings. Expansion joint fittings shall be used only where necessary to provide for expansion and contraction of the pipes. Expansion joint fittings shall be of the typical material suitable for use with the type of piping in which such fittings are installed.

308.9 Parallel water distribution systems. Piping bundles for manifold systems shall be supported in accordance with Table 308.5. Support at changes in direction shall be in accordance with the manufacturer's instructions. Where hot water piping is bundled with cold or hot water piping, each hot water pipe shall be insulated.

SECTION 309
FLOOD HAZARD RESISTANCE

309.1 General. Plumbing systems and equipment in structures erected in flood hazard areas shall be constructed in accordance with the requirements of this section and the *International Building Code*.

[B] 309.2 Flood hazard. For structures located in flood hazard areas, the following systems and equipment shall be located and installed as required by Section 1612 of the *International Building Code*.

> **Exception:** The following systems are permitted to be located below the elevation required by Section 1612 of the *International Building Code* for utilities and attendant equipment provided that the systems are designed and installed to prevent water from entering or accumulating within their components and the systems are constructed to resist hydrostatic and hydrodynamic loads and stresses, including the effects of buoyancy, during the occurrence of flooding to up to such elevation.
>
> 1. All water service pipes.
> 2. Pump seals in individual water supply systems where the pump is located below the *design flood elevation*.
> 3. Covers on potable water wells shall be sealed, except where the top of the casing well or pipe sleeve is elevated to not less than 1 foot (305 mm) above the *design flood elevation*.
> 4. All sanitary drainage piping.
> 5. All storm drainage piping.
> 6. Manhole covers shall be sealed, except where elevated to or above the *design flood elevation*.
> 7. All other plumbing fixtures, faucets, fixture fittings, piping systems and equipment.
> 8. Water heaters.
> 9. Vents and vent systems.

[B] 309.3 Flood hazard areas subject to high-velocity wave action. Structures located in flood hazard areas subject to high-velocity wave action shall meet the requirements of Section 309.2. The plumbing systems, pipes and fixtures shall not be mounted on or penetrate through walls intended to break away under flood loads.

SECTION 310
WASHROOM AND TOILET ROOM REQUIREMENTS

310.1 Light and ventilation. Washrooms and toilet rooms shall be illuminated and ventilated in accordance with the *International Building Code* and *International Mechanical Code*.

310.2 Location of fixtures and compartments. The location of plumbing fixtures and the requirements for compartments and partitions shall be in accordance with Section 405.3.

310.3 Interior finish. Interior finish surfaces of toilet rooms shall comply with the *International Building Code*.

SECTION 311
TOILET FACILITIES FOR WORKERS

311.1 General. Toilet facilities shall be provided for construction workers and such facilities shall be maintained in a sanitary condition. Construction worker toilet facilities of the nonsewer type shall conform to ANSI Z4.3.

SECTION 312
TESTS AND INSPECTIONS

312.1 Required tests. The permit holder shall make the applicable tests prescribed in Sections 312.2 through 312.10 to determine compliance with the provisions of this code. The permit holder shall give reasonable advance notice to the code official when the plumbing work is ready for tests. The equipment, material, power and labor necessary for the inspection and test shall be furnished by the permit holder and the permit holder shall be responsible for determining that the work will withstand the test pressure prescribed in the following tests. All plumbing system piping shall be tested with either water or, for piping systems other than plastic, by air. After the plumbing fixtures have been set and their traps filled with water, the entire drainage system shall be submitted to final tests. The code official shall require the removal of any cleanouts if necessary to ascertain whether the pressure has reached all parts of the system.

312.1.1 Test gauges. Gauges used for testing shall be as follows:

1. Tests requiring a pressure of 10 pounds per square inch (psi) (69 kPa) or less shall utilize a testing gauge having increments of 0.10 psi (0.69 kPa) or less.
2. Tests requiring a pressure of greater than 10 psi (69 kPa) but less than or equal to 100 psi (689 kPa) shall utilize a testing gauge having increments of 1 psi (6.9 kPa) or less.
3. Tests requiring a pressure of greater than 100 psi (689 kPa) shall utilize a testing gauge having increments of 2 psi (14 kPa) or less.

312.2 Drainage and vent water test. A water test shall be applied to the drainage system either in its entirety or in sections. If applied to the entire system, all openings in the piping shall be tightly closed, except the highest opening, and the system shall be filled with water to the point of overflow.

If the system is tested in sections, each opening shall be tightly plugged except the highest openings of the section under test, and each section shall be filled with water, but no section shall be tested with less than a 10-foot (3048 mm) head of water. In testing successive sections, at least the upper 10 feet (3048 mm) of the next preceding section shall be tested so that no joint or pipe in the building, except the uppermost 10 feet (3048 mm) of the system, shall have been submitted to a test of less than a 10-foot (3048 mm) head of water. This pressure shall be held for not less than 15 minutes. The system shall then be tight at all points.

312.3 Drainage and vent air test. Plastic piping shall not be tested using air. An air test shall be made by forcing air into the system until there is a uniform gauge pressure of 5 psi (34.5 kPa) or sufficient to balance a 10-inch (254 mm) column of mercury. This pressure shall be held for a test period of not less than 15 minutes. Any adjustments to the test pressure required because of changes in ambient temperatures or the seating of gaskets shall be made prior to the beginning of the test period.

312.4 Drainage and vent final test. The final test of the completed drainage and vent systems shall be visual and in sufficient detail to determine compliance with the provisions of this code. Where a smoke test is utilized, it shall be made by filling all traps with water and then introducing into the entire system a pungent, thick smoke produced by one or more smoke machines. When the smoke appears at *stack* openings on the roof, the *stack* openings shall be closed and a pressure equivalent to a 1-inch water column (248.8 Pa) shall be held for a test period of not less than 15 minutes.

312.5 Water supply system test. Upon completion of a section of or the entire water supply system, the system, or portion completed, shall be tested and proved tight under a water pressure not less than the working pressure of the system; or, for piping systems other than plastic, by an air test of not less than 50 psi (344 kPa). This pressure shall be held for not less than 15 minutes. The water utilized for tests shall be obtained from a potable source of supply. The required tests shall be performed in accordance with this section and Section 107.

312.6 Gravity sewer test. Gravity *sewer* tests shall consist of plugging the end of the *building sewer* at the point of connection with the *public sewer*, filling the *building sewer* with water, testing with not less than a 10-foot (3048 mm) head of water and maintaining such pressure for 15 minutes.

312.7 Forced sewer test. Forced *sewer* tests shall consist of plugging the end of the *building sewer* at the point of connection with the *public sewer* and applying a pressure of 5 psi (34.5 kPa) greater than the pump rating, and maintaining such pressure for 15 minutes.

312.8 Storm drainage system test. *Storm drain* systems within a building shall be tested by water or air in accordance with Section 312.2 or 312.3.

312.9 Shower liner test. Where shower floors and receptors are made water-tight by the application of materials required by Section 417.5.2, the completed liner installation shall be tested. The pipe from the shower drain shall be plugged water tight for the test. The floor and receptor area shall be filled with potable water to a depth of not less than 2 inches (51

mm) measured at the threshold. Where a threshold of at least 2 inches (51 mm) high does not exist, a temporary threshold shall be constructed to retain the test water in the lined floor or receptor area to a level not less than 2 inches (51 mm) deep measured at the threshold. The water shall be retained for a test period of not less than 15 minutes, and there shall not be evidence of leakage.

312.10 Inspection and testing of backflow prevention assemblies. Inspection and testing shall comply with Sections 312.10.1 and 312.10.2.

312.10.1 Inspections. Annual inspections shall be made of all backflow prevention assemblies and air gaps to determine whether they are operable.

312.10.2 Testing. Reduced pressure principle, double check, pressure vacuum breaker, reduced pressure detector fire protection, double check detector fire protection, and spill-resistant vacuum breaker backflow preventer assemblies and hose connection backflow preventers shall be tested at the time of installation, immediately after repairs or relocation and at least annually. The testing procedure shall be performed in accordance with one of the following standards: ASSE 5013, ASSE 5015, ASSE 5020, ASSE 5047, ASSE 5048, ASSE 5052, ASSE 5056, CSA B64.10 or CSA B64.10.1.

SECTION 313
EQUIPMENT EFFICIENCIES

313.1 General. Equipment efficiencies shall be in accordance with the *International Energy Conservation Code*.

SECTION 314
CONDENSATE DISPOSAL

[M] 314.1 Fuel-burning appliances. Liquid combustion by-products of condensing appliances shall be collected and discharged to an *approved* plumbing fixture or disposal area in accordance with the manufacturer's instructions. Condensate piping shall be of *approved* corrosion-resistant material and shall not be smaller than the drain connection on the appliance. Such piping shall maintain a horizontal slope in the direction of discharge of not less than one-eighth unit vertical in 12 units horizontal (1-percent slope).

[M] 314.2 Evaporators and cooling coils. Condensate drain systems shall be provided for equipment and appliances containing evaporators or cooling coils. Condensate drain systems shall be designed, constructed and installed in accordance with Sections 314.2.1 through 314.2.4.

[M] 314.2.1 Condensate disposal. Condensate from all cooling coils and evaporators shall be conveyed from the drain pan outlet to an *approved* place of disposal. Such piping shall maintain a horizontal slope in the direction of discharge of not less than one-eighth unit vertical in 12 units horizontal (1-percent slope). Condensate shall not discharge into a street, alley or other areas so as to cause a nuisance.

[M] 314.2.2 Drain pipe materials and sizes. Components of the condensate disposal system shall be cast iron, galvanized steel, copper, cross-linked polyethylene, polybutylene, polyethylene, ABS, CPVC or PVC pipe or tubing. All components shall be selected for the pressure and temperature rating of the installation. Joints and connections shall be made in accordance with the applicable provisions of Chapter 7 relative to the material type. Condensate waste and drain line size shall be not less than $^3/_4$-inch (19 mm) internal diameter and shall not decrease in size from the drain pan connection to the place of condensate disposal. Where the drain pipes from more than one unit are manifolded together for condensate drainage, the pipe or tubing shall be sized in accordance with Table 314.2.2.

[M] TABLE 314.2.2
CONDENSATE DRAIN SIZING

EQUIPMENT CAPACITY	MINIMUM CONDENSATE PIPE DIAMETER
Up to 20 tons of refrigeration	$^3/_4$ inch
Over 20 tons to 40 tons of refrigeration	1 inch
Over 40 tons to 90 tons of refrigeration	$1^1/_4$ inch
Over 90 tons to 125 tons of refrigeration	$1^1/_2$ inch
Over 125 tons to 250 tons of refrigeration	2 inch

For SI: 1 inch = 25.4 mm, 1 ton of capacity = 3.517 kW.

[M] 314.2.3 Auxiliary and secondary drain systems. In addition to the requirements of Section 314.2.1, where damage to any building components could occur as a result of overflow from the equipment primary condensate removal system, one of the following auxiliary protection methods shall be provided for each cooling coil or fuel-fired appliance that produces condensate:

1. An auxiliary drain pan with a separate drain shall be provided under the coils on which condensation will occur. The auxiliary pan drain shall discharge to a conspicuous point of disposal to alert occupants in the event of a stoppage of the primary drain. The pan shall have a depth of not less than $1^1/_2$ inches (38 mm), shall be not less than 3 inches (76 mm) larger than the unit or the coil dimensions in width and length and shall be constructed of corrosion-resistant material. Galvanized sheet metal pans shall have a thickness of not less than 0.0236-inch (0.6010 mm) (No. 24 gage) galvanized sheet metal. Nonmetallic pans shall have a thickness of not less than 0.0625 inch (1.6 mm).

2. A separate overflow drain line shall be connected to the drain pan provided with the equipment. Such overflow drain shall discharge to a conspicuous point of disposal to alert occupants in the event of a stoppage of the primary drain. The overflow drain line shall connect to the drain pan at a higher level than the primary drain connection.

3. An auxiliary drain pan without a separate drain line shall be provided under the coils on which condensate will occur. Such pan shall be equipped with a water-level detection device conforming to UL 508 that will shut off the equipment served prior to overflow of the pan. The auxiliary drain pan shall be

constructed in accordance with Item 1 of this section.

4. A water-level detection device conforming to UL 508 shall be provided that will shut off the equipment served in the event that the primary drain is blocked. The device shall be installed in the primary drain line, the overflow drain line, or in the equipment-supplied drain pan, located at a point higher than the primary drain line connection and below the overflow rim of such pan.

   Exception: Fuel-fired appliances that automatically shut down operation in the event of a stoppage in the condensate drainage system.

[M] 314.2.3.1 Water-level monitoring devices. On down-flow units and all other coils that do not have a secondary drain or provisions to install a secondary or auxiliary drain pan, a water-level monitoring device shall be installed inside the primary drain pan. This device shall shut off the equipment served in the event that the primary drain becomes restricted. Devices installed in the drain line shall not be permitted.

[M] 314.2.3.2 Appliance, equipment and insulation in pans. Where appliances, equipment or insulation are subject to water damage when auxiliary drain pans fill such portions of the appliances, equipment and insulation shall be installed above the *flood level rim* of the pan. Supports located inside of the pan to support the appliance or equipment shall be water resistant and *approved*.

[M] 314.2.4 Traps. Condensate drains shall be trapped as required by the equipment or appliance manufacturer.

SECTION 315
PENETRATIONS

315.1 Sealing of annular spaces. The annular space between the outside of a pipe and the inside of a pipe sleeve or between the outside of a pipe and an opening in a building envelope wall, floor, or ceiling assembly penetrated by a pipe shall be sealed in an approved manner with caulking material, foam sealant or closed with a gasketing system. The caulking material, foam sealant or gasketing system shall be designed for the conditions at the penetration location and shall be compatible with the pipe, sleeve and building materials in contact with the sealing materials. Annular spaces created by pipes penetrating fire-resistance-rated assemblies or membranes of such assemblies shall be sealed or closed in accordance with Section 713 of the *International Building Code.*

SECTION 316
ALTERNATIVE ENGINEERED DESIGN

316.1 Alternative engineered design. The design, documentation, inspection, testing and approval of an *alternative engineered design* plumbing system shall comply with Sections 316.1.1 through 316.1.6.

316.1.1 Design criteria. An *alternative engineered design* shall conform to the intent of the provisions of this code and shall provide an equivalent level of quality, strength, effectiveness, fire resistance, durability and safety. Material, equipment or components shall be designed and installed in accordance with the manufacturer's instructions.

✳✳ **316.1.2 Submittal.** The registered design professional shall indicate on the permit application that the plumbing system is an *alternative engineered design*. The permit and permanent permit records shall indicate that an *alternative engineered design* was part of the *approved* installation.

✳✳ **316.1.3 Technical data.** The registered design professional shall submit sufficient technical data to substantiate the proposed *alternative engineered design* and to prove that the performance meets the intent of this code.

✳✳ **316.1.4 Construction documents.** The registered design professional shall submit to the code official two complete sets of signed and sealed construction documents for the *alternative engineering design* shall include floor plans and a riser diagram of the work. Where appropriate, the construction documents shall indicate the direction of flow, all pipe sizes, grade of horizontal piping, loading, and location of fixtures and appliances.

✳✳ **316.1.5 Design approval.** Where the code official determines that the *alternative engineered design* conforms to the intent of this code, the plumbing system shall be *approved*. If the *alternative engineered design* is not *approved*, the code official shall notify the registered design professional in writing, stating the reasons thereof.

✳✳ **316.1.6 Inspection and testing.** The *alternative engineered design* shall be tested and inspected in accordance with the requirements of Sections 107 and 312.

CHAPTER 4

FIXTURES, FAUCETS AND FIXTURE FITTINGS

SECTION 401
GENERAL

401.1 Scope. This chapter shall govern the materials, design and installation of plumbing fixtures, faucets and fixture fittings in accordance with the type of *occupancy*, and shall provide for the minimum number of fixtures for various types of occupancies.

401.2 Prohibited fixtures and connections. Water closets having a concealed trap seal or an unventilated space or having walls that are not thoroughly washed at each discharge in accordance with ASME A112.19.2/CSA B45.1 shall be prohibited. Any water closet that permits siphonage of the contents of the bowl back into the tank shall be prohibited. Trough urinals shall be prohibited.

401.3 Water conservation. The maximum water flow rates and flush volume for plumbing fixtures and fixture fittings shall comply with Section 604.4.

SECTION 402
FIXTURE MATERIALS

402.1 Quality of fixtures. Plumbing fixtures shall be constructed of *approved* materials, with smooth, impervious surfaces, free from defects and concealed fouling surfaces, and shall conform to standards cited in this code. All porcelain enameled surfaces on plumbing fixtures shall be acid resistant.

402.2 Materials for specialty fixtures. Materials for specialty fixtures not otherwise covered in this code shall be of stainless steel, soapstone, chemical stoneware or plastic, or shall be lined with lead, copper-base alloy, nickel-copper alloy, corrosion-resistant steel or other material especially suited to the application for which the fixture is intended.

402.3 Sheet copper. Sheet copper for general applications shall conform to ASTM B 152 and shall not weigh less than 12 ounces per square foot (3.7 kg/m^2).

402.4 Sheet lead. Sheet lead for pans shall not weigh less than 4 pounds per square foot (19.5 kg/m^2) coated with an asphalt paint or other *approved* coating.

SECTION 403
MINIMUM PLUMBING FACILITIES

403.1 Minimum number of fixtures. Plumbing fixtures shall be provided for the type of *occupancy* and in the minimum number shown in Table 403.1. Types of occupancies not shown in Table 403.1 shall be considered individually by the code official. The number of occupants shall be determined by the *International Building Code*. *Occupancy* classification shall be determined in accordance with the *International Building Code*.

TABLE 403.1
MINIMUM NUMBER OF REQUIRED PLUMBING FIXTURES[a]
(See Sections 403.2 and 403.3)

NO.	CLASSIFICATION	OCCUPANCY	DESCRIPTION	WATER CLOSETS (URINALS SEE SECTION 419.2)		LAVATORIES		BATHTUBS/ SHOWERS	DRINKING FOUNTAIN[e,f] (SEE SECTION 410.1)	OTHER
				MALE	FEMALE	MALE	FEMALE			
1	Assembly	A-1[d]	Theaters and other buildings for the performing arts and motion pictures	1 per 125	1 per 65	1 per 200		—	1 per 500	1 service sink
		A-2[d]	Nightclubs, bars, taverns, dance halls and buildings for similar purposes	1 per 40	1 per 40	1 per 75		—	1 per 500	1 service sink
			Restaurants, banquet halls and food courts	1 per 75	1 per 75	1 per 200		—	1 per 500	1 service sink
		A-3[d]	Auditoriums without permanent seating, art galleries, exhibition halls, museums, lecture halls, libraries, arcades and gymnasiums	1 per 125	1 per 65	1 per 200		—	1 per 500	1 service sink
			Passenger terminals and transportation facilities	1 per 500	1 per 500	1 per 750		—	1 per 1,000	1 service sink
			Places of worship and other religious services.	1 per 150	1 per 75	1 per 200		—	1 per 1,000	1 service sink

(continued)

TABLE 403.1 —continued
MINIMUM NUMBER OF REQUIRED PLUMBING FIXTURES[a]
(See Sections 403.2 and 403.3)

NO.	CLASSIFICATION	OCCUPANCY	DESCRIPTION	WATER CLOSETS (URINALS SEE SECTION 419.2)		LAVATORIES		BATHTUBS/ SHOWERS	DRINKING FOUNTAIN[e,f] (SEE SECTION 410.1)	OTHER
				MALE	FEMALE	MALE	FEMALE			
1 (cont.)	Assembly	A-4	Coliseums, arenas, skating rinks, pools and tennis courts for indoor sporting events and activities	1 per 75 for the first 1,500 and 1 per 120 for the remainder exceeding 1,500	1 per 40 for the first 1,520 and 1 per 60 for the remainder exceeding 1,520	1 per 200	1 per 150	—	1 per 1,000	1 service sink
		A-5	Stadiums, amusement parks, bleachers and grandstands for outdoor sporting events and activities	1 per 75 for the first 1,500 and 1 per 120 for the remainder exceeding 1,500	1 per 40 for the first 1,520 and 1 per 60 for the remainder exceeding 1,520	1 per 200	1 per 150	—	1 per 1,000	1 service sink
2	Business	B	Buildings for the transaction of business, professional services, other services involving merchandise, office buildings, banks, light industrial and similar uses	1 per 25 for the first 50 and 1 per 50 for the remainder exceeding 50		1 per 40 for the first 80 and 1 per 80 for the remainder exceeding 80		—	1 per 100	1 service sink[g]
3	Educational	E	Educational facilities	1 per 50		1 per 50		—	1 per 100	1 service sink
4	Factory and industrial	F-1 and F-2	Structures in which occupants are engaged in work fabricating, assembly or processing of products or materials	1 per 100		1 per 100		(see Section 411)	1 per 400	1 service sink
5	Institutional	I-1	Residential care	1 per 10		1 per 10		1 per 8	1 per 100	1 service sink
		I-2	Hospitals, ambulatory nursing home care recipient	1 per room[c]		1 per room[c]		1 per 15	1 per 100	1 service sink per floor
			Employees, other than residential care[b]	1 per 25		1 per 35		—	1 per 100	—
			Visitors, other than residential care	1 per 75		1 per 100		—	1 per 500	—
		I-3	Prisons[b]	1 per cell		1 per cell		1 per 15	1 per 100	1 service sink
			Reformitories, detention centers, and correctional centers[b]	1 per 15		1 per 15		1 per 15	1 per 100	1 service sink
			Employees[b]	1 per 25		1 per 35		—	1 per 100	—
		I-4	Adult day care and child day care	1 per 15		1 per 15		1	1 per 100	1 service sink

(continued)

TABLE 403.1 —continued
MINIMUM NUMBER OF REQUIRED PLUMBING FIXTURES[a]
(See Sections 403.2 and 403.3)

NO.	CLASSIFICATION	OCCUPANCY	DESCRIPTION	WATER CLOSETS (URINALS SEE SECTION 419.2)		LAVATORIES		BATHTUBS/ SHOWERS	DRINKING FOUNTAIN[e, f] (SEE SECTION 410.1)	OTHER
				MALE	FEMALE	MALE	FEMALE			
6	Mercantile	M	Retail stores, service stations, shops, salesrooms, markets and shopping centers	1 per 500		1 per 750		—	1 per 1,000	1 service sink[g]
7	Residential	R-1	Hotels, motels, boarding houses (transient)	1 per sleeping unit		1 per sleeping unit		1 per sleeping unit	—	1 service sink
		R-2	Dormitories, fraternities, sororities and boarding houses (not transient)	1 per 10		1 per 10		1 per 8	1 per 100	1 service sink
		R-2	Apartment house	1 per dwelling unit		1 per dwelling unit		1 per dwelling unit	—	1 kitchen sink per dwelling unit; 1 automatic clothes washer connection per 20 dwelling units
		R-3	Congregate living facilities with 16 or fewer persons	1 per 10		1 per 10		1 per 8	1 per 100	1 service sink
		R-3	One- and two-family dwellings	1 per dwelling unit		1 per dwelling unit		1 per dwelling unit	—	1 kitchen sink per dwelling unit; 1 automatic clothes washer connection per dwelling unit
		R-4	Congregate living facilities with 16 or fewer persons	1 per 10		1 per 10		1 per 8	1 per 100	1 service sink
8	Storage	S-1 S-2	Structures for the storage of goods, warehouses, storehouse and freight depots. Low and Moderate Hazard.	1 per 100		1 per 100		See Section411	1 per 1,000	1 service sink

a. The fixtures shown are based on one fixture being the minimum required for the number of persons indicated or any fraction of the number of persons indicated. The number of occupants shall be determined by the *International Building Code*.
b. Toilet facilities for employees shall be separate from facilities for inmates or care recipients.

(continued)

TABLE 403.1 —continued
MINIMUM NUMBER OF REQUIRED PLUMBING FIXTURES[a]
(See Sections 403.2 and 403.3)

c. A single-occupant toilet room with one water closet and one lavatory serving not more than two adjacent patient sleeping units shall be permitted where such room is provided with direct access from each patient sleeping unit and with provisions for privacy.

d. The occupant load for seasonal outdoor seating and entertainment areas shall be included when determining the minimum number of facilities required.

e. The minimum number of required drinking fountains shall comply with Table 403.1 and Chapter 11 of the *International Building Code.*

f. Drinking fountains are not required for an occupant load of 15 or fewer.

g. For business and mercantile occupancies with an occupant load of 15 or fewer, service sinks shall not be required.

403.1.1 Fixture calculations. To determine the occupant load of each sex, the total occupant load shall be divided in half. To determine the required number of fixtures, the fixture ratio or ratios for each fixture type shall be applied to the occupant load of each sex in accordance with Table 403.1. Fractional numbers resulting from applying the fixture ratios of Table 403.1 shall be rounded up to the next whole number. For calculations involving multiple *occupancies*, such fractional numbers for each *occupancy* shall first be summed and then rounded up to the next whole number.

Exception: The total occupant load shall not be required to be divided in half where *approved* statistical data indicates a distribution of the sexes of other than 50 percent of each sex.

403.1.2 Family or assisted-use toilet and bath fixtures. Fixtures located within family or assisted-use toilet and bathing rooms required by Section 1109.2.1 of the *International Building Code* are permitted to be included in the number of required fixtures for either the male or female occupants in assembly and mercantile *occupancies.*

403.2 Separate facilities. Where plumbing fixtures are required, separate facilities shall be provided for each sex.

Exceptions:

1. Separate facilities shall not be required for dwelling units and sleeping units.

2. Separate facilities shall not be required in structures or tenant spaces with a total occupant load, including both employees and customers, of 15 or fewer.

3. Separate facilities shall not be required in mercantile *occupancies* in which the maximum occupant load is 100 or fewer.

403.2.1 Family or assisted-use toilet facilities serving as separate facilities. Where a building or tenant space requires a separate toilet facility for each sex and each toilet facility is required to have only one water closet, two family/assisted-use toilet facilities shall be permitted to serve as the required separate facilities. Family or assisted-use toilet facilities shall not be required to be identified for exclusive use by either sex as required by Section 403.4.

403.3 Required public toilet facilities. Customers, patrons and visitors shall be provided with *public* toilet facilities in structures and tenant spaces intended for public utilization. The number of plumbing fixtures located within the required toilet facilities shall be provided in accordance with Section 403 for all users. Employees shall be provided with toilet facilities in all *occupancies.* Employee toilet facilities shall be either separate or combined employee and *public* toilet facilities.

Exception: Public toilet facilities shall not be required in open or enclosed parking garages. Toilet facilities shall not be required in parking garages where there are no parking attendants.

403.3.1 Access. The route to the *public* toilet facilities required by Section 403.3 shall not pass through kitchens, storage rooms or closets. Access to the required facilities shall be from within the building or from the exterior of the building. All routes shall comply with the accessibility requirements of the *International Building Code.* The public shall have access to the required toilet facilities at all times that the building is occupied.

[B] 403.3.2 Toilet room location. Toilet rooms shall not open directly into a room used for the preparation of food for service to the public.

403.3.3 Location of toilet facilities in occupancies other than malls. In occupancies other than covered and open mall buildings, the required *public* and employee toilet facilities shall be located not more than one story above or below the space required to be provided with toilet facilities, and the path of travel to such facilities shall not exceed a distance of 500 feet (152 m).

Exception: The location and maximum travel distances to required employee facilities in factory and industrial *occupancies* are permitted to exceed that required by this section, provided that the location and maximum travel distance are *approved.*

403.3.4 Location of toilet facilities in malls. In covered and open mall buildings, the required *public* and employee toilet facilities shall be located not more than one story above or below the space required to be provided with toilet facilities, and the path of travel to such facilities shall not exceed a distance of 300 feet (91 440 mm). In mall buildings, the required facilities shall be based on total square footage within a covered mall building or within the perimeter line of an open mall building, and facilities shall be installed in each individual store or in a central toilet area located in accordance with this section. The maximum travel distance to central toilet facilities in mall buildings shall be measured from the main entrance of any store or tenant space. In mall buildings, where employees' toilet facilities are not provided in the individual store, the maximum travel distance shall be measured from the employees' work area of the store or tenant space.

403.3.5 Pay facilities. Where pay facilities are installed, such facilities shall be in excess of the required minimum facilities. Required facilities shall be free of charge.

403.3.6 Door locking. Where a toilet room is provided for the use of multiple occupants, the egress door for the room shall not be lockable from the inside of the room. This section does not apply to family or assisted-use toilet rooms.

403.4 Signage. Required *public* facilities shall be designated by a legible sign for each sex. Signs shall be readily visible and located near the entrance to each toilet facility. Signs for accessible toilet facilities shall comply with Section 1110 of the *International Building Code*.

403.4.1 Directional signage. Directional signage indicating the route to the *public* facilities shall be posted in accordance with Section 3107 of the *International Building Code*. Such signage shall be located in a corridor or aisle, at the entrance to the facilities for customers and visitors.

403.5 Drinking fountain location. Drinking fountains shall not be required to be located in individual tenant spaces provided that public drinking fountains are located within a travel distance of 500 feet (152 mm) of the most remote location in the tenant space and not more than one story above or below the tenant space. Where the tenant space is in a covered or open mall, such distance shall not exceed 300 feet (91 440 mm). Drinking fountains shall be located on an accessible route.

SECTION 404
ACCESSIBLE PLUMBING FACILITIES

404.1 Where required. Accessible plumbing facilities and fixtures shall be provided in accordance with the *International Building Code*.

SECTION 405
INSTALLATION OF FIXTURES

405.1 Water supply protection. The supply lines and fittings for every plumbing fixture shall be installed so as to prevent backflow.

405.2 Access for cleaning. Plumbing fixtures shall be installed so as to afford easy *access* for cleaning both the fixture and the area around the fixture.

405.3 Setting. Fixtures shall be set level and in proper alignment with reference to adjacent walls.

405.3.1 Water closets, urinals, lavatories and bidets. A water closet, urinal, lavatory or bidet shall not be set closer than 15 inches (381 mm) from its center to any side wall, partition, vanity or other obstruction, or closer than 30 inches (762 mm) center to center between adjacent fixtures. There shall be not less than a 21-inch (533 mm) clearance in front of the water closet, urinal, lavatory or bidet to any wall, fixture or door. Water closet compartments shall be not less than 30 inches (762 mm) in width and not less than 60 inches (1524 mm) in depth for floor-

mounted water closets and not less than 30 inches (762 mm) in width and 56 inches (1422 mm) in depth for wall-hung water closets.

405.3.2 Public lavatories. In employee and *public* toilet rooms, the required lavatory shall be located in the same room as the required water closet.

405.3.3 Location of fixtures and piping. Piping, fixtures or equipment shall not be located in such a manner as to interfere with the normal operation of windows, doors or other means of egress openings.

405.3.4 Water closet compartment. Each water closet utilized by the *public* or employees shall occupy a separate compartment with walls or partitions and a door enclosing the fixtures to ensure privacy.

Exceptions:

1. Water closet compartments shall not be required in a single-occupant toilet room with a lockable door.

2. Toilet rooms located in child day care facilities and containing two or more water closets shall be permitted to have one water closet without an enclosing compartment.

3. This provision is not applicable to toilet areas located within Group I-3 housing areas.

405.3.5 Urinal partitions. Each urinal utilized by the *public* or employees shall occupy a separate area with walls or partitions to provide privacy. The walls or partitions shall begin at a height not greater than 12 inches (305 mm) from and extend not less than 60 inches (1524 mm) above the finished floor surface. The walls or partitions shall extend from the wall surface at each side of the urinal not less than 18 inches (457 mm) or to a point not less than 6 inches (152 mm) beyond the outermost front lip of the urinal measured from the finished backwall surface, whichever is greater.

Exceptions:

1. Urinal partitions shall not be required in a single occupant or family/assisted-use toilet room with a lockable door.

2. Toilet rooms located in child day care facilities and containing two or more urinals shall be permitted to have one urinal without partitions.

405.4 Floor and wall drainage connections. Connections between the drain and floor outlet plumbing fixtures shall be made with a floor flange or a waste connector and sealing gasket. The waste connector and sealing gasket joint shall comply with the joint tightness test of ASME A112.4.3 and shall be installed in accordance with the manufacturer's instructions. The flange shall be attached to the drain and anchored to the structure. Connections between the drain and wall-hung water closets shall be made with an *approved* extension nipple or horn adaptor. The water closet shall be bolted to the hanger with corrosion-resistant bolts or screws. Joints shall be sealed with an *approved* elastomeric gasket, flange-to-fixture connection complying with ASME A112.4.3 or an *approved* setting compound.

405.4.1 Floor flanges. Floor flanges for water closets or similar fixtures shall not be less than 0.125 inch (3.2 mm) thick for brass, 0.25 inch (6.4 mm) thick for plastic, and 0.25 inch (6.4 mm) thick and not less than a 2-inch (51 mm) caulking depth for cast-iron or galvanized malleable iron.

Floor flanges of hard lead shall weigh not less than 1 pound, 9 ounces (0.7 kg) and shall be composed of lead alloy with not less than 7.75-percent antimony by weight. Closet screws and bolts shall be of brass. Flanges shall be secured to the building structure with corrosion-resistant screws or bolts.

405.4.2 Securing floor outlet fixtures. Floor outlet fixtures shall be secured to the floor or floor flanges by screws or bolts of corrosion-resistant material.

405.4.3 Securing wall-hung water closet bowls. Wall-hung water closet bowls shall be supported by a concealed metal carrier that is attached to the building structural members so that strain is not transmitted to the closet connector or any other part of the plumbing system. The carrier shall conform to ASME A112.6.1M or ASME A112.6.2.

405.5 Water-tight joints. Joints formed where fixtures come in contact with walls or floors shall be sealed.

405.6 Plumbing in mental health centers. In mental health centers, pipes or traps shall not be exposed, and fixtures shall be bolted through walls.

405.7 Design of overflows. Where any fixture is provided with an overflow, the waste shall be designed and installed so that standing water in the fixture will not rise in the overflow when the stopper is closed, and no water will remain in the overflow when the fixture is empty.

405.7.1 Connection of overflows. The overflow from any fixture shall discharge into the drainage system on the inlet or fixture side of the trap.

Exception: The overflow from a flush tank serving a water closet or urinal shall discharge into the fixture served.

405.8 Slip joint connections. Slip joints shall be made with an *approved* elastomeric gasket and shall only be installed on the trap outlet, trap inlet and within the trap seal. Fixtures with concealed slip-joint connections shall be provided with an *access* panel or utility space not less than 12 inches (305 mm) in its smallest dimension or other *approved* arrangement so as to provide *access* to the slip joint connections for inspection and repair.

405.9 Design and installation of plumbing fixtures. Integral fixture fitting mounting surfaces on manufactured plumbing fixtures or plumbing fixtures constructed on site, shall meet the design requirements of ASME A112.19.2/CSA B45.1 or ASME A112.19.3/CSA B45.4.

SECTION 406
AUTOMATIC CLOTHES WASHERS

406.1 Water connection. The water supply to an automatic clothes washer shall be protected against backflow by an air gap installed integrally within the machine or with the installation of a backflow preventer in accordance with Section 608.

406.2 Waste connection. The waste from an automatic clothes washer shall discharge through an air break into a standpipe in accordance with Section 802.4 or into a laundry sink. The trap and fixture drain for an automatic clothes washer standpipe shall be not less than 2 inches (51 mm) in diameter. The fixture drain for the standpipe serving an automatic clothes washer shall connect to a 3-inch (76 mm) or larger diameter fixture *branch* or *stack*. Automatic clothes washers that discharge by gravity shall be permitted to drain to a waste receptor or an approved trench drain.

SECTION 407
BATHTUBS

407.1 Approval. Bathtubs shall conform to ANSI Z124.1.2, ASME A112.19.1/CSA B45.2, ASME A112.19.2/CSA B45.1, ASME A112.19.3/CSA B45.4 or CSA B45.5.

407.2 Bathtub waste outlets and overflows. Bathtubs shall be equipped with a waste outlet and an overflow outlet. The outlets shall be connected to waste tubing or piping not less than $1^1/_2$ inches (38 mm) in diameter. The waste outlet shall be equipped with a water-tight stopper.

407.3 Glazing. Windows and doors within a bathtub enclosure shall conform to the safety glazing requirements of the *International Building Code*.

407.4 Bathtub enclosure. Doors within a bathtub enclosure shall conform to ASME A112.19.15.

SECTION 408
BIDETS

408.1 Approval. Bidets shall conform to ASME A112.19.2/CSA B45.1.

408.2 Water connection. The water supply to a bidet shall be protected against backflow by an *air gap* or backflow preventer in accordance with Section 608.13.1, 608.13.2, 608.13.3, 608.13.5, 608.13.6 or 608.13.8.

408.3 Bidet water temperature. The discharge water temperature from a bidet fitting shall be limited to a maximum temperature of 110°F (43°C) by a water temperature limiting device conforming to ASSE 1070 or CSA B125.3.

SECTION 409
DISHWASHING MACHINES

409.1 Approval. Commercial dishwashing machines shall conform to ASSE 1004 and NSF 3.

409.2 Water connection. The water supply to a dishwashing machine shall be protected against backflow by an *air gap* or backflow preventer in accordance with Section 608.

409.3 Waste connection. The waste connection of a dishwashing machine shall comply with Section 802.1.6 or 802.1.7, as applicable.

SECTION 410
DRINKING FOUNTAINS

410.1 Approval. Drinking fountains shall conform to ASME A112.19.1/CSA B45.2 or ASME A112.19.2/CSA B45.1 and water coolers shall conform to ARI 1010. Drinking fountains and water coolers shall conform to NSF 61, Section 9.

[B] 410.2 Minimum number. Where drinking fountains are required, not fewer than two drinking fountains shall be provided. One drinking fountain shall comply with the requirements for people who use a wheelchair and one drinking fountain shall comply with the requirements for standing persons.

> **Exception:** A single drinking fountain that complies with the requirements for people who use a wheelchair and standing persons shall be permitted to be substituted for two separate drinking fountains.

410.3 Substitution. Where restaurants provide drinking water in a container free of charge, drinking fountains shall not be required in those restaurants. In other occupancies, where drinking fountains are required, water coolers or bottled water dispensers shall be permitted to be substituted for not more than 50 percent of the required number of drinking fountains.

410.4 Prohibited location. Drinking fountains, water coolers and bottle water dispensers shall not be installed in *public* restrooms.

SECTION 411
EMERGENCY SHOWERS AND EYEWASH STATIONS

411.1 Approval. Emergency showers and eyewash stations shall conform to ISEA Z358.1.

411.2 Waste connection. Waste connections shall not be required for emergency showers and eyewash stations.

SECTION 412
FLOOR AND TRENCH DRAINS

412.1 Approval. Floor drains shall conform to ASME A112.3.1, ASME A112.6.3 or CSA B79. Trench drains shall comply with ASME A112.6.3.

412.2 Floor drains. Floor drains shall have removable strainers. The floor drain shall be constructed so that the drain is capable of being cleaned. *Access* shall be provided to the drain inlet. Ready *access* shall be provided to floor drains.

> **Exception:** Floor drains serving refrigerated display cases shall be provided with *access*.

412.3 Size of floor drains. Floor drains shall have a drain outlet not less than 2 inches (51 mm) in diameter.

412.4 Public laundries and central washing facilities. In public coin-operated laundries and in the central washing facilities of multiple family dwellings, the rooms containing automatic clothes washers shall be provided with floor drains located to readily drain the entire floor area. Such drains shall have a outlet of not less than 3 inches (76 mm) in diameter.

SECTION 413
FOOD WASTE GRINDER UNITS

413.1 Approval. Domestic food waste grinders shall conform to ASSE 1008. Food waste grinders shall not increase the drainage fixture unit load on the sanitary drainage system.

413.2 Domestic food waste grinder waste outlets. Domestic food waste grinders shall be connected to a drain of not less than $1^1/_2$ inches (38 mm) in diameter.

413.3 Commercial food waste grinder waste outlets. Commercial food waste grinders shall be connected to a drain not less than $1^1/_2$ inches (38 mm) in diameter. Commercial food waste grinders shall be connected and trapped separately from any other fixtures or sink compartments.

413.4 Water supply required. All food waste grinders shall be provided with a supply of cold water. The water supply shall be protected against backflow by an *air gap* or backflow preventer in accordance with Section 608.

SECTION 414
GARBAGE CAN WASHERS

414.1 Water connection. The water supply to a garbage can washer shall be protected against backflow by an *air gap* or a backflow preventer in accordance with Section 608.13.1, 608.13.2, 608.13.3, 608.13.5, 608.13.6 or 608.13.8.

414.2 Waste connection. Garbage can washers shall be trapped separately. The receptacle receiving the waste from the washer shall have a removable basket or strainer to prevent the discharge of large particles into the drainage system.

SECTION 415
LAUNDRY TRAYS

415.1 Approval. Laundry trays shall conform to ANSI Z124.6, ASME A112.19.1/CSA B45.2, ASME A112.19.2/ CSA B45.1, ASME A112.19.3/CSA B45.4.

415.2 Waste outlet. Each compartment of a laundry tray shall be provided with a waste outlet not less than $1^1/_2$ inches (38 mm) in diameter and a strainer or crossbar to restrict the clear opening of the waste outlet.

SECTION 416
LAVATORIES

416.1 Approval. Lavatories shall conform to ANSI Z124.3, ASME A112.19.1/CSA B45.2, ASME A112.19.2/CSA B45.1, or ASME A112.19.3/CSA B45.4. Group wash-up equipment shall conform to the requirements of Section 402. Every 20 inches (508 mm) of rim space shall be considered as one lavatory.

416.2 Cultured marble lavatories. Cultured marble vanity tops with an integral lavatory shall conform to ANSI Z124.3 or CSA B45.5.

416.3 Lavatory waste outlets. Lavatories shall have waste outlets not less than $1^1/_4$ inches (32 mm) in diameter. A strainer, pop-up stopper, crossbar or other device shall be provided to restrict the clear opening of the waste outlet.

416.4 Moveable lavatory systems. Moveable lavatory systems shall comply with ASME A112.19.12.

416.5 Tempered water for public hand-washing facilities. *Tempered water* shall be delivered from lavatories and group wash fixtures located in public toilet facilities provided for customers, patrons and visitors. *Tempered water* shall be delivered through an *approved* water-temperature limiting device that conforms to ASSE 1070 or CSA B125.3.

SECTION 417
SHOWERS

417.1 Approval. Prefabricated showers and shower compartments shall conform to ANSI Z124.1.2, ASME A112.19.2/ CSA B45.1 or CSA B45.5. Shower valves for individual showers shall conform to the requirements of Section 424.3.

417.2 Water supply riser. Water supply risers from the shower valve to the shower head outlet, whether exposed or concealed, shall be attached to the structure. The attachment to the structure shall be made by the use of support devices designed for use with the specific piping material or by fittings anchored with screws.

417.3 Shower waste outlet. Waste outlets serving showers shall be not less than $1^1/_2$ inches (38 mm) in diameter and, for other than waste outlets in bathtubs, shall have removable strainers not less than 3 inches (76 mm) in diameter with strainer openings not less than $^1/_4$ inch (6.4 mm) in least dimension. Where each shower space is not provided with an individual waste outlet, the waste outlet shall be located and the floor pitched so that waste from one shower does not flow over the floor area serving another shower. Waste outlets shall be fastened to the waste pipe in an *approved* manner.

417.4 Shower compartments. Shower compartments shall be not less than 900 square inches (0.58 m²) in interior cross-sectional area. Shower compartments shall be not less than 30 inches (762 mm) in least dimension as measured from the finished interior dimension of the compartment, exclusive of fixture valves, showerheads, soap dishes, and safety grab bars or rails. Except as required in Section 404, the minimum required area and dimension shall be measured from the finished interior dimension at a height equal to the top of the threshold and at a point tangent to its centerline and shall be continued to a height not less than 70 inches (1778 mm) above the shower drain outlet.

Exception: Shower compartments having not less than 25 inches (635 mm) in minimum dimension measured from the finished interior dimension of the compartment, provided that the shower compartment has not less than of 1,300 square inches (.838 m²) of cross-sectional area.

417.4.1 Wall area. The wall area above built-in tubs with installed shower heads and in shower compartments shall be constructed of smooth, noncorrosive and nonabsorbent waterproof materials to a height not less than 6 feet (1829 mm) above the room floor level, and not less than 70 inches (1778 mm) where measured from the compartment floor at the drain. Such walls shall form a water-tight joint with each other and with either the tub, receptor or shower floor.

417.4.2 Access. The shower compartment access and egress opening shall have a clear and unobstructed finished width of not less than 22 inches (559 mm). Shower compartments required to be designed in conformance to accessibility provisions shall comply with Section 404.1.

417.5 Shower floors or receptors. Floor surfaces shall be constructed of impervious, noncorrosive, nonabsorbent and waterproof materials.

417.5.1 Support. Floors or receptors under shower compartments shall be laid on, and supported by, a smooth and structurally sound base.

417.5.2 Shower lining. Floors under shower compartments, except where prefabricated receptors have been provided, shall be lined and made water tight utilizing material complying with Sections 417.5.2.1 through 417.5.2.6. Such liners shall turn up on all sides not less than 2 inches (51 mm) above the finished threshold level. Liners shall be recessed and fastened to an *approved* backing so as not to occupy the space required for wall covering, and shall not be nailed or perforated at any point less than 1 inch (25 mm) above the finished threshold. Liners shall be pitched one-fourth unit vertical in 12 units horizontal (2-percent slope) and shall be sloped toward the fixture drains and be securely fastened to the waste outlet at the seepage entrance, making a water-tight joint between the liner and the outlet. The completed liner shall be tested in accordance with Section 312.9.

Exceptions:

1. Floor surfaces under shower heads provided for rinsing laid directly on the ground are not required to comply with this section.

2. Where a sheet-applied, load-bearing, bonded, waterproof membrane is installed as the shower lining, the membrane shall not be required to be recessed.

417.5.2.1 PVC sheets. Plasticized polyvinyl chloride (PVC) sheets shall meet the requirements of ASTM D 4551. Sheets shall be joined by solvent welding in accordance with the manufacturer's installation instructions.

417.5.2.2 Chlorinated polyethylene (CPE) sheets. Nonplasticized chlorinated polyethylene sheet shall meet the requirements of ASTM D 4068. The liner shall be joined in accordance with the manufacturer's installation instructions.

417.5.2.3 Sheet lead. Sheet lead shall weigh not less than 4 pounds per square foot (19.5 kg/m²) and shall be coated with an asphalt paint or other *approved* coating. The lead sheet shall be insulated from conducting substances other than the connecting drain by 15-pound (6.80 kg) asphalt felt or an equivalent. Sheet lead shall be joined by burning.

417.5.2.4 Sheet copper. Sheet copper shall conform to ASTM B 152 and shall weigh not less than 12 ounces per square foot (3.7 kg/m²). The copper sheet shall be insulated from conducting substances other than the connecting drain by 15-pound (6.80 kg) asphalt felt or an equivalent. Sheet copper shall be joined by brazing or soldering.

417.5.2.5 Sheet-applied, load-bearing, bonded, waterproof membranes. Sheet-applied, load-bearing, bonded, waterproof membranes shall meet requirements of ANSI A118.10 and shall be applied in accordance with the manufacturer's installation instructions.

417.5.2.6 Liquid-type, trowel-applied, load-bearing, bonded waterproof materials. Liquid-type, trowel-applied, load-bearing, bonded waterproof materials shall meet the requirements of ANSI A118.10 and shall be applied in accordance with the manufacturer's instructions.

417.6 Glazing. Windows and doors within a shower enclosure shall conform to the safety glazing requirements of the *International Building Code*.

SECTION 418
SINKS

418.1 Approval. Sinks shall conform to ANSI Z124.6, ASME A112.19.1/CSA B45.2, ASME A112.19.2/CSA B45.1 or ASME A112.19.3/CSA B45.4.

418.2 Sink waste outlets. Sinks shall be provided with waste outlets having a diameter not less than $1^1/_2$ inches (38 mm). A strainer or crossbar shall be provided to restrict the clear opening of the waste outlet.

418.3 Moveable sink systems. Moveable sink systems shall comply with ASME A112.19.12.

SECTION 419
URINALS

419.1 Approval. Urinals shall conform to ANSI Z124.9, ASME A112.19.2/CSA B45.1, ASME A112.19.19 or CSA B45.5. Urinals shall conform to the water consumption requirements of Section 604.4. Water-supplied urinals shall conform to the hydraulic performance requirements of ASME A112.19.2/CSA B45.1 or CSA B45.5.

419.2 Substitution for water closets. In each bathroom or toilet room, urinals shall not be substituted for more than 67 percent of the required water closets in assembly and educational *occupancies*. Urinals shall not be substituted for more than 50 percent of the required water closets in all other *occupancies*.

[B] 419.3 Surrounding material. Wall and floor space to a point 2 feet (610 mm) in front of a urinal lip and 4 feet (1219 mm) above the floor and at least 2 feet (610 mm) to each side of the urinal shall be waterproofed with a smooth, readily cleanable, nonabsorbent material.

SECTION 420
WATER CLOSETS

420.1 Approval. Water closets shall conform to the water consumption requirements of Section 604.4 and shall conform to ANSI Z124.4, ASME A112.19.2/CSA B45.1, ASME A112.19.3/CSA B45.4 or CSA B45.5. Water closets shall conform to the hydraulic performance requirements of ASME A112.19.2/CSA B45.1. Water closet tanks shall conform to ANSI Z124.4, ASME A112.19.2/CSA B45.1, ASME A112.19.3/CSA B45.4 or CSA B45.5. Electro-hydraulic water closets shall comply with ASME A112.19.2/CSA B45.1.

420.2 Water closets for public or employee toilet facilities. Water closet bowls for *public* or employee toilet facilities shall be of the elongated type.

420.3 Water closet seats. Water closets shall be equipped with seats of smooth, nonabsorbent material. All seats of water closets provided for *public* or employee toilet facilities shall be of the hinged open-front type. Integral water closet seats shall be of the same material as the fixture. Water closet seats shall be sized for the water closet bowl type.

420.4 Water closet connections. A 4-inch by 3-inch (102 mm by 76 mm) closet bend shall be acceptable. Where a 3-inch (76 mm) bend is utilized on water closets, a 4-inch by 3-inch (102 mm by 76 mm) flange shall be installed to receive the fixture horn.

SECTION 421
WHIRLPOOL BATHTUBS

421.1 Approval. Whirlpool bathtubs shall comply with ASME A112.19.7/CSA B45.10.

421.2 Installation. Whirlpool bathtubs shall be installed and tested in accordance with the manufacturer's installation instructions. The pump shall be located above the weir of the fixture trap.

421.3 Drain. The pump drain and circulation piping shall be sloped to drain the water in the volute and the circulation piping when the whirlpool bathtub is empty.

421.4 Suction fittings. Suction fittings for whirlpool bathtubs shall comply with ASME A112.19.7/CSA B45.10.

421.5 Access to pump. *Access* shall be provided to circulation pumps in accordance with the fixture or pump manufacturer's installation instructions. Where the manufacturer's instructions do not specify the location and minimum size of field-fabricated *access* openings, an opening not less than 12-inches by 12-inches (305 mm by 305 mm) shall be installed to provide *access* to the circulation pump. Where pumps are located more than 2 feet (609 mm) from the *access* opening, an opening not less than 18-inches by 18-inches (457 mm by 457 mm) shall be installed. A door or panel shall be permitted to close the opening. In all cases, the *access* opening shall be unobstructed and of the size necessary to permit the removal and replacement of the circulation pump.

421.6 Whirlpool enclosure. Doors within a whirlpool enclosure shall conform to ASME A112.19.15.

SECTION 422
HEALTH CARE FIXTURES AND EQUIPMENT

422.1 Scope. This section shall govern those aspects of health care plumbing systems that differ from plumbing systems in other structures. Health care plumbing systems shall conform to the requirements of this section in addition to the other requirements of this code. The provisions of this section shall apply to the special devices and equipment installed and maintained in the following *occupancies*: nursing homes, homes for the aged, orphanages, infirmaries, first aid stations, psychiatric facilities, clinics, professional offices of dentists and doctors, mortuaries, educational facilities, surgery, dentistry, research and testing laboratories, establishments manufacturing pharmaceutical drugs and medicines, and other structures with similar apparatus and equipment classified as plumbing.

422.2 Approval. All special plumbing fixtures, equipment, devices and apparatus shall be of an *approved* type.

422.3 Protection. All devices, appurtenances, appliances and apparatus intended to serve some special function, such as sterilization, distillation, processing, cooling, or storage of ice or foods, and that connect to either the water supply or drainage system, shall be provided with protection against backflow, flooding, fouling, contamination of the water supply system and stoppage of the drain.

422.4 Materials. Fixtures designed for therapy, special cleansing or disposal of waste materials, combinations of such purposes, or any other special purpose, shall be of smooth, impervious, corrosion-resistant materials and, where subjected to temperatures in excess of 180°F (82°C), shall be capable of withstanding, without damage, higher temperatures.

422.5 Access. *Access* shall be provided to concealed piping in connection with special fixtures where such piping contains steam traps, valves, relief valves, check valves, vacuum breakers or other similar items that require periodic inspection, servicing, maintenance or repair. *Access* shall be provided to concealed piping that requires periodic inspection, maintenance or repair.

422.6 Clinical sink. A clinical sink shall have an integral trap in which the upper portion of a visible trap seal provides a water surface. The fixture shall be designed so as to permit complete removal of the contents by siphonic or blowout action and to reseal the trap. A flushing rim shall provide water to cleanse the interior surface. The fixture shall have the flushing and cleansing characteristics of a water closet.

422.7 Prohibited usage of clinical sinks and service sinks. A clinical sink serving a soiled utility room shall not be considered as a substitute for, or be utilized as, a service sink. A service sink shall not be utilized for the disposal of urine, fecal matter or other human waste.

422.8 Ice prohibited in soiled utility room. Machines for manufacturing ice, or any device for the handling or storage of ice, shall not be located in a soiled utility room.

422.9 Sterilizer equipment requirements. The approval and installation of all sterilizers shall conform to the requirements of the *International Mechanical Code*.

422.9.1 Sterilizer piping. *Access* for the purposes of inspection and maintenance shall be provided to all sterilizer piping and devices necessary for the operation of sterilizers.

422.9.2 Steam supply. Steam supplies to sterilizers, including those connected by pipes from overhead mains or branches, shall be drained to prevent any moisture from reaching the sterilizer. The condensate drainage from the steam supply shall be discharged by gravity.

422.9.3 Steam condensate return. Steam condensate returns from sterilizers shall be a gravity return system.

422.9.4 Condensers. Pressure sterilizers shall be equipped with a means of condensing and cooling the exhaust steam vapors. Nonpressure sterilizers shall be equipped with a device that will automatically control the vapor, confining the vapors within the vessel.

422.10 Special elevations. Control valves, vacuum outlets and devices protruding from a wall of an operating, emergency, recovery, examining or delivery room, or in a corridor or other location where patients are transported on a wheeled stretcher, shall be located at an elevation that prevents bumping the patient or stretcher against the device.

SECTION 423
SPECIALTY PLUMBING FIXTURES

423.1 Water connections. Baptisteries, ornamental and lily pools, aquariums, ornamental fountain basins, swimming pools, and similar constructions, where provided with water supplies, shall be protected against backflow in accordance with Section 608.

423.2 Approval. Specialties requiring water and waste connections shall be submitted for approval.

SECTION 424
FAUCETS AND OTHER FIXTURE FITTINGS

424.1 Approval. Faucets and fixture fittings shall conform to ASME A112.18.1/CSA B125.1. Faucets and fixture fittings that supply drinking water for human ingestion shall conform to the requirements of NSF 61, Section 9. Flexible water connectors exposed to continuous pressure shall conform to the requirements of Section 605.6.

424.1.1 Faucets and supply fittings. Faucets and supply fittings shall conform to the water consumption requirements of Section 604.4.

424.1.2 Waste fittings. Waste fittings shall conform to ASME A112.18.2/CSA B125.2, ASTM F 409 or to one of the standards listed in Tables 702.1 and 702.4 for aboveground drainage and vent pipe and fittings.

424.2 Hand showers. Hand-held showers shall conform to ASME A112.18.1/CSA B125.1. Hand-held showers shall provide backflow protection in accordance with ASME

A112.18.1/ CSA B125.1 or shall be protected against backflow by a device complying with ASME A112.18.3.

424.3 Individual shower valves. Individual shower and tub-shower combination valves shall be balanced-pressure, thermostatic or combination balanced-pressure/thermostatic valves that conform to the requirements of ASSE 1016 or ASME A112.18.1/CSA B125.1 and shall be installed at the point of use. Shower and tub-shower combination valves required by this section shall be equipped with a means to limit the maximum setting of the valve to 120°F (49°C), which shall be field adjusted in accordance with the manufacturer's instructions. In-line thermostatic valves shall not be utilized for compliance with this section.

424.4 Multiple (gang) showers. Multiple (gang) showers supplied with a single-tempered water supply pipe shall have the water supply for such showers controlled by an *approved* automatic temperature control mixing valve that conforms to ASSE 1069 or CSA B125.3, or each shower head shall be individually controlled by a balanced-pressure, thermostatic or combination balanced-pressure/thermostatic valve that conforms to ASSE1016 or ASME A112.18.1/CSA B125.1 and is installed at the point of use. Such valves shall be equipped with a means to limit the maximum setting of the valve to 120°F (49°C), which shall be field adjusted in accordance with the manufacturers' instructions.

424.5 Bathtub and whirlpool bathtub valves. The *hot water* supplied to bathtubs and whirlpool bathtubs shall be limited to a maximum temperature of 120°F (49°C) by a water-temperature limiting device that conforms to ASSE 1070 or CSA B125.3, except where such protection is otherwise provided by a combination tub/shower valve in accordance with Section 424.3.

424.6 Hose-connected outlets. Faucets and fixture fittings with hose connected outlets shall conform to ASME A112.18.3 or ASME A112.18.1/CSA B125.1.

424.7 Temperature-actuated, flow reduction valves for individual fixture fittings. Temperature-actuated, flow reduction devices, where installed for individual fixture fittings, shall conform to ASSE 1062. Such valves shall not be used alone as a substitute for the balanced pressure, thermostatic or combination shower valves required in Section 424.3.

424.8 Transfer valves. Deck-mounted bath/shower transfer valves containing an integral atmospheric vacuum breaker shall conform to the requirements of ASME A112.18.7.

424.9 Water closet personal hygiene devices. Personal hygiene devices integral to water closets or water closet seats shall conform to the requirements of ASME A112.4.2.

SECTION 425
FLUSHING DEVICES FOR
WATER CLOSETS AND URINALS

425.1 Flushing devices required. Each water closet, urinal, clinical sink and any plumbing fixture that depends on trap siphonage to discharge the fixture contents to the drainage system shall be provided with a flushometer valve, flushometer tank or a flush tank designed and installed to supply water in quantity and rate of flow to flush the contents of the fixture, cleanse the fixture and refill the fixture trap.

425.1.1 Separate for each fixture. A flushing device shall not serve more than one fixture.

425.2 Flushometer valves and tanks. Flushometer valves and tanks shall comply with ASSE 1037 or CSA B125.3. Vacuum breakers on flushometer valves shall conform to the performance requirements of ASSE 1001 or CSA B64.1.1. *Access* shall be provided to vacuum breakers. Flushometer valves shall be of the water conservation type and shall not be used where the water pressure is lower than the minimum required for normal operation. When operated, the valve shall automatically complete the cycle of operation, opening fully and closing positively under the water supply pressure. Each flushometer valve shall be provided with a means for regulating the flow through the valve. The trap seal to the fixture shall be automatically refilled after each flushing cycle.

425.3 Flush tanks. Flush tanks equipped for manual flushing shall be controlled by a device designed to refill the tank after each discharge and to shut off completely the water flow to the tank when the tank is filled to operational capacity. The trap seal to the fixture shall be automatically refilled after each flushing. The water supply to flush tanks equipped for automatic flushing shall be controlled with a timing device or sensor control devices.

425.3.1 Fill valves. All flush tanks shall be equipped with an antisiphon fill valve conforming to ASSE 1002 or CSA B125.3. The fill valve backflow preventer shall be located not less than 1 inch (25 mm) above the full opening of the overflow pipe.

425.3.2 Overflows in flush tanks. Flush tanks shall be provided with overflows discharging to the water closet or urinal connected thereto and shall be sized to prevent flooding the tank at the maximum rate at which the tanks are supplied with water according to the manufacturer's design conditions. The opening of the overflow pipe shall be located above the flood level rim of the water closet or urinal or above a secondary overflow in the flush tank.

425.3.3 Sheet copper. Sheet copper utilized for flush tank linings shall conform to ASTM B 152 and shall not weigh less than 10 ounces per square foot (0.03 kg/m²).

425.3.4 Access required. All parts in a flush tank shall be accessible for repair and replacement.

425.4 Flush pipes and fittings. Flush pipes and fittings shall be of nonferrous material and shall conform to ASME A112.19.5/CSA B45.15.

SECTION 426
MANUAL FOOD AND BEVERAGE
DISPENSING EQUIPMENT

426.1 Approval. Manual food and beverage dispensing equipment shall conform to the requirements of NSF 18.

SECTION 427
FLOOR SINKS

427.1 Approval. Sanitary floor sinks shall conform to the requirements of ASME A112.6.7.

CHAPTER 5
WATER HEATERS

SECTION 501
GENERAL

501.1 Scope. The provisions of this chapter shall govern the materials, design and installation of water heaters and the related safety devices and appurtenances.

501.2 Water heater as space heater. Where a combination potable water heating and space heating system requires water for space heating at temperatures higher than 140°F (60°C), a master thermostatic mixing valve complying with ASSE 1017 shall be provided to limit the water supplied to the potable *hot water* distribution system to a temperature of 140°F (60°C) or less. The potability of the water shall be maintained throughout the system.

501.3 Drain valves. Drain valves for emptying shall be installed at the bottom of each tank-type water heater and *hot water* storage tank. Drain valves shall conform to ASSE 1005.

501.4 Location. Water heaters and storage tanks shall be located and connected so as to provide *access* for observation, maintenance, servicing and replacement.

501.5 Water heater labeling. All water heaters shall be third-party certified.

501.6 Water temperature control in piping from tankless heaters. The temperature of water from tankless water heaters shall be not greater than 140°F (60°C) where intended for domestic uses. This provision shall not supersede the requirement for protective shower valves in accordance with Section 424.3.

501.7 Pressure marking of storage tanks. Storage tanks and water heaters installed for domestic hot water shall have the maximum allowable working pressure clearly and indelibly stamped in the metal or marked on a plate welded thereto or otherwise permanently attached. Such markings shall be in an accessible position outside of the tank so as to make inspection or reinspection readily possible.

501.8 Temperature controls. Hot water supply systems shall be equipped with automatic temperature controls capable of adjustments from the lowest to the highest acceptable temperature settings for the intended temperature operating range.

SECTION 502
INSTALLATION

502.1 General. Water heaters shall be installed in accordance with the manufacturer's installation instructions. Oil-fired water heaters shall conform to the requirements of this code and the *International Mechanical Code*. Electric water heaters shall conform to the requirements of this code and provisions of NFPA 70. Gas-fired water heaters shall conform to the requirements of the *International Fuel Gas Code*.

502.1.1 Elevation and protection. Elevation of water heater ignition sources and mechanical damage protection requirements for water heaters shall be in accordance with the *International Mechanical Code* and the *International Fuel Gas Code*.

502.2 Rooms used as a plenum. Water heaters using solid, liquid or gas fuel shall not be installed in a room containing air-handling machinery where such room is used as a plenum.

502.3 Water heaters installed in attics. Attics containing a water heater shall be provided with an opening and unobstructed passageway large enough to allow removal of the water heater. The passageway shall be not less than 30 inches (762 mm) in height and 22 inches (559 mm) in width and not more than 20 feet (6096 mm) in length when measured along the centerline of the passageway from the opening to the water heater. The passageway shall have continuous solid flooring not less than 24 inches (610 mm) in width. A level service space not less than 30 inches (762 mm) in length and 30 inches (762 mm) in width shall be present at the front or service side of the water heater. The clear *access* opening dimensions shall be not less than 20 inches by 30 inches (508 mm by 762 mm) where such dimensions are large enough to allow removal of the water heater.

502.4 Seismic supports. Where earthquake loads are applicable in accordance with the *International Building Code*, water heater supports shall be designed and installed for the seismic forces in accordance with the *International Building Code*.

502.5 Clearances for maintenance and replacement. Appliances shall be provided with *access* for inspection, service, repair and replacement without disabling the function of a fire-resistance-rated assembly or removing permanent construction, other appliances or any other piping or ducts not connected to the appliance being inspected, serviced, repaired or replaced. A level working space not less than 30 inches in length and 30 inches in width (762 mm by 762 mm) shall be provided in front of the control side to service an appliance.

SECTION 503
CONNECTIONS

503.1 Cold water line valve. The cold water *branch* line from the main water supply line to each hot water storage tank or water heater shall be provided with a valve, located near the equipment and serving only the hot water storage tank or water heater. The valve shall not interfere or cause a disruption of the cold water supply to the remainder of the cold water system. The valve shall be provided with *access* on the same floor level as the water heater served.

503.2 Water circulation. The method of connecting a circulating water heater to the tank shall provide proper circulation of water through the water heater. The pipe or tubes required for the installation of appliances that will draw from the water heater or storage tank shall comply with the provisions of this code for material and installation.

SECTION 504
SAFETY DEVICES

504.1 Antisiphon devices. An *approved* means, such as a cold water "dip" tube with a hole at the top or a vacuum relief valve installed in the cold water supply line above the top of the heater or tank, shall be provided to prevent siphoning of any storage water heater or tank.

504.2 Vacuum relief valve. Bottom fed water heaters and bottom fed tanks connected to water heaters shall have a vacuum relief valve installed. The vacuum relief valve shall comply with ANSI Z21.22.

504.3 Shutdown. A means for disconnecting an electric hot water supply system from its energy supply shall be provided in accordance with NFPA 70. A separate valve shall be provided to shut off the energy fuel supply to all other types of hot water supply systems.

504.4 Relief valve. Storage water heaters operating above atmospheric pressure shall be provided with an *approved*, self-closing (levered) pressure relief valve and temperature relief valve or combination thereof. The relief valve shall conform to ANSI Z21.22. The relief valve shall not be used as a means of controlling thermal expansion.

> **504.4.1 Installation.** Such valves shall be installed in the shell of the water heater tank. Temperature relief valves shall be so located in the tank as to be actuated by the water in the top 6 inches (152 mm) of the tank served. For installations with separate storage tanks, the *approved*, self-closing (levered) pressure relief valve and temperature relief valve or combination thereof conforming to ANSI Z21.22 valves shall be installed on both the storage water heater and storage tank. There shall not be a check valve or shutoff valve between a relief valve and the heater or tank served.

504.5 Relief valve approval. Temperature and pressure relief valves, or combinations thereof, and energy cutoff devices shall bear the label of an *approved* agency and shall have a temperature setting of not more than 210°F (99°C) and a pressure setting not exceeding the tank or water heater manufacturer's rated working pressure or 150 psi (1035 kPa), whichever is less. The relieving capacity of each pressure relief valve and each temperature relief valve shall equal or exceed the heat input to the water heater or storage tank.

504.6 Requirements for discharge piping. The discharge piping serving a pressure relief valve, temperature relief valve or combination thereof shall:

1. Not be directly connected to the drainage system.

2. Discharge through an *air gap* located in the same room as the water heater.

3. Not be smaller than the diameter of the outlet of the valve served and shall discharge full size to the *air gap*.

4. Serve a single relief device and shall not connect to piping serving any other relief device or equipment.

5. Discharge to the floor, to the pan serving the water heater or storage tank, to a waste receptor or to the outdoors.

6. Discharge in a manner that does not cause personal injury or structural damage.

7. Discharge to a termination point that is readily observable by the building occupants.

8. Not be trapped.

9. Be installed so as to flow by gravity.

10. Not terminate more than 6 inches (152 mm) above the floor or waste receptor.

11. Not have a threaded connection at the end of such piping.

12. Not have valves or tee fittings.

13. Be constructed of those materials listed in Section 605.4 or materials tested, rated and *approved* for such use in accordance with ASME A112.4.1.

504.7 Required pan. Where a storage tank-type water heater or a hot water storage tank is installed in a location where water leakage from the tank will cause damage, the tank shall be installed in a galvanized steel pan having a material thickness of not less than 0.0236 inch (0.6010mm) (No. 24 gage), or other pans approved for such use.

> **504.7.1 Pan size and drain.** The pan shall be not less than $1^1/_2$ inches (38 mm) in depth and shall be of sufficient size and shape to receive all dripping or condensate from the tank or water heater. The pan shall be drained by an indirect waste pipe having a diameter of not less than $^3/_4$ inch (19 mm). Piping for safety pan drains shall be of those materials listed in Table 605.4.

> **504.7.2 Pan drain termination.** The pan drain shall extend full-size and terminate over a suitably located indirect waste receptor or floor drain or extend to the exterior of the building and terminate not less than 6 inches (152 mm) and not more than 24 inches (610 mm) above the adjacent ground surface.

SECTION 505
INSULATION

[E] 505.1 Unfired vessel insulation. Unfired hot water storage tanks shall be insulated to R-12.5 (h • ft² • °F)/Btu (R-2.2 m² • K/W).

CHAPTER 6
WATER SUPPLY AND DISTRIBUTION

SECTION 601
GENERAL

601.1 Scope. This chapter shall govern the materials, design and installation of water supply systems, both hot and cold, for utilization in connection with human occupancy and habitation and shall govern the installation of individual water supply systems.

601.2 Solar energy utilization. Solar energy systems used for heating potable water or using an independent medium for heating potable water shall comply with the applicable requirements of this code. The use of solar energy shall not compromise the requirements for cross connection or protection of the potable water supply system required by this code.

601.3 Existing piping used for grounding. Existing metallic water service piping used for electrical grounding shall not be replaced with nonmetallic pipe or tubing until other *approved* means of grounding is provided.

601.4 Tests. The potable water distribution system shall be tested in accordance with Section 312.5.

SECTION 602
WATER REQUIRED

602.1 General. Structures equipped with plumbing fixtures and utilized for human occupancy or habitation shall be provided with a potable supply of water in the amounts and at the pressures specified in this chapter.

602.2 Potable water required. Only potable water shall be supplied to plumbing fixtures that provide water for drinking, bathing or culinary purposes, or for the processing of food, medical or pharmaceutical products. Unless otherwise provided in this code, potable water shall be supplied to all plumbing fixtures.

602.3 Individual water supply. Where a potable public water supply is not available, individual sources of potable water supply shall be utilized.

602.3.1 Sources. Dependent on geological and soil conditions and the amount of rainfall, individual water supplies are of the following types: drilled well, driven well, dug well, bored well, spring, stream or cistern. Surface bodies of water and land cisterns shall not be sources of individual water supply unless properly treated by *approved* means to prevent contamination.

602.3.2 Minimum quantity. The combined capacity of the source and storage in an individual water supply system shall supply the fixtures with water at rates and pressures as required by this chapter.

602.3.3 Water quality. Water from an individual water supply shall be *approved* as potable by the authority having jurisdiction prior to connection to the plumbing system.

602.3.4 Disinfection of system. After construction or major repair, the individual water supply system shall be purged of deleterious matter and disinfected in accordance with Section 610.

602.3.5 Pumps. Pumps shall be rated for the transport of potable water. Pumps in an individual water supply system shall be constructed and installed so as to prevent contamination from entering a potable water supply through the pump units. Pumps shall be sealed to the well casing or covered with a water-tight seal. Pumps shall be designed to maintain a prime and installed such that ready *access* is provided to the pump parts of the entire assembly for repairs.

602.3.5.1 Pump enclosure. The pump room or enclosure around a well pump shall be drained and protected from freezing by heating or other *approved* means. Where pumps are installed in basements, such pumps shall be mounted on a block or shelf not less than 18 inches (457 mm) above the basement floor. Well pits shall be prohibited.

SECTION 603
WATER SERVICE

603.1 Size of water service pipe. The water service pipe shall be sized to supply water to the structure in the quantities and at the pressures required in this code. The water service pipe shall be not less than $^3/_4$ inch (19.1 mm) in diameter.

603.2 Separation of water service and building sewer. Water service pipe and the *building sewer* shall be separated by not less than 5 feet (1524 mm) of undisturbed or compacted earth.

Exceptions:

1. The required separation distance shall not apply where the bottom of the water service pipe within 5 feet (1524 mm) of the *sewer* is not less than 12 inches (305 mm) above the top of the highest point of the *sewer* and the pipe materials conform to Table 702.3.

2. Water service pipe is permitted to be located in the same trench with a *building sewer*, provided such *sewer* is constructed of materials listed in Table 702.2.

3. The required separation distance shall not apply where a water service pipe crosses a *sewer* pipe, provided the water service pipe is sleeved to a point not less than 5 feet (1524 mm) horizontally from the *sewer* pipe centerline on both sides of such crossing with pipe materials listed in Table 605.3, 702.2 or 702.3.

603.2.1 Water service near sources of pollution. Potable water service pipes shall not be located in, under or above cesspools, septic tanks, septic tank drainage fields or seepage pits (see Section 605.1 for soil and groundwater conditions).

SECTION 604
DESIGN OF BUILDING WATER DISTRIBUTION SYSTEM

604.1 General. The design of the water distribution system shall conform to *accepted engineering practice*. Methods utilized to determine pipe sizes shall be *approved*.

604.2 System interconnection. At the points of interconnection between the hot and cold water supply piping systems and the individual fixtures, appliances or devices, provisions shall be made to prevent flow between such piping systems.

604.3 Water distribution system design criteria. The water distribution system shall be designed, and pipe sizes shall be selected such that under conditions of peak demand, the capacities at the fixture supply pipe outlets shall not be less than shown in Table 604.3. The minimum flow rate and flow pressure provided to fixtures and appliances not listed in Table 604.3 shall be in accordance with the manufacturer's installation instructions.

TABLE 604.3
WATER DISTRIBUTION SYSTEM DESIGN CRITERIA REQUIRED CAPACITY AT FIXTURE SUPPLY PIPE OUTLETS

FIXTURE SUPPLY OUTLET SERVING	FLOW RATE[a] (gpm)	FLOW PRESSURE (psi)
Bathtub, balanced-pressure, thermostatic or combination balanced-pressure/thermo-static mixing valve	4	20
Bidet, thermostatic mixing valve	2	20
Combination fixture	4	8
Dishwasher, residential	2.75	8
Drinking fountain	0.75	8
Laundry tray	4	8
Lavatory	2	8
Shower	3	8
Shower, balanced-pressure, thermostatic or combination balanced-pressure/thermo-static mixing valve	3	20
Sillcock, hose bibb	5	8
Sink, residential	2.5	8
Sink, service	3	8
Urinal, valve	12	25
Water closet, blow out, flushometer valve	25	45
Water closet, flushometer tank	1.6	20
Water closet, siphonic, flushometer valve	25	35
Water closet, tank, close coupled	3	20
Water closet, tank, one piece	6	20

For SI: 1 pound per square inch = 6.895 kPa,
     1 gallon per minute = 3.785 L/m.

a. For additional requirements for flow rates and quantities, see Section 604.4.

604.4 Maximum flow and water consumption. The maximum water consumption flow rates and quantities for all plumbing fixtures and fixture fittings shall be in accordance with Table 604.4.

Exceptions:

1. Blowout design water closets having a water consumption not greater than $3^1/_2$ gallons (13 L) per flushing cycle.

2. Vegetable sprays.

3. Clinical sinks having a water consumption not greater than $4^1/_2$ gallons (17 L) per flushing cycle.

4. Service sinks.

5. Emergency showers.

TABLE 604.4
MAXIMUM FLOW RATES AND CONSUMPTION FOR PLUMBING FIXTURES AND FIXTURE FITTINGS

PLUMBING FIXTURE OR FIXTURE FITTING	MAXIMUM FLOW RATE OR QUANTITY[b]
Lavatory, private	2.2 gpm at 60 psi
Lavatory, public (metering)	0.25 gallon per metering cycle
Lavatory, public (other than metering)	0.5 gpm at 60 psi
Shower head[a]	2.5 gpm at 80 psi
Sink faucet	2.2 gpm at 60 psi
Urinal	1.0 gallon per flushing cycle
Water closet	1.6 gallons per flushing cycle

For SI: 1 gallon = 3.785 L, 1 gallon per minute = 3.785 L/m,
     1 pound per square inch = 6.895 kPa.

a. A hand-held shower spray is a shower head.
b. Consumption tolerances shall be determined from referenced standards.

604.5 Size of fixture supply. The minimum size of a fixture supply pipe shall be as shown in Table 604.5. The fixture supply pipe shall terminate not more than 30 inches (762 mm) from the point of connection to the fixture. A reduced-size flexible water connector installed between the supply pipe and the fixture shall be of an *approved* type. The supply pipe shall extend to the floor or wall adjacent to the fixture. The minimum size of individual distribution lines utilized in gridded or parallel water distribution systems shall be as shown in Table 604.5.

604.6 Variable street pressures. Where street water main pressures fluctuate, the building water distribution system shall be designed for the minimum pressure available.

604.7 Inadequate water pressure. Wherever water pressure from the street main or other source of supply is insufficient to provide flow pressures at fixture outlets as required under Table 604.3, a water pressure booster system conforming to Section 606.5 shall be installed on the building water supply system.

604.8 Water pressure reducing valve or regulator. Where water pressure within a building exceeds 80 psi (552 kPa) static, an *approved* water-pressure reducing valve conforming to ASSE 1003 or CSA B356 with strainer shall be

TABLE 604.5
MINIMUM SIZES OF FIXTURE WATER SUPPLY PIPES

FIXTURE	MINIMUM PIPE SIZE (inch)
Bathtubs[a] (60″ × 32″ and smaller)	$^1/_2$
Bathtubs[a] (larger than 60″ × 32″)	$^1/_2$
Bidet	$^3/_8$
Combination sink and tray	$^1/_2$
Dishwasher, domestic[a]	$^1/_2$
Drinking fountain	$^3/_8$
Hose bibbs	$^1/_2$
Kitchen sink[a]	$^1/_2$
Laundry, 1, 2 or 3 compartments[a]	$^1/_2$
Lavatory	$^3/_8$
Shower, single head[a]	$^1/_2$
Sinks, flushing rim	$^3/_4$
Sinks, service	$^1/_2$
Urinal, flush tank	$^1/_2$
Urinal, flushometer valve	$^3/_4$
Wall hydrant	$^1/_2$
Water closet, flush tank	$^3/_8$
Water closet, flushometer valve	1
Water closet, flushometer tank	$^3/_8$
Water closet, one piece[a]	$^1/_2$

For SI: 1 inch = 25.4 mm, 1 foot = 304.8 mm,
1 pound per square inch = 6.895 kPa.

a. Where the developed length of the distribution line is 60 feet or less, and the available pressure at the meter is 35 psi or greater, the minimum size of an individual distribution line supplied from a manifold and installed as part of a parallel water distribution system shall be one nominal tube size smaller than the sizes indicated.

installed to reduce the pressure in the building water distribution piping to not greater than 80 psi (552 kPa) static.

Exception: Service lines to sill cocks and outside hydrants, and main supply risers where pressure from the mains is reduced to 80 psi (552 kPa) or less at individual fixtures.

604.8.1 Valve design. The pressure-reducing valve shall be designed to remain open to permit uninterrupted water flow in case of valve failure.

604.8.2 Repair and removal. Water-pressure reducing valves, regulators and strainers shall be so constructed and installed as to permit repair or removal of parts without breaking a pipeline or removing the valve and strainer from the pipeline.

604.9 Water hammer. The flow velocity of the water distribution system shall be controlled to reduce the possibility of water hammer. A water-hammer arrestor shall be installed where quick-closing valves are utilized. Water-hammer arrestors shall be installed in accordance with the manufacturer's instructions. Water-hammer arrestors shall conform to ASSE 1010.

604.10 Gridded and parallel water distribution system manifolds. Hot water and cold water manifolds installed with gridded or parallel connected individual distribution lines to each fixture or fixture fitting shall be designed in accordance with Sections 604.10.1 through 604.10.3.

604.10.1 Manifold sizing. Hot water and cold water manifolds shall be sized in accordance with Table 604.10.1. The total gallons per minute is the demand of all outlets supplied.

TABLE 604.10.1
MANIFOLD SIZING

NOMINAL SIZE INTERNAL DIAMETER (inches)	MAXIMUM DEMAND (gpm)	
	Velocity at 4 feet per second	Velocity at 8 feet per second
$^1/_2$	2	5
$^3/_4$	6	11
1	10	20
$1^1/_4$	15	31
$1^1/_2$	22	44

For SI: 1 inch = 25.4 mm, 1 gallon per minute = 3.785 L/m,
1 foot per second = 0.305 m/s.

604.10.2 Valves. Individual fixture shutoff valves installed at the manifold shall be identified as to the fixture being supplied.

604.10.3 Access. *Access* shall be provided to manifolds with integral factory- or field-installed valves.

604.11 Individual pressure balancing in-line valves for individual fixture fittings. Where individual pressure balancing in-line valves for individual fixture fittings are installed, such valves shall comply with ASSE 1066. Such valves shall be installed in an accessible location and shall not be utilized alone as a substitute for the balanced pressure, thermostatic or combination shower valves required in Section 424.3.

SECTION 605
MATERIALS, JOINTS AND CONNECTIONS

605.1 Soil and ground water. The installation of a water service or water distribution pipe shall be prohibited in soil and ground water contaminated with solvents, fuels, organic compounds or other detrimental materials causing permeation, corrosion, degradation or structural failure of the piping material. Where detrimental conditions are suspected, a chemical analysis of the soil and ground water conditions shall be required to ascertain the acceptability of the water service or water distribution piping material for the specific installation. Where detrimental conditions exist, *approved* alternative materials or routing shall be required.

605.2 Lead content of water supply pipe and fittings. Pipe and pipe fittings, including valves and faucets, utilized in the water supply system shall have a maximum of 8-percent lead content.

605.3 Water service pipe. Water service pipe shall conform to NSF 61 and shall conform to one of the standards listed in Table 605.3. Water service pipe or tubing, installed underground and outside of the structure, shall have a working pressure rating of not less than 160 psi (1100 kPa) at 73.4°F (23°C). Where the water pressure exceeds 160 psi (1100 kPa), piping material shall have a working pressure rating not less than the highest available pressure. Water service piping materials not third-party certified for water distribution shall terminate at or before the full open valve located at the entrance to the structure. All ductile iron water service piping shall be cement mortar lined in accordance with AWWA C104.

605.3.1 Dual check-valve-type backflow preventer. Dual check-valve backflow preventers installed on the water supply system shall comply with ASSE 1024 or CSA B64.6.

605.4 Water distribution pipe. Water distribution pipe shall conform to NSF 61 and shall conform to one of the standards listed in Table 605.4. Hot water distribution pipe and tubing shall have a pressure rating of not less than 100 psi (690 kPa) at 180°F (82°C).

605.5 Fittings. Pipe fittings shall be *approved* for installation with the piping material installed and shall comply with the applicable standards listed in Table 605.5. Pipe fittings utilized in water supply systems shall also comply with NSF 61. Ductile and gray iron pipe fittings shall be cement mortar lined in accordance with AWWA C104.

605.5.1 Mechanically formed tee fittings. Mechanically extracted outlets shall have a height not less than three times the thickness of the branch tube wall.

605.5.1.1 Full flow assurance. Branch tubes shall not restrict the flow in the run tube. A dimple/depth stop shall be formed in the branch tube to ensure that penetration into the collar is of the correct depth. For inspection purposes, a second dimple shall be placed $^1/_4$ inch (6.4 mm) above the first dimple. Dimples shall be aligned with the tube run.

605.5.1.2 Brazed joints. Mechanically formed tee fittings shall be brazed in accordance with Section 605.14.1.

605.6 Flexible water connectors. Flexible water connectors exposed to continuous pressure shall conform to ASME A112.18.6/CSA B125.6. *Access* shall be provided to all flexible water connectors.

605.7 Valves. All valves shall be of an *approved* type and compatible with the type of piping material installed in the system. Ball valves, gate valves, globe valves and plug valves intended to supply drinking water shall meet the requirements of NSF 61.

605.8 Manufactured pipe nipples. Manufactured pipe nipples shall conform to one of the standards listed in Table 605.8.

TABLE 605.3
WATER SERVICE PIPE

MATERIAL	STANDARD
Acrylonitrile butadiene styrene (ABS) plastic pipe	ASTM D 1527; ASTM D 2282
Asbestos-cement pipe	ASTM C 296
Brass pipe	ASTM B 43
Chlorinated polyvinyl chloride (CPVC) plastic pipe	ASTM D 2846; ASTM F 441; ASTM F 442; CSA B137.6
Copper or copper-alloy pipe	ASTM B 42; ASTM B 302
Copper or copper-alloy tubing (Type K, WK, L, WL, M or WM)	ASTM B 75; ASTM B 88; ASTM B 251; ASTM B 447
Cross-linked polyethylene (PEX) plastic pipe and tubing	ASTM F 876; ASTM F 877; AWWA C904; CSA B137.5
Cross-linked polyethylene/aluminum/cross-linked polyethylene (PEX-AL-PEX) pipe	ASTM F 1281; ASTM F 2262; CSA B137.10M
Cross-linked polyethylene/aluminum/high-density polyethylene (PEX-AL-HDPE)	ASTM F 1986
Ductile iron water pipe	AWWA C151/A21.51; AWWA C115/A21.15
Galvanized steel pipe	ASTM A 53
Polyethylene (PE) plastic pipe	ASTM D 2239; ASTM D 3035; AWWA C901; CSA B137.1
Polyethylene (PE) plastic tubing	ASTM D 2737; AWWA C901; CSA B137.1
Polyethylene/aluminum/polethylene (PE-AL-PE) pipe	ASTM F 1282; CSA B137.9
Polyethylene of raised temperature (PE-RT) plastic tubing	ASTM F 2769
Polypropylene (PP) plastic pipe or tubing	ASTM F 2389; CSA B137.11
Polyvinyl chloride (PVC) plastic pipe	ASTM D 1785; ASTM D 2241; ASTM D 2672; CSA B137.3
Stainless steel pipe (Type 304/304L)	ASTM A 312; ASTM A 778
Stainless steel pipe (Type 316/316L)	ASTM A 312; ASTM A 778

TABLE 605.4
WATER DISTRIBUTION PIPE

MATERIAL	STANDARD
Brass pipe	ASTM B 43
Chlorinated polyvinyl chloride (CPVC) plastic pipe and tubing	ASTM D 2846; ASTM F 441; ASTM F 442; CSA B137.6
Copper or copper-alloy pipe	ASTM B 42; ASTM B 302
Copper or copper-alloy tubing (Type K, WK, L, WL, M or WM)	ASTM B 75; ASTM B 88; ASTM B 251; ASTM B 447
Cross-linked polyethylene (PEX) plastic tubing	ASTM F 876; ASTM F 877; CSA B137.5
Cross-linked polyethylene/aluminum/cross-linked polyethylene (PEX-AL-PEX) pipe	ASTM F 1281; ASTM F 2262; CSA B137.10M
Cross-linked polyethylene/aluminum/high-density polyethylene (PEX-AL-HDPE)	ASTM F 1986
Ductile iron pipe	AWWA C151/A21.51; AWWA C115/A21.15
Galvanized steel pipe	ASTM A 53
Polyethylene/aluminum/polyethylene (PE-AL-PE) composite pipe	ASTM F 1282
Polyethylene of raised temperature (PE-RT) plastic tubing	ASTM F 2769
Polypropylene (PP) plastic pipe or tubing	ASTM F 2389; CSA B137.11
Stainless steel pipe (Type 304/304L)	ASTM A 312; ASTM A 778
Stainless steel pipe (Type 316/316L)	ASTM A 312; ASTM A 778

TABLE 605.5
PIPE FITTINGS

MATERIAL	STANDARD
Acrylonitrile butadiene styrene (ABS) plastic	ASTM D 2468
Cast-iron	ASME B16.4; ASME B16.12
Chlorinated polyvinyl chloride (CPVC) plastic	ASSE 1061; ASTM D 2846; ASTM F 437; ASTM F 438; ASTM F 439; CSA B137.6
Copper or copper alloy	ASSE 1061; ASME B16.15; ASME B16.18; ASME B16.22; ASME B16.23; ASME B16.26; ASME B16.29
Cross-linked polyethylene/aluminum/high-density polyethylene (PEX-AL-HDPE)	ASTM F 1986
Fittings for cross-linked polyethylene (PEX) plastic tubing	ASSE 1061, ASTM F 877; ASTM F 1807; ASTM F 1960; ASTM F 2080; ASTM F 2098, ASTM F 2159; ASTM F 2434; ASTM F 2735; CSA B137.5
Fittings for polyethylene of raised temperature (PE-RT) plastic tubing	ASTM F 1807; ASTM F 2098; ASTM F 2159; ASTM F 2735
Gray iron and ductile iron	AWWA C110/A21.10; AWWA C153/A21.53
Insert fittings for polyethylene/aluminum/polyethylene (PE-AL-PE) and cross-linked polyethylene/aluminum/cross-linked polyethylene (PEX-AL-PEX)	ASTM F 1974; ASTM F 1281; ASTM F 1282; CSA B137.9; CSA B137.10M
Malleable iron	ASME B16.3
Metal (brass) insert fittings for polyethylene/aluminum/polyethylene (PE-AL-PE) and cross-linked polyethylene/aluminum/cross-linked polyethylene (PEX-AL-PEX)	ASTM F 1974
Polyethylene (PE) plastic pipe	ASTM D 2609; ASTM D 2683; ASTM D 3261; ASTM F 1055; CSA B137.1
Polypropylene (PP) plastic pipe or tubing	ASTM F 2389; CSA B137.11
Polyvinyl chloride (PVC) plastic	ASTM D 2464; ASTM D 2466; ASTM D 2467; CSA B137.2; CSA B137.3
Stainless steel (Type 304/304L)	ASTM A 312; ASTM A 778
Stainless steel (Type 316/316L)	ASTM A 312; ASTM A 778
Steel	ASME B16.9; ASME B16.11; ASME B16.28

**TABLE 605.8
MANUFACTURED PIPE NIPPLES**

MATERIAL	STANDARD
Brass-, copper-, chromium-plated	ASTM B 687
Steel	ASTM A 733

605.9 Prohibited joints and connections. The following types of joints and connections shall be prohibited:

1. Cement or concrete joints.

2. Joints made with fittings not *approved* for the specific installation.

3. Solvent-cement joints between different types of plastic pipe.

4. Saddle-type fittings.

605.10 ABS plastic. Joints between ABS plastic pipe or fittings shall comply with Sections 605.10.1 through 605.10.3.

605.10.1 Mechanical joints. Mechanical joints on water pipes shall be made with an elastomeric seal conforming to ASTM D 3139. Mechanical joints shall only be installed in underground systems, unless otherwise *approved*. Joints shall be installed only in accordance with the manufacturer's instructions.

605.10.2 Solvent cementing. Joint surfaces shall be clean and free from moisture. Solvent cement that conforms to ASTM D 2235 shall be applied to all joint surfaces. The joint shall be made while the cement is wet. Joints shall be made in accordance with ASTM D 2235. Solvent-cement joints shall be permitted above or below ground.

605.10.3 Threaded joints. Threads shall conform to ASME B1.20.1. Schedule 80 or heavier pipe shall be permitted to be threaded with dies specifically designed for plastic pipe. *Approved* thread lubricant or tape shall be applied on the male threads only.

605.11 Asbestos-cement. Joints between asbestos-cement pipe or fittings shall be made with a sleeve coupling of the same composition as the pipe, sealed with an elastomeric ring conforming to ASTM D 1869.

605.12 Brass. Joints between brass pipe and fittings shall comply with Sections 605.12.1 through 605.12.4.

605.12.1 Brazed joints. All joint surfaces shall be cleaned. An *approved* flux shall be applied where required. The joint shall be brazed with a filler metal conforming to AWS A5.8.

605.12.2 Mechanical joints. Mechanical joints shall be installed in accordance with the manufacturer's instructions.

605.12.3 Threaded joints. Threads shall conform to ASME B1.20.1. Pipe-joint compound or tape shall be applied on the male threads only.

605.12.4 Welded joints. All joint surfaces shall be cleaned. The joint shall be welded with an *approved* filler metal.

605.13 Gray iron and ductile iron joints. Joints for gray and ductile iron pipe and fittings shall comply with AWWA C111 and shall be installed in accordance with the manufacturer's instructions.

605.14 Copper pipe. Joints between copper or copper-alloy pipe or fittings shall comply with Sections 605.14.1 through 605.14.5.

605.14.1 Brazed joints. All joint surfaces shall be cleaned. An *approved* flux shall be applied where required. The joint shall be brazed with a filler metal conforming to AWS A5.8.

605.14.2 Mechanical joints. Mechanical joints shall be installed in accordance with the manufacturer's instructions.

605.14.3 Soldered joints. Solder joints shall be made in accordance with the methods of ASTM B 828. Cut tube ends shall be reamed to the full inside diameter of the tube end. Joint surfaces shall be cleaned. A flux conforming to ASTM B 813 shall be applied. The joint shall be soldered with a solder conforming to ASTM B 32. The joining of water supply piping shall be made with lead-free solder and fluxes. "Lead free" shall mean a chemical composition equal to or less than 0.2-percent lead.

605.14.4 Threaded joints. Threads shall conform to ASME B1.20.1. Pipe-joint compound or tape shall be applied on the male threads only.

605.14.5 Welded joints. Joint surfaces shall be cleaned. The joint shall be welded with an *approved* filler metal.

605.15 Copper tubing. Joints between copper or copper-alloy tubing or fittings shall comply with Sections 605.15.1 through 605.15.4.

605.15.1 Brazed joints. Joint surfaces shall be cleaned. An *approved* flux shall be applied where required. The joint shall be brazed with a filler metal conforming to AWS A5.8.

605.15.2 Flared joints. Flared joints for water pipe shall be made by a tool designed for that operation.

605.15.3 Mechanical joints. Mechanical joints shall be installed in accordance with the manufacturer's instructions.

605.15.4 Soldered joints. Solder joints shall be made in accordance with the methods of ASTM B 828. All cut tube ends shall be reamed to the full inside diameter of the tube end. All joint surfaces shall be cleaned. A flux conforming to ASTM B 813 shall be applied. The joint shall be soldered with a solder conforming to ASTM B 32. The joining of water supply piping shall be made with lead-free solders and fluxes. "Lead free" shall mean a chemical composition equal to or less than 0.2-percent lead.

605.16 CPVC plastic. Joints between CPVC plastic pipe or fittings shall comply with Sections 605.16.1 through 605.16.3.

605.16.1 Mechanical joints. Mechanical joints shall be installed in accordance with the manufacturer's instructions.

605.16.2 Solvent cementing. Joint surfaces shall be clean and free from moisture, and an *approved* primer shall be applied. Solvent cement, orange in color and conforming to ASTM F 493, shall be applied to joint surfaces. The joint shall be made while the cement is wet, and in accordance with ASTM D 2846 or ASTM F 493. Solvent-cement joints shall be permitted above or below ground.

> **Exception:** A primer is not required where all of the following conditions apply:
>
> 1. The solvent cement used is third-party certified as conforming to ASTM F 493.
> 2. The solvent cement used is yellow in color.
> 3. The solvent cement is used only for joining $^1/_2$ inch (12.7 mm) through 2 inch (51 mm) diameter CPVC pipe and fittings.
> 4. The CPVC pipe and fittings are manufactured in accordance with ASTM D 2846.

605.16.3 Threaded joints. Threads shall conform to ASME B1.20.1. Schedule 80 or heavier pipe shall be permitted to be threaded with dies specifically designed for plastic pipe, but the pressure rating of the pipe shall be reduced by 50 percent. Thread by socket molded fittings shall be permitted. *Approved* thread lubricant or tape shall be applied on the male threads only.

605.17 Cross-linked polyethylene plastic. Joints between cross-linked polyethylene plastic tubing or fittings shall comply with Sections 605.17.1 and 605.17.2.

605.17.1 Flared joints. Flared pipe ends shall be made by a tool designed for that operation.

605.17.2 Mechanical joints. Mechanical joints shall be installed in accordance with the manufacturer's instructions. Fittings for cross-linked polyethylene (PEX) plastic tubing shall comply with the applicable standards listed in Table 605.5 and shall be installed in accordance with the manufacturer's instructions. PEX tubing shall be factory marked with the appropriate standards for the fittings that the PEX manufacturer specifies for use with the tubing.

605.18 Steel. Joints between galvanized steel pipe or fittings shall comply with Sections 605.18.1 and 605.18.2.

605.18.1 Threaded joints. Threads shall conform to ASME B1.20.1. Pipe-joint compound or tape shall be applied on the male threads only.

605.18.2 Mechanical joints. Joints shall be made with an *approved* elastomeric seal. Mechanical joints shall be installed in accordance with the manufacturer's instructions.

605.19 Polyethylene plastic. Joints between polyethylene plastic pipe and tubing or fittings shall comply with Sections 605.19.1 through 605.19.4.

605.19.1 Flared joints. Flared joints shall be permitted where so indicated by the pipe manufacturer. Flared joints shall be made by a tool designed for that operation.

605.19.2 Heat-fusion joints. Joint surfaces shall be clean and free from moisture. All joint surfaces shall be heated to melt temperature and joined. The joint shall be undisturbed until cool. Joints shall be made in accordance with ASTM D 2657.

605.19.3 Mechanical joints. Mechanical joints shall be installed in accordance with the manufacturer's instructions.

605.19.4 Installation. Polyethylene pipe shall be cut square, with a cutter designed for plastic pipe. Except where joined by heat fusion, pipe ends shall be chamfered to remove sharp edges. Kinked pipe shall not be installed. The minimum pipe bending radius shall not be less than 30 pipe diameters, or the minimum coil radius, whichever is greater. Piping shall not be bent beyond straightening of the curvature of the coil. Bends shall not be permitted within 10 pipe diameters of any fitting or valve. Stiffener inserts installed with compression-type couplings and fittings shall not extend beyond the clamp or nut of the coupling or fitting.

605.20 Polypropylene (PP) plastic. Joints between PP plastic pipe and fittings shall comply with Section 605.20.1 or 605.20.2.

605.20.1 Heat-fusion joints. Heat-fusion joints for polypropylene pipe and tubing joints shall be installed with socket-type heat-fused polypropylene fittings, butt-fusion polypropylene fittings or electrofusion polypropylene fittings. Joint surfaces shall be clean and free from moisture. The joint shall be undisturbed until cool. Joints shall be made in accordance with ASTM F 2389.

605.20.2 Mechanical and compression sleeve joints. Mechanical and compression sleeve joints shall be installed in accordance with the manufacturer's instructions.

605.21 Polyethylene/aluminum/polyethylene (PE-AL-PE) and cross-linked polyethylene/aluminum/cross-linked polyethylene (PEX-AL-PEX). Joints between PE-AL-PE and PEX-AL-PEX pipe and fittings shall comply with Section 605.21.1.

605.21.1 Mechanical joints. Mechanical joints shall be installed in accordance with the manufacturer's instructions. Fittings for PE-AL-PE and PEX-AL-PEX as described in ASTM F 1974, ASTM F 1281, ASTM F 1282, CSA B137.9 and CSA B137.10M shall be installed in accordance with the manufacturer's instructions.

605.22 PVC plastic. Joints between PVC plastic pipe or fittings shall comply with Sections 605.22.1 through 605.22.3.

605.22.1 Mechanical joints. Mechanical joints on water pipe shall be made with an elastomeric seal conforming to ASTM D 3139. Mechanical joints shall not be installed in above-ground systems unless otherwise *approved*. Joints shall be installed in accordance with the manufacturer's instructions.

605.22.2 Solvent cementing. Joint surfaces shall be clean and free from moisture. A purple primer that conforms to ASTM F 656 shall be applied. Solvent cement not purple in color and conforming to ASTM D 2564 or CSA B137.3 shall be applied to all joint surfaces. The joint shall be made while the cement is wet and shall be in accordance with ASTM D 2855. Solvent-cement joints shall be permitted above or below ground.

605.22.3 Threaded joints. Threads shall conform to ASME B1.20.1. Schedule 80 or heavier pipe shall be permitted to be threaded with dies specifically designed for plastic pipe, but the pressure rating of the pipe shall be reduced by 50 percent. Thread by socket molded fittings shall be permitted. *Approved* thread lubricant or tape shall be applied on the male threads only.

605.23 Stainless steel. Joints between stainless steel pipe and fittings shall comply with Sections 605.23.1 and 605.23.2.

605.23.1 Mechanical joints. Mechanical joints shall be installed in accordance with the manufacturer's instructions.

605.23.2 Welded joints. All joint surfaces shall be cleaned. The joint shall be welded autogenously or with an *approved* filler metal as referenced in ASTM A 312.

605.24 Joints between different materials. Joints between different piping materials shall be made with a mechanical joint of the compression or mechanical-sealing type, or as permitted in Sections 605.24.1, 605.24.2 and 605.24.3. Connectors or adapters shall have an elastomeric seal conforming to ASTM D 1869 or ASTM F 477. Joints shall be installed in accordance with the manufacturer's instructions.

605.24.1 Copper or copper-alloy tubing to galvanized steel pipe. Joints between copper or copper-alloy tubing and galvanized steel pipe shall be made with a brass fitting or dielectric fitting or a dielectric union conforming to ASSE 1079. The copper tubing shall be soldered to the fitting in an *approved* manner, and the fitting shall be screwed to the threaded pipe.

605.24.2 Plastic pipe or tubing to other piping material. Joints between different grades of plastic pipe or between plastic pipe and other piping material shall be made with an *approved* adapter fitting.

605.24.3 Stainless steel. Joints between stainless steel and different piping materials shall be made with a mechanical joint of the compression or mechanical sealing type or a dielectric fitting or a dielectric union conforming to ASSE 1079.

605.25 Polyethylene of raised temperature plastic. Joints between polyethylene of raised temperature plastic tubing and fittings shall be in accordance with Sections 605.25.1 and 605.25.2.

605.25.1 Flared joints. Flared pipe ends shall be made by a tool designed for that operation.

605.25.2 Mechanical joints. Mechanical joints shall be installed in accordance with the manufacturer's instructions. Fittings for polyethylene of raised temperature plastic tubing shall comply with the applicable standards listed in Table 605.5 and shall be installed in accordance with the manufacturer's instructions. Polyethylene of raised temperature plastic tubing shall be factory marked with the applicable standards for the fittings that the manufacturer of the tubing specifies for use with the tubing.

SECTION 606
INSTALLATION OF THE BUILDING WATER DISTRIBUTION SYSTEM

606.1 Location of full-open valves. Full-open valves shall be installed in the following locations:

1. On the building water service pipe from the public water supply near the curb.

2. On the water distribution supply pipe at the entrance into the structure.

3. On the discharge side of every water meter.

4. On the base of every water riser pipe in occupancies other than multiple-family residential *occupancies* that are two stories or less in height and in one- and two-family residential *occupancies*.

5. On the top of every water down-feed pipe in *occupancies* other than one- and two-family residential *occupancies*.

6. On the entrance to every water supply pipe to a dwelling unit, except where supplying a single fixture equipped with individual stops.

7. On the water supply pipe to a gravity or pressurized water tank.

8. On the water supply pipe to every water heater.

606.2 Location of shutoff valves. Shutoff valves shall be installed in the following locations:

1. On the fixture supply to each plumbing fixture other than bathtubs and showers in one- and two-family residential *occupancies*, and other than in individual sleeping units that are provided with unit shutoff valves in hotels, motels, boarding houses and similar *occupancies*.

2. On the water supply pipe to each sillcock.

3. On the water supply pipe to each appliance or mechanical equipment.

606.3 Access to valves. *Access* shall be provided to all full-open valves and shutoff valves.

606.4 Valve identification. Service and hose bibb valves shall be identified. All other valves installed in locations that are not adjacent to the fixture or appliance shall be identified, indicating the fixture or appliance served.

606.5 Water pressure booster systems. Water pressure booster systems shall be provided as required by Sections 606.5.1 through 606.5.10.

606.5.1 Water pressure booster systems required. Where the water pressure in the public water main or individual water supply system is insufficient to supply the minimum pressures and quantities specified in this code, the supply shall be supplemented by an elevated water tank, a hydropneumatic pressure booster system or a water pressure booster pump installed in accordance with Section 606.5.5.

606.5.2 Support. All water supply tanks shall be supported in accordance with the *International Building Code*.

606.5.3 Covers. All water supply tanks shall be covered to keep out unauthorized persons, dirt and vermin. The covers of gravity tanks shall be vented with a return bend vent pipe with an area not less than the area of the down-feed riser pipe, and the vent shall be screened with a corrosion-resistant screen of not less than 16 by 20 mesh per inch (630 by 787 mesh per m).

606.5.4 Overflows for water supply tanks. A gravity or suction water supply tank shall be provided with an overflow with a diameter not less than that shown in Table 606.5.4. The overflow outlet shall discharge at a point not less than 6 inches (152 mm) above the roof or roof drain; floor or floor drain; or over an open water-supplied fixture. The overflow outlet shall be covered with a corrosion-resistant screen of not less than 16 by 20 mesh per inch (630 by 787 mesh per m) and by $^1/_4$-inch (6.4 mm) hardware cloth or shall terminate in a horizontal angle seat check valve. Drainage from overflow pipes shall be directed so as not to freeze on roof walks.

TABLE 606.5.4
SIZES FOR OVERFLOW PIPES FOR WATER SUPPLY TANKS

MAXIMUM CAPACITY OF WATER SUPPLY LINE TO TANK (gpm)	DIAMETER OF OVERFLOW PIPE (inches)
0 – 50	2
50 – 150	$2^1/_2$
150 – 200	3
200 – 400	4
400 – 700	5
700 – 1,000	6
Over 1,000	8

For SI: 1 inch = 25.4 mm, 1 gallon per minute = 3.785 L/m.

606.5.5 Low-pressure cutoff required on booster pumps. A low-pressure cutoff shall be installed on all booster pumps in a water pressure booster system to prevent creation of a vacuum or negative pressure on the suction side of the pump when a positive pressure of 10 psi (68.94 kPa) or less occurs on the suction side of the pump.

606.5.6 Potable water inlet control and location. Potable water inlets to gravity tanks shall be controlled by a fill valve or other automatic supply valve installed so as to prevent the tank from overflowing. The inlet shall be terminated so as to provide an *air gap* not less than 4 inches (102 mm) above the overflow.

606.5.7 Tank drain pipes. A valved pipe shall be provided at the lowest point of each tank to permit emptying of the tank. The tank drain pipe shall discharge as required for overflow pipes and shall not be smaller in size than specified in Table 606.5.7.

TABLE 606.5.7
SIZE OF DRAIN PIPES FOR WATER TANKS

TANK CAPACITY (gallons)	DRAIN PIPE (inches)
Up to 750	1
751 to 1,500	$1^1/_2$
1,501 to 3,000	2
3,001 to 5,000	$2^1/_2$
5,000 to 7,500	3
Over 7,500	4

For SI: 1 inch = 25.4 mm, 1 gallon = 3.785 L.

606.5.8 Prohibited location of potable supply tanks. Potable water gravity tanks or manholes of potable water pressure tanks shall not be located directly under any soil or waste piping or any source of contamination.

606.5.9 Pressure tanks, vacuum relief. All water pressure tanks shall be provided with a vacuum relief valve at the top of the tank that will operate up to a maximum water pressure of 200 psi (1380 kPa) and up to a maximum temperature of 200°F (93°C). The size of such vacuum relief valve shall be not less than $^1/_2$ inch (12.7 mm).

Exception: This section shall not apply to pressurized captive air diaphragm/bladder tanks.

606.5.10 Pressure relief for tanks. Every pressure tank in a hydropneumatic pressure booster system shall be protected with a pressure relief valve. The pressure relief valve shall be set at a maximum pressure equal to the rating of the tank. The relief valve shall be installed on the supply pipe to the tank or on the tank. The relief valve shall discharge by gravity to a safe place of disposal.

606.6 Water supply system test. Upon completion of a section of or the entire water supply system, the system, or portion completed, shall be tested in accordance with Section 312.

606.7 Labeling of water distribution pipes in bundles. Where water distribution piping is bundled at installation, each pipe in the bundle shall be indentified using stenciling or commercially available pipe labels. The identification shall indicate the pipe contents and the direction of flow in the pipe. The interval of the identification markings on the pipe shall not exceed 25 feet (7620 mm). There shall be not less than one identification label on each pipe in each room, space or story.

SECTION 607
HOT WATER SUPPLY SYSTEM

607.1 Where required. In residential *occupancies*, *hot water* shall be supplied to plumbing fixtures and equipment utilized for bathing, washing, culinary purposes, cleansing, laundry or building maintenance. In nonresidential *occupancies*, *hot water* shall be supplied for culinary purposes, cleansing, laundry or building maintenance purposes. In nonresidential *occupancies*, *hot water* or *tempered water* shall be supplied for bathing and washing purposes.

607.1.1 Temperature limiting means. A thermostat control for a water heater shall not serve as the temperature limiting means for the purposes of complying with the requirements of this code for maximum allowable hot or tempered water delivery temperature at fixtures.

607.1.2 Tempered water temperature control. *Tempered water* shall be supplied through a water temperature limiting device that conforms to ASSE 1070 and shall limit the *tempered water* to a maximum of 110°F (43°C). This provision shall not supersede the requirement for protective shower valves in accordance with Section 424.3.

607.2 Hot or tempered water supply to fixtures. The developed length of hot or tempered water piping, from the source of hot water to the fixtures that require hot or tempered water, shall not exceed 50 feet (15 240 mm). Recirculating system piping and heat-traced piping shall be considered to be sources of hot or tempered water.

[E] 607.2.1 Hot water system controls. Automatic circulating hot water system pumps or heat trace shall be arranged to be conveniently turned off, automatically or manually, when the hot water system is not in operation.

607.2.2 Recirculating pump. Where a thermostatic mixing valve is used in a system with a hot water recirculating pump, the *hot water* or *tempered water* return line shall be routed to the cold water inlet pipe of the water heater and the cold water inlet pipe or the hot water return connection of the thermostatic mixing valve.

607.3 Thermal expansion control. A means of controlling increased pressure caused by thermal expansion shall be provided where required in accordance with Sections 607.3.1 and 607.3.2.

607.3.1 Pressure-reducing valve. For water service system sizes up to and including 2 inches (51 mm), a device for controlling pressure shall be installed where, because of thermal expansion, the pressure on the downstream side of a pressure-reducing valve exceeds the pressure-reducing valve setting.

607.3.2 Backflow prevention device or check valve. Where a backflow prevention device, check valve or other device is installed on a water supply system utilizing storage water heating equipment such that thermal expansion causes an increase in pressure, a device for controlling pressure shall be installed.

607.4 Flow of hot water to fixtures. Fixture fittings, faucets and diverters shall be installed and adjusted so that the flow of hot water from the fittings corresponds to the left-hand side of the fixture fitting.

Exception: Shower and tub/shower mixing valves conforming to ASSE 1016 or ASME A112.18.1/CSA B125.1, where the flow of hot water corresponds to the markings on the device.

[E] 607.5 Pipe insulation. Hot water piping in automatic temperature maintenance systems shall be insulated with 1 inch (25 mm) of insulation having a conductivity not exceeding 0.27 Btu per inch/h • ft² • °F (1.53 W per 25 mm/m² • K). The first 8 feet (2438 mm) of hot water piping from a *hot water* source that does not have heat traps shall be insulated with 0.5 inch (12.7 mm) of material having a conductivity not exceeding 0.27 Btu per inch/h • ft² • °F (1.53 W per 25 mm/ m² • K).

SECTION 608
PROTECTION OF POTABLE WATER SUPPLY

608.1 General. A potable water supply system shall be designed, installed and maintained in such a manner so as to prevent contamination from nonpotable liquids, solids or gases being introduced into the potable water supply through cross-connections or any other piping connections to the system. Backflow preventer applications shall conform to Table 608.1, except as specifically stated in Sections 608.2 through 608.16.10.

608.2 Plumbing fixtures. The supply lines and fittings for plumbing fixtures shall be installed so as to prevent backflow. Plumbing fixture fittings shall provide backflow protection in accordance with ASME A112.18.1/CSA B125.1.

608.3 Devices, appurtenances, appliances and apparatus. Devices, appurtenances, appliances and apparatus intended to serve some special function, such as sterilization, distillation, processing, cooling, or storage of ice or foods, and that connect to the water supply system, shall be provided with protection against backflow and contamination of the water supply system. Water pumps, filters, softeners, tanks and other appliances and devices that handle or treat potable water shall be protected against contamination.

608.3.1 Special equipment, water supply protection. The water supply for hospital fixtures shall be protected against backflow with a reduced pressure principle backflow prevention assembly, an atmospheric or spill-resistant vacuum breaker assembly, or an air gap. Vacuum breakers for bedpan washer hoses shall not be located less than 5 feet (1524 mm) above the floor. Vacuum breakers for hose connections in health care or laboratory areas shall not be less than 6 feet (1829 mm) above the floor.

608.4 Water service piping. Water service piping shall be protected in accordance with Sections 603.2 and 603.2.1.

608.5 Chemicals and other substances. Chemicals and other substances that produce either toxic conditions, taste, odor or discoloration in a potable water system shall not be introduced into, or utilized in, such systems.

608.6 Cross-connection control. Cross connections shall be prohibited, except where approved methods are installed to protect the potable water supply.

608.6.1 Private water supplies. Cross connections between a private water supply and a potable public supply shall be prohibited.

608.7 Valves and outlets prohibited below grade. Potable water outlets and combination stop-and-waste valves shall not be installed underground or below grade. Freezeproof yard hydrants that drain the riser into the ground are considered to be stop-and-waste valves.

Exception: Freezeproof yard hydrants that drain the riser into the ground shall be permitted to be installed, provided that the potable water supply to such hydrants is protected upstream of the hydrants in accordance with Section 608 and the hydrants are permanently identified as nonpotable outlets by *approved* signage that reads as follows: "Caution, Nonpotable Water. Do Not Drink."

608.8 Identification of nonpotable water. Where nonpotable water systems are installed, the piping conveying the nonpotable water shall be identified either by color marking or metal tags in accordance with Sections 608.8.1 through 608.8.3. All nonpotable water outlets such as hose connections, open ended pipes, and faucets shall be identified at the point of use for each outlet with the words, "Nonpotable—not safe for drinking." The words shall be indelibly printed on a tag or sign constructed of corrosion-resistant waterproof material or shall be indelibly printed on the fixture. The letters of the words shall be not less than 0.5 inches (12.7 mm) in height and in colors in contrast to the background on which they are applied.

608.8.1 Information. Pipe identification shall include the contents of the piping system and an arrow indicating the direction of flow. Hazardous piping systems shall also contain information addressing the nature of the hazard. Pipe identification shall be repeated at intervals not exceeding 25 feet (7620 mm) and at each point where the piping passes through a wall, floor or roof. Lettering shall be readily observable within the room or space where the piping is located.

608.8.2 Color. The color of the pipe identification shall be discernable and consistent throughout the building. The color purple shall be used to identify reclaimed, rain and gray water distribution systems.

608.8.3 Size. The size of the background color field and lettering shall comply with Table 608.8.3.

TABLE 608.8.3
SIZE OF PIPE IDENTIFICATION

PIPE DIAMETER (inches)	LENGTH BACKGROUND COLOR FIELD (inches)	SIZE OF LETTERS (inches)
$^3/_4$ to $1^1/_4$	8	0.5
$1^1/_2$ to 2	8	0.75
$2^1/_2$ to 6	12	1.25
8 to 10	24	2.5
over 10	32	3.5

For SI   1 inch = 25.4 mm.

608.9 Reutilization prohibited. Water utilized for the cooling of equipment or other processes shall not be returned to the potable water system. Such water shall be discharged into a drainage system through an *air gap* or shall be utilized for nonpotable purposes.

608.10 Reuse of piping. Piping that has been utilized for any purpose other than conveying potable water shall not be utilized for conveying potable water.

608.11 Painting of water tanks. The interior surface of a potable water tank shall not be lined, painted or repaired with any material that changes the taste, odor, color or potability of the water supply when the tank is placed in, or returned to, service.

608.12 Pumps and other appliances. Water pumps, filters, softeners, tanks and other devices that handle or treat potable water shall be protected against contamination.

608.13 Backflow protection. Means of protection against backflow shall be provided in accordance with Sections 608.13.1 through 608.13.9.

608.13.1 Air gap. The minimum required *air gap* shall be measured vertically from the lowest end of a potable water outlet to the *flood level rim* of the fixture or receptacle into which such potable water outlet discharges. Air gaps shall comply with ASME A112.1.2 and *air gap* fittings shall comply with ASME A112.1.3.

608.13.2 Reduced pressure principle backflow prevention assemblies Reduced pressure principle backflow prevention assemblies shall conform to ASSE 1013, AWWA C511, CSA B64.4 or CSA B64.4.1. Reduced pressure detector assembly backflow preventers shall conform to ASSE 1047. These devices shall be permitted to be installed where subject to continuous pressure conditions. The relief opening shall discharge by *air gap* and shall be prevented from being submerged.

608.13.3 Backflow preventer with intermediate atmospheric vent. Backflow preventers with intermediate atmospheric vents shall conform to ASSE 1012 or CSA B64.3. These devices shall be permitted to be installed where subject to continuous pressure conditions. The relief opening shall discharge by *air gap* and shall be prevented from being submerged.

608.13.4 Barometric loop. Barometric loops shall precede the point of connection and shall extend vertically to a height of 35 feet (10 668 mm). A barometric loop shall only be utilized as an atmospheric-type or pressure-type vacuum breaker.

608.13.5 Pressure vacuum breaker assemblies. Pressure vacuum breaker assemblies shall conform to ASSE 1020 or CSA B64.1.2. Spill-resistant vacuum breaker assemblies shall comply with ASSE 1056. These assemblies are designed for installation under continuous pressure conditions where the critical level is installed at the required height. Pressure vacuum breaker assemblies shall not be installed in locations where spillage could cause damage to the structure.

608.13.6 Atmospheric-type vacuum breakers. Pipe-applied atmospheric-type vacuum breakers shall conform to ASSE 1001 or CSA B64.1.1. Hose-connection vacuum breakers shall conform to ASSE 1011, ASSE 1019, ASSE 1035, ASSE 1052, CSA B64.2, CSA B64.2.1, CSA B64.2.1.1, CSA B64.2.2 or CSA B64.7. These devices shall operate under normal atmospheric pressure when the critical level is installed at the required height.

608.13.7 Double check-valve assemblies. Double check-valve assemblies shall conform to ASSE 1015, CSA B64.5, CSA B64.5.1 or AWWA C510. Double-detector check-valve assemblies shall conform to ASSE 1048. These devices shall be capable of operating under continuous pressure conditions.

TABLE 608.1
APPLICATION OF BACKFLOW PREVENTERS

DEVICE	DEGREE OF HAZARD[a]	APPLICATION[b]	APPLICABLE STANDARDS
Air gap	High or low hazard	Backsiphonage or backpressure	ASME A112.1.2
Air gap fittings for use with plumbing fixtures, appliances and appurtenances	High or low hazard	Backsiphonage or backpressure	ASME A112.1.3
Antisiphon-type fill valves for gravity water closet flush tanks	High hazard	Backsiphonage only	ASSE 1002, CSA B125.3
Backflow preventer for carbonated beverage machines	Low hazard	Backpressure or backsiphonage Sizes $^1/_4''$–$^3/_8''$	ASSE 1022
Backflow preventer with intermediate atmospheric vents	Low hazard	Backpressure or backsiphonage Sizes $^1/_4''$–$^3/_4''$	ASSE 1012, CSA B64.3
Barometric loop	High or low hazard	Backsiphonage only	(See Section 608.13.4)
Double check backflow prevention assembly and double check fire protection backflow prevention assembly	Low hazard	Backpressure or backsiphonage Sizes $^3/_8''$–16″	ASSE 1015, AWWA C510, CSA B64.5, CSA B64.5.1
Double check detector fire protection backflow prevention assemblies	Low hazard	Backpressure or backsiphonage (Fire sprinkler systems) Sizes 2″–16″	ASSE 1048
Dual-check-valve-type backflow preventer	Low hazard	Backpressure or backsiphonage Sizes $^1/_4''$–1″	ASSE 1024, CSA B64.6
Hose connection backflow preventer	High or low hazard	Low head backpressure, rated working pressure, backpressure or backsiphonage Sizes $^1/_2''$–1″	ASSE 1052, CSA B64.2.1.1
Hose connection vacuum breaker	High or low hazard	Low head backpressure or backsiphonage Sizes $^1/_2''$, $^3/_4''$, 1″	ASSE 1011, CSA B64.2, CSA B64.2.1
Laboratory faucet backflow preventer	High or low hazard	Low head backpressure and backsiphonage	ASSE 1035, CSA B64.7
Pipe-applied atmospheric-type vacuum breaker	High or low hazard	Backsiphonage only Sizes $^1/_4''$–4″	ASSE 1001, CSA B64.1.1
Pressure vacuum breaker assembly	High or low hazard	Backsiphonage only Sizes $^1/_2''$–2″	ASSE 1020, CSA B64.1.2
Reduced pressure principle backflow prevention assembly and reduced pressure principle fire protection backflow prevention assembly	High or low hazard	Backpressure or backsiphonage Sizes $^3/_8''$–16″	ASSE 1013, AWWA C511, CSA B64.4, CSA B64.4.1
Reduced pressure detector fire protection backflow prevention assemblies	High or low hazard	Backsiphonage or backpressure (Fire sprinkler systems)	ASSE 1047
Spill-resistant vacuum breaker assembly	High or low hazard	Backsiphonage only Sizes $^1/_4''$–2″	ASSE 1056
Vacuum breaker wall hydrants, frost-resistant, automatic draining type	High or low hazard	Low head backpressure or backsiphonage Sizes $^3/_4''$, 1″	ASSE 1019, CSA B64.2.2

For SI: 1 inch = 25.4 mm.

a. Low hazard—See Pollution (Section 202).
   High hazard—See Contamination (Section 202).
b. See Backpressure (Section 202).
   See Backpressure, low head (Section 202).
   See Backsiphonage (Section 202).

608.13.8 Spill-resistant pressure vacuum breaker assemblies. Spill-resistant pressure vacuum breaker assemblies shall conform to ASSE 1056 or CSA B64.1.3. These assemblies are designed for installation under continuous-pressure conditions where the critical level is installed at the required height.

608.13.9 Chemical dispenser backflow devices. Backflow devices for chemical dispensers shall comply with ASSE 1055 or shall be equipped with an *air gap* fitting.

608.14 Location of backflow preventers. Access shall be provided to backflow preventers as specified by the manufacturer's instructions.

608.14.1 Outdoor enclosures for backflow prevention devices. Outdoor enclosures for backflow prevention devices shall comply with ASSE 1060.

608.14.2 Protection of backflow preventers. Backflow preventers shall not be located in areas subject to freezing except where they can be removed by means of unions or are protected from freezing by heat, insulation or both.

608.14.2.1 Relief port piping. The termination of the piping from the relief port or *air gap* fitting of a backflow preventer shall discharge to an *approved* indirect waste receptor or to the outdoors where it will not cause damage or create a nuisance.

608.15 Protection of potable water outlets. All potable water openings and outlets shall be protected against backflow in accordance with Section 608.15.1, 608.15.2, 608.15.3, 608.15.4, 608.15.4.1 or 608.15.4.2.

608.15.1 Protection by air gap. Openings and outlets shall be protected by an *air gap* between the opening and the fixture *flood level rim* as specified in Table 608.15.1. Openings and outlets equipped for hose connection shall be protected by means other than an *air gap*.

608.15.2 Protection by reduced pressure principle backflow prevention assembly. Openings and outlets shall be protected by a reduced pressure principle backflow prevention assembly or a reduced pressure principle fire protection backflow prevention assembly on potable water supplies.

608.15.3 Protection by a backflow preventer with intermediate atmospheric vent. Openings and outlets shall be protected by a backflow preventer with an intermediate atmospheric vent.

608.15.4 Protection by a vacuum breaker. Openings and outlets shall be protected by atmospheric-type or pressure-type vacuum breakers. The critical level of the vacuum breaker shall be set not less than 6 inches (152 mm) above the *flood level rim* of the fixture or device. Fill valves shall be set in accordance with Section 425.3.1. Vacuum breakers shall not be installed under exhaust hoods or similar locations that will contain toxic fumes or vapors. Pipe-applied vacuum breakers shall be installed not less than 6 inches (152 mm) above the *flood level rim* of the fixture, receptor or device served.

608.15.4.1 Deck-mounted and integral vacuum breakers. *Approved* deck-mounted or equipment-mounted vacuum breakers and faucets with integral atmospheric vacuum breakers or spill-resistant vacuum breaker assemblies shall be installed in accordance with the manufacturer's instructions and the requirements for labeling with the critical level not less than 1 inch (25 mm) above the *flood level rim*.

608.15.4.2 Hose connections. Sillcocks, hose bibbs, wall hydrants and other openings with a hose connection shall be protected by an atmospheric-type or pressure-type vacuum breaker or a permanently attached hose connection vacuum breaker.

Exceptions:

1. This section shall not apply to water heater and boiler drain valves that are provided with hose connection threads and that are intended only for tank or vessel draining.

2. This section shall not apply to water supply valves intended for connection of clothes washing machines where backflow prevention is otherwise provided or is integral with the machine.

608.16 Connections to the potable water system. Connections to the potable water system shall conform to Sections 608.16.1 through 608.16.10.

608.16.1 Beverage dispensers. The water supply connection to beverage dispensers shall be protected against backflow by a backflow preventer conforming to ASSE 1022 or by an *air gap*. The portion of the backflow preventer device downstream from the second check valve and the piping downstream therefrom shall not be affected by carbon dioxide gas.

608.16.2 Connections to boilers. The potable supply to the boiler shall be equipped with a backflow preventer with an intermediate atmospheric vent complying with ASSE 1012 or CSA B64.3. Where conditioning chemicals are introduced into the system, the potable water connection shall be protected by an *air gap* or a reduced pressure principle backflow preventer, complying with ASSE 1013, CSA B64.4 or AWWA C511.

608.16.3 Heat exchangers. Heat exchangers utilizing an essentially toxic transfer fluid shall be separated from the potable water by double-wall construction. An *air gap* open to the atmosphere shall be provided between the two walls. Heat exchangers utilizing an essentially nontoxic transfer fluid shall be permitted to be of single-wall construction.

608.16.4 Connections to automatic fire sprinkler systems and standpipe systems. The potable water supply to automatic fire sprinkler and standpipe systems shall be protected against backflow by a double check backflow prevention assembly, a double check fire protection backflow prevention assembly or a reduced pressure principle fire protection backflow prevention assembly.

Exceptions:

1. Where systems are installed as a portion of the water distribution system in accordance with the

requirements of this code and are not provided with a fire department connection, isolation of the water supply system shall not be required.

2. Isolation of the water distribution system is not required for deluge, preaction or dry pipe systems.

608.16.4.1 Additives or nonpotable source. Where systems under continuous pressure contain chemical additives or antifreeze, or where systems are connected to a nonpotable secondary water supply, the potable water supply shall be protected against backflow by a reduced pressure principle backflow prevention assembly or a reduced pressure principle fire protection backflow prevention assembly. Where chemical additives or antifreeze are added to only a portion of an automatic fire sprinkler or standpipe system, the reduced pressure principle backflow prevention assembly or the reduced pressure principle fire protection backflow prevention assembly shall be permitted to be located so as to isolate that portion of the system. Where systems are not under continuous pressure, the potable water supply shall be protected against backflow by an air gap or an atmospheric vacuum breaker conforming to ASSE 1001 or CSA B64.1.1.

608.16.5 Connections to lawn irrigation systems. The potable water supply to lawn irrigation systems shall be protected against backflow by an atmospheric vacuum breaker, a pressure vacuum breaker assembly or a reduced pressure principle backflow prevention assembly. Valves shall not be installed downstream from an atmospheric vacuum breaker. Where chemicals are introduced into the system, the potable water supply shall be protected against backflow by a reduced pressure principle backflow prevention assembly.

608.16.6 Connections subject to backpressure. Where a potable water connection is made to a nonpotable line, fixture, tank, vat, pump or other equipment subject to high-hazard back-pressure, the potable water connection shall be protected by a reduced pressure principle backflow prevention assembly.

608.16.7 Chemical dispensers. Where chemical dispensers connect to the potable water distribution system, the water supply system shall be protected against backflow in accordance with Section 608.13.1, 608.13.2, 608.13.5, 608.13.6, 608.13.8 or 608.13.9.

608.16.8 Portable cleaning equipment. Where the portable cleaning equipment connects to the water distribution system, the water supply system shall be protected against backflow in accordance with Section 608.13.1, 608.13.2, 608.13.3, 608.13.7 or 608.13.8.

608.16.9 Dental pump equipment. Where dental pumping equipment connects to the water distribution system, the water supply system shall be protected against backflow in accordance with Section 608.13.1, 608.13.2, 608.13.5, 608.13.6 or 608.13.8.

608.16.10 Coffee machines and noncarbonated beverage dispensers. The water supply connection to coffee machines and noncarbonated beverage dispensers shall be protected against backflow by a backflow preventer conforming to ASSE 1022 or by an *air gap*.

608.17 Protection of individual water supplies. An individual water supply shall be located and constructed so as to be safeguarded against contamination in accordance with Sections 608.17.1 through 608.17.8.

608.17.1 Well locations. A potable ground water source or pump suction line shall not be located closer to potential sources of contamination than the distances shown in Table 608.17.1. In the event the underlying rock structure

TABLE 608.15.1
MINIMUM REQUIRED AIR GAPS

FIXTURE	MINIMUM AIR GAP	
	Away from a wall[a] (inches)	Close to a wall (inches)
Lavatories and other fixtures with effective opening not greater than $^1/_2$ inch in diameter	1	$1^1/_2$
Sink, laundry trays, gooseneck back faucets and other fixtures with effective openings not greater than $^3/_4$ inch in diameter	$1^1/_2$	$2^1/_2$
Over-rim bath fillers and other fixtures with effective openings not greater than 1 inch in diameter	2	3
Drinking water fountains, single orifice not greater than $^7/_{16}$ inch in diameter or multiple orifices with a total area of 0.150 square inch (area of circle $^7/_{16}$ inch in diameter)	1	$1^1/_2$
Effective openings greater than 1 inch	Two times the diameter of the effective opening	Three times the diameter of the effective opening

For SI: 1 inch = 25.4 mm.

a. Applicable where walls or obstructions are spaced from the nearest inside-edge of the spout opening a distance greater than three times the diameter of the effective opening for a single wall, or a distance greater than four times the diameter of the effective opening for two intersecting walls.

is limestone or fragmented shale, the local or state health department shall be consulted on well site location. The distances in Table 608.17.1 constitute minimum separation and shall be increased in areas of creviced rock or limestone, or where the direction of movement of the ground water is from sources of contamination toward the well.

TABLE 608.17.1
DISTANCE FROM CONTAMINATION TO
PRIVATE WATER SUPPLIES AND PUMP SUCTION LINES

SOURCE OF CONTAMINATION	DISTANCE (feet)
Barnyard	100
Farm silo	25
Pasture	100
Pumphouse floor drain of cast iron draining to ground surface	2
Seepage pits	50
Septic tank	25
Sewer	10
Subsurface disposal fields	50
Subsurface pits	50

For SI: 1 foot = 304.8 mm.

608.17.2 Elevation. Well sites shall be positively drained and shall be at higher elevations than potential sources of contamination.

608.17.3 Depth. Private potable well supplies shall not be developed from a water table less than 10 feet (3048 mm) below the ground surface.

608.17.4 Water-tight casings. Each well shall be provided with a water-tight casing extending to not less than 10 feet (3048 mm) below the ground surface. Casings shall extend not less than 6 inches (152 mm) above the well platform. Casings shall be large enough to permit installation of a separate drop pipe. Casings shall be sealed at the bottom in an impermeable stratum or extend several feet into the water-bearing stratum.

608.17.5 Drilled or driven well casings. Drilled or driven well casings shall be of steel or other *approved* material. Where drilled wells extend into a rock formation, the well casing shall extend to and set firmly in the formation. The annular space between the earth and the outside of the casing shall be filled with cement grout to a depth of not less than 10 feet (3048 mm) below the ground surface. In an instance of casing to rock installation, the grout shall extend to the rock surface.

608.17.6 Dug or bored well casings. Dug or bored well casings shall be of water-tight concrete, tile, or galvanized or corrugated metal pipe extending to not less than 10 feet (3048 mm) below the ground surface. Where the water table is more than 10 feet (3048 mm) below the ground surface, the water-tight casing shall extend below the table surface. Well casings for dug wells or bored wells constructed with sections of concrete, tile, or galvanized or corrugated metal pipe shall be surrounded by 6 inches (152 mm) of grout poured into the hole between the out-

side of the casing and the ground and extending not less than 10 feet (3048 mm) below the ground surface.

608.17.7 Cover. Potable water wells shall be equipped with an overlapping water-tight cover at the top of the well casing or pipe sleeve such that contaminated water or other substances are prevented from entering the well through the annular opening at the top of the well casing, wall or pipe sleeve. Covers shall extend downward not less than 2 inches (51 mm) over the outside of the well casing or wall. A dug well cover shall be provided with a pipe sleeve permitting the withdrawal of the pump suction pipe, cylinder or jet body without disturbing the cover. Where pump sections or discharge pipes enter or leave a well through the side of the casing, the circle of contact shall be water tight.

608.17.8 Drainage. Potable water wells and springs shall be constructed such that surface drainage will be diverted away from the well or spring.

SECTION 609
HEALTH CARE PLUMBING

609.1 Scope. This section shall govern those aspects of health care plumbing systems that differ from plumbing systems in other structures. Health care plumbing systems shall conform to the requirements of this section in addition to the other requirements of this code. The provisions of this section shall apply to the special devices and equipment installed and maintained in the following occupancies: nursing homes, homes for the aged, orphanages, infirmaries, first aid stations, psychiatric facilities, clinics, professional offices of dentists and doctors, mortuaries, educational facilities, surgery, dentistry, research and testing laboratories, establishments manufacturing pharmaceutical drugs and medicines, and other structures with similar apparatus and equipment classified as plumbing.

609.2 Water service. Hospitals shall have two water service pipes installed in such a manner so as to minimize the potential for an interruption of the supply of water in the event of a water main or water service pipe failure.

609.3 Hot water. *Hot water* shall be provided to supply all of the hospital fixture, kitchen and laundry requirements. Special fixtures and equipment shall have hot water supplied at a temperature specified by the manufacturer. The hot water system shall be installed in accordance with Section 607.

609.4 Vacuum breaker installation. Vacuum breakers shall be installed not less than 6 inches (152 mm) above the *flood level rim* of the fixture or device in accordance with Section 608. The *flood level rim* of hose connections shall be the maximum height at which any hose is utilized.

609.5 Prohibited water closet and clinical sink supply. Jet- or water-supplied orifices, except those supplied by the flush connections, shall not be located in or connected with a water closet bowl or clinical sink. This section shall not prohibit an *approved* bidet installation.

609.6 Clinical, hydrotherapeutic and radiological equipment. Clinical, hydrotherapeutic, radiological or any equipment that is supplied with water or that discharges to the waste system shall conform to the requirements of this section and Section 608.

609.7 Condensate drain trap seal. A water supply shall be provided for cleaning, flushing and resealing the condensate trap, and the trap shall discharge through an *air gap* in accordance with Section 608.

609.8 Valve leakage diverter. Each water sterilizer filled with water through directly connected piping shall be equipped with an *approved* leakage diverter or bleed line on the water supply control valve to indicate and conduct any leakage of unsterile water away from the sterile zone.

SECTION 610
DISINFECTION OF POTABLE WATER SYSTEM

610.1 General. New or repaired potable water systems shall be purged of deleterious matter and disinfected prior to utilization. The method to be followed shall be that prescribed by the health authority or water purveyor having jurisdiction or, in the absence of a prescribed method, the procedure described in either AWWA C651 or AWWA C652, or as described in this section. This requirement shall apply to "onsite" or "in-plant" fabrication of a system or to a modular portion of a system.

1. The pipe system shall be flushed with clean, potable water until dirty water does not appear at the points of outlet.

2. The system or part thereof shall be filled with a water/chlorine solution containing not less than 50 parts per million (50 mg/L) of chlorine, and the system or part thereof shall be valved off and allowed to stand for 24 hours; or the system or part thereof shall be filled with a water/chlorine solution containing not less than 200 parts per million (200 mg/L) of chlorine and allowed to stand for 3 hours.

3. Following the required standing time, the system shall be flushed with clean potable water until the chlorine is purged from the system.

4. The procedure shall be repeated where shown by a bacteriological examination that contamination remains present in the system.

SECTION 611
DRINKING WATER TREATMENT UNITS

611.1 Design. Drinking water treatment units shall meet the requirements of NSF42, NSF 44, NSF 53, NSF 62 or CSA B483.1.

611.2 Reverse osmosis systems. The discharge from a reverse osmosis drinking water treatment unit shall enter the drainage system through an *air gap* or an *air gap* device that meets the requirements of NSF 58 or CSA B483.1.

611.3 Connection tubing. The tubing to and from drinking water treatment units shall be of a size and material as recommended by the manufacturer. The tubing shall comply with NSF 14, NSF 42, NSF 44, NSF 53, NSF 58 or NSF 61.

SECTION 612
SOLAR SYSTEMS

612.1 Solar systems. The construction, installation, alterations and repair of systems, equipment and appliances intended to utilize solar energy for space heating or cooling, domestic hot water heating, swimming pool heating or process heating shall be in accordance with the *International Mechanical Code*.

SECTION 613
TEMPERATURE CONTROL DEVICES AND VALVES

613.1 Temperature-actuated mixing valves. Temperature-actuated mixing valves, which are installed to reduce water temperatures to defined limits, shall comply with ASSE 1017.

CHAPTER 7

SANITARY DRAINAGE

SECTION 701
GENERAL

701.1 Scope. The provisions of this chapter shall govern the materials, design, construction and installation of sanitary drainage systems.

701.2 Sewer required. Buildings in which plumbing fixtures are installed and premises having drainage piping shall be connected to a *public sewer*, where available, or an *approved private* sewage disposal system in accordance with the *International Private Sewage Disposal Code*.

701.3 Separate sewer connection. A building having plumbing fixtures installed and intended for human habitation, occupancy or use on premises abutting on a street, alley or easement in which there is a *public sewer* shall have a separate connection with the *sewer*. Where located on the same lot, multiple buildings shall not be prohibited from connecting to a common *building sewer* that connects to the *public sewer*.

701.4 Sewage treatment. Sewage or other waste from a plumbing system that is deleterious to surface or subsurface waters shall not be discharged into the ground or into any waterway unless it has first been rendered innocuous through subjection to an *approved* form of treatment.

701.5 Damage to drainage system or public sewer. Wastes detrimental to the *public sewer* system or to the functioning of the sewage-treatment plant shall be treated and disposed of in accordance with Section 1003 as directed by the code official.

701.6 Tests. The sanitary drainage system shall be tested in accordance with Section 312.

701.7 Connections. Direct connection of a steam exhaust, blowoff or drip pipe shall not be made with the building drainage system. Waste water where discharged into the building drainage system shall be at a temperature not greater than 140°F (60°C). Where higher temperatures exist, *approved* cooling methods shall be provided.

701.8 Engineered systems. Engineered sanitary drainage systems shall conform to the provisions of Sections 316 and 714.

701.9 Drainage piping in food service areas. Exposed soil or waste piping shall not be installed above any working, storage or eating surfaces in food service establishments.

SECTION 702
MATERIALS

702.1 Above-ground sanitary drainage and vent pipe. Above-ground soil, waste and vent pipe shall conform to one of the standards listed in Table 702.1.

TABLE 702.1
ABOVE-GROUND DRAINAGE AND VENT PIPE

MATERIAL	STANDARD
Acrylonitrile butadiene styrene (ABS) plastic pipe in IPS diameters, including Schedule 40, DR 22 (PS 200) and DR 24 (PS 140); with a solid, cellular core or composite wall	ASTM D 2661; ASTM F 628; ASTM F 1488; CSA B181.1
Brass pipe	ASTM B 43
Cast-iron pipe	ASTM A 74; ASTM A 888; CISPI 301
Copper or copper-alloy pipe	ASTM B 42; ASTM B 302
Copper or copper-alloy tubing (Type K, L, M or DWV)	ASTM B 75; ASTM B 88; ASTM B 251; ASTM B 306
Galvanized steel pipe	ASTM A 53
Glass pipe	ASTM C 1053
Polyolefin pipe	ASTM F 1412; CSA B181.3
Polyvinyl chloride (PVC) plastic pipe in IPS diameters, including schedule 40, DR 22 (PS 200), and DR 24 (PS 140); with a solid, cellular core or composite wall	ASTM D 2665; ASTM F 891; ASTM F 1488; CSA B181.2
Polyvinyl chloride (PVC) plastic pipe with a 3.25-inch O.D. and a solid, cellular core or composite wall	ASTM D 2949, ASTM F 1488
Polyvinylidene fluoride (PVDF) plastic pipe	ASTM F 1673; CSA B181.3
Stainless steel drainage systems, Types 304 and 316L	ASME A112.3.1

702.2 Underground building sanitary drainage and vent pipe. Underground building sanitary drainage and vent pipe shall conform to one of the standards listed in Table 702.2.

702.3 Building sewer pipe. *Building sewer* pipe shall conform to one of the standards listed in Table 702.3.

702.4 Fittings. Pipe fittings shall be *approved* for installation with the piping material installed and shall comply with the applicable standards listed in Table 702.4.

702.5 Chemical waste system. A chemical waste system shall be completely separated from the sanitary drainage system. The chemical waste shall be treated in accordance with Section 803.2 before discharging to the sanitary drainage system. Separate drainage systems for chemical wastes and vent pipes shall be of an *approved* material that is resistant to corrosion and degradation for the concentrations of chemicals involved.

702.6 Lead bends and traps. Lead bends and traps shall not be less than $^1/_8$ inch (3.2 mm) wall thickness.

TABLE 702.2
UNDERGROUND BUILDING DRAINAGE AND VENT PIPE

MATERIAL	STANDARD
Acrylonitrile butadiene styrene (ABS) plastic pipe in IPS diameters, including schedule 40, DR 22 (PS 200) and DR 24 (PS 140); with a solid, cellular core, or composite wall	ASTM D 2661; ASTM F 628; ASTM F 1488; CSA B181.1
Asbestos-cement pipe	ASTM C 428
Cast-iron pipe	ASTM A 74; ASTM A 888; CISPI 301
Copper or copper-alloy tubing (Type K, L, M or DWV)	ASTM B 75; ASTM B 88; ASTM B 251; ASTM B 306
Polyolefin pipe	ASTM F 1412; CSA B181.3
Polyvinyl chloride (PVC) plastic pipe in IPS diameters, including schedule 40, DR 22 (PS 200) and DR 24 (PS 140); with a solid, cellular core, or composite wall	ASTM D 2665; ASTM F 891; ASTM F 1488; CSA B181.2
Polyvinyl chloride (PVC) plastic pipe with a 3.25-inch O.D. and a solid, cellular core, or composite wall	ASTM D 2949, ASTM F 1488
Polyvinylidene fluoride (PVDF) plastic pipe	ASTM F 1673; CSA B181.3
Stainless steel drainage systems, Type 316L	ASME A 112.3.1

TABLE 702.3
BUILDING SEWER PIPE

MATERIAL	STANDARD
Acrylonitrile butadiene styrene (ABS) plastic pipe in IPS diameters, including schedule 40, DR 22 (PS 200) and DR 24 (PS 140); with a solid, cellular core or composite wall	ASTM D 2661; ASTM F 628; ASTM F 1488; CSA B181.1
Acrylonitrile butadiene styrene (ABS) plastic pipe in sewer and drain diameters, including SDR 42 (PS 20), PS 35, SDR 35 (PS 45), PS 50, PS 100, PS 140, SDR 23.5 (PS 150) and PS 200; with a solid, cellular core or composite wall	ASTM F 1488; ASTM D 2751
Asbestos-cement pipe	ASTM C 428
Cast-iron pipe	ASTM A 74; ASTM A 888; CISPI 301
Concrete pipe	ASTM C 14; ASTM C 76; CSA A257.1M; CSA A257.2M
Copper or copper-alloy tubing (Type K or L)	ASTM B 75; ASTM B 88; ASTM B 251
Polyethylene (PE) plastic pipe (SDR-PR)	ASTM F 714
Polyvinyl chloride (PVC) plastic pipe in IPS diameters, including schedule 40, DR 22 (PS 200) and DR 24 (PS 140); with a solid, cellular core or composite wall	ASTM D 2665; ASTM F 891; ASTM F 1488
Polyvinyl chloride (PVC) plastic pipe in sewer and drain diameters, including PS 25, SDR 41 (PS 28), PS 35, SDR 35 (PS 46), PS 50, PS 100, SDR 26 (PS 115), PS 140 and PS 200; with a solid, cellular core or composite wall	ASTM F 891; ASTM F 1488; ASTM D 3034; CSA B182.2; CSA B182.4
Polyvinyl chloride (PVC) plastic pipe with a 3.25-inch O.D. and a solid, cellular core or composite wall.	ASTM D 2949, ASTM F 1488
Polyvinylidene fluoride (PVDF) plastic pipe	ASTM F 1673; CSA B181.3
Stainless steel drainage systems, Types 304 and 316L	ASME A112.3.1
Vitrified clay pipe	ASTM C 4; ASTM C 700

TABLE 702.4
PIPE FITTINGS

MATERIAL	STANDARD
Acrylonitrile butadiene styrene (ABS) plastic pipe in IPS diameters	ASTM D 2661; ASTM F 628; CSA B181.1
Acrylonotrile butadiene styrene (ABS) plastic pipe in sewer and drain diameters	ASTM D 2751
Asbestos cement	ASTM C 428
Cast iron	ASME B 16.4; ASME B 16.12; ASTM A 74; ASTM A 888; CISPI 301
Copper or copper alloy	ASME B 16.15; ASME B 16.18; ASME B 16.22; ASME B 16.23; ASME B 16.26; ASME B 16.29
Glass	ASTM C 1053
Gray iron and ductile iron	AWWA C 110/A21.10
Malleable iron	ASME B 16.3
Polyolefin	ASTM F 1412; CSA B181.3
Polyvinyl chloride (PVC) plastic in IPS diameters	ASTM D 2665; ASTM F 1866
Polyvinyl chloride (PVC) plastic pipe in sewer and drain diameters	ASTM D 3034
Polyvinyl chloride (PVC) plastic pipe with a 3.25-inch O.D.	ASTM D 2949
Polyvinylidene fluoride (PVDF) plastic pipe	ASTM F 1673; CSA B181.3
Stainless steel drainage systems, Types 304 and 316L	ASME A 112.3.1
Steel	ASME B 16.9; ASME B 16.11; ASME B 16.28
Vitrified clay	ASTM C 700

For SI: 1 inch = 25.4 mm, 1 inch per foot = 83.3 mm/m.

SECTION 703
BUILDING SEWER

703.1 Building sewer pipe near the water service. Where the *building sewer* is installed within 5 feet (1524 mm) of the water service, the installation shall comply with the provisions of Section 603.2.

703.2 Drainage pipe in filled ground. Where a *building sewer* or *building drain* is installed on filled or unstable ground, the drainage pipe shall conform to one of the standards for ABS plastic pipe, cast-iron pipe, copper or copper-alloy tubing, or PVC plastic pipe listed in Table 702.3.

703.3 Sanitary and storm sewers. Where separate systems of sanitary drainage and storm drainage are installed in the same property, the sanitary and storm building sewers or drains shall be permitted to be laid side by side in one trench.

703.4 Existing building sewers and drains. Existing building sewers and drains shall connect with new *building sewer* and drainage systems only where found by examination and test to conform to the new system in quality of material. The code official shall notify the owner to make the changes necessary to conform to this code.

703.5 Cleanouts on building sewers. Cleanouts on building sewers shall be located as set forth in Section 708.

SECTION 704
DRAINAGE PIPING INSTALLATION

704.1 Slope of horizontal drainage piping. Horizontal drainage piping shall be installed in uniform alignment at uniform slopes. The slope of a horizontal drainage pipe shall be not less than that indicated in Table 704.1.

TABLE 704.1
SLOPE OF HORIZONTAL DRAINAGE PIPE

SIZE (inches)	MINIMUM SLOPE (inch per foot)
$2^1/_2$ or less	$^1/_4$
3 to 6	$^1/_8$
8 or larger	$^1/_{16}$

704.2 Change in size. The size of the drainage piping shall not be reduced in size in the direction of the flow. A 4-inch by 3-inch (102 mm by 76 mm) water closet connection shall not be considered as a reduction in size.

704.3 Connections to offsets and bases of stacks. Horizontal branches shall connect to the bases of stacks at a point located not less than 10 times the diameter of the drainage *stack* downstream from the *stack*. Horizontal branches shall connect to horizontal *stack* offsets at a point located not less than 10 times the diameter of the drainage *stack* downstream from the upper *stack*.

704.4 Future fixtures. Drainage piping for future fixtures shall terminate with an *approved* cap or plug.

SECTION 705
JOINTS

705.1 General. This section contains provisions applicable to joints specific to sanitary drainage piping.

705.2 ABS plastic. Joints between ABS plastic pipe or fittings shall comply with Sections 705.2.1 through 705.2.3.

705.2.1 Mechanical joints. Mechanical joints on drainage pipes shall be made with an elastomeric seal conforming to ASTM C 1173, ASTM D 3212 or CSA B602. Mechanical joints shall be installed only in underground systems unless otherwise *approved*. Joints shall be installed in accordance with the manufacturer's instructions.

705.2.2 Solvent cementing. Joint surfaces shall be clean and free from moisture. Solvent cement that conforms to ASTM D 2235 or CSA B181.1 shall be applied to all joint surfaces. The joint shall be made while the cement is wet. Joints shall be made in accordance with ASTM D 2235,

ASTM D 2661, ASTM F 628 or CSA B181.1. Solvent-cement joints shall be permitted above or below ground.

705.2.3 Threaded joints. Threads shall conform to ASME B1.20.1. Schedule 80 or heavier pipe shall be permitted to be threaded with dies specifically designed for plastic pipe. *Approved* thread lubricant or tape shall be applied on the male threads only.

705.3 Asbestos cement. Joints between asbestos-cement pipe or fittings shall be made with a sleeve coupling of the same composition as the pipe, sealed with an elastomeric ring conforming to ASTM D 1869.

705.4 Brass. Joints between brass pipe or fittings shall comply with Sections 705.4.1 through 705.4.4.

705.4.1 Brazed joints. All joint surfaces shall be cleaned. An *approved* flux shall be applied where required. The joint shall be brazed with a filler metal conforming to AWS A5.8.

705.4.2 Mechanical joints. Mechanical joints shall be installed in accordance with the manufacturer's instructions.

705.4.3 Threaded joints. Threads shall conform to ASME B1.20.1. Pipe-joint compound or tape shall be applied on the male threads only.

705.4.4 Welded joints. All joint surfaces shall be cleaned. The joint shall be welded with an *approved* filler metal.

705.5 Cast iron. Joints between cast-iron pipe or fittings shall comply with Sections 705.5.1 through 705.5.3.

705.5.1 Caulked joints. Joints for hub and spigot pipe shall be firmly packed with oakum or hemp. Molten lead shall be poured in one operation to a depth of not less than 1 inch (25 mm). The lead shall not recede more than $^{1}/_{8}$ inch (3.2 mm) below the rim of the hub and shall be caulked tight. Paint, varnish or other coatings shall not be permitted on the jointing material until after the joint has been tested and *approved*. Lead shall be run in one pouring and shall be caulked tight. Acid-resistant rope and acid-proof cement shall be permitted.

705.5.2 Compression gasket joints. Compression gaskets for hub and spigot pipe and fittings shall conform to ASTM C 564 and shall be tested to ASTM C 1563. Gaskets shall be compressed when the pipe is fully inserted.

705.5.3 Mechanical joint coupling. Mechanical joint couplings for hubless pipe and fittings shall comply with CISPI 310, ASTM C 1277 or ASTM C 1540. The elastomeric sealing sleeve shall conform to ASTM C 564 or CSA B602 and shall be provided with a center stop. Mechanical joint couplings shall be installed in accordance with the manufacturer's installation instructions.

705.6 Concrete joints. Joints between concrete pipe and fittings shall be made with an elastomeric seal conforming to ASTM C 443, ASTM C 1173, CSA A257.3M or CSA B602.

705.7 Coextruded composite ABS pipe, joints. Joints between coextruded composite pipe with an ABS outer layer or ABS fittings shall comply with Sections 705.7.1 and 705.7.2.

705.7.1 Mechanical joints. Mechanical joints on drainage pipe shall be made with an elastomeric seal conforming to ASTM C1173, ASTM D 3212 or CSA B602. Mechanical joints shall not be installed in above-ground systems, unless otherwise *approved*. Joints shall be installed in accordance with the manufacturer's instructions.

705.7.2 Solvent cementing. Joint surfaces shall be clean and free from moisture. Solvent cement that conforms to ASTM D 2235 or CSA B181.1 shall be applied to all joint surfaces. The joint shall be made while the cement is wet. Joints shall be made in accordance with ASTM D 2235, ASTM D 2661, ASTM F 628 or CSA B181.1. Solvent-cement joints shall be permitted above or below ground.

705.8 Coextruded composite PVC pipe. Joints between coextruded composite pipe with a PVC outer layer or PVC fittings shall comply with Sections 705.8.1 and 705.8.2.

705.8.1 Mechanical joints. Mechanical joints on drainage pipe shall be made with an elastomeric seal conforming to ASTM D 3212. Mechanical joints shall not be installed in above-ground systems, unless otherwise *approved*. Joints shall be installed in accordance with the manufacturer's instructions.

705.8.2 Solvent cementing. Joint surfaces shall be clean and free from moisture. A purple primer that conforms to ASTM F 656 shall be applied. Solvent cement not purple in color and conforming to ASTM D 2564, CSA B137.3, CSA B181.2 or CSA B182.1 shall be applied to all joint surfaces. The joint shall be made while the cement is wet and shall be in accordance with ASTM D 2855. Solvent-cement joints shall be permitted above or below ground.

705.9 Copper pipe. Joints between copper or copper-alloy pipe or fittings shall comply with Sections 705.9.1 through 705.9.5.

705.9.1 Brazed joints. All joint surfaces shall be cleaned. An *approved* flux shall be applied where required. The joint shall be brazed with a filler metal conforming to AWS A5.8.

705.9.2 Mechanical joints. Mechanical joints shall be installed in accordance with the manufacturer's instructions.

705.9.3 Soldered joints. Solder joints shall be made in accordance with the methods of ASTM B 828. All cut tube ends shall be reamed to the full inside diameter of the tube end. All joint surfaces shall be cleaned. A flux conforming to ASTM B 813 shall be applied. The joint shall be soldered with a solder conforming to ASTM B 32.

705.9.4 Threaded joints. Threads shall conform to ASME B1.20.1. Pipe-joint compound or tape shall be applied on the male threads only.

705.9.5 Welded joints. All joint surfaces shall be cleaned. The joint shall be welded with an *approved* filler metal.

705.10 Copper tubing. Joints between copper or copper-alloy tubing or fittings shall comply with Sections 705.10.1 through 705.10.3.

705.10.1 Brazed joints. All joint surfaces shall be cleaned. An *approved* flux shall be applied where required. The joint shall be brazed with a filler metal conforming to AWS A5.8.

705.10.2 Mechanical joints. Mechanical joints shall be installed in accordance with the manufacturer's instructions.

705.10.3 Soldered joints. Solder joints shall be made in accordance with the methods of ASTM B 828. All cut tube ends shall be reamed to the full inside diameter of the tube end. All joint surfaces shall be cleaned. A flux conforming to ASTM B 813 shall be applied. The joint shall be soldered with a solder conforming to ASTM B 32.

705.11 Borosilicate glass joints. Glass-to-glass connections shall be made with a bolted compression-type stainless steel (300 series) coupling with contoured acid-resistant elastomeric compression ring and a fluorocarbon polymer inner seal ring; or with caulked joints in accordance with Section 705.11.1.

705.11.1 Caulked joints. Lead-caulked joints for hub and spigot soil pipe shall be firmly packed with oakum or hemp and filled with molten lead not less than 1 inch (25 mm) in depth and not to recede more than $^1/_8$ inch (3.2 mm) below the rim of the hub. Paint, varnish or other coatings shall not be permitted on the jointing material until after the joint has been tested and *approved*. Lead shall be run in one pouring and shall be caulked tight. Acid-resistant rope and acidproof cement shall be permitted.

705.12 Steel. Joints between galvanized steel pipe or fittings shall comply with Sections 705.12.1 and 705.12.2.

705.12.1 Threaded joints. Threads shall conform to ASME B1.20.1. Pipe-joint compound or tape shall be applied on the male threads only.

705.12.2 Mechanical joints. Joints shall be made with an *approved* elastomeric seal. Mechanical joints shall be installed in accordance with the manufacturer's instructions.

705.13 Lead. Joints between lead pipe or fittings shall comply with Sections 705.13.1 and 705.13.2.

705.13.1 Burned. Burned joints shall be uniformly fused together into one continuous piece. The thickness of the joint shall be at least as thick as the lead being joined. The filler metal shall be of the same material as the pipe.

705.13.2 Wiped. Joints shall be fully wiped, with an exposed surface on each side of the joint not less than $^3/_4$ inch (19.1 mm). The joint shall be not less than $^3/_8$ inch (9.5 mm) thick at the thickest point.

705.14 PVC plastic. Joints between PVC plastic pipe or fittings shall comply with Sections 705.14.1 through 705.14.3.

705.14.1 Mechanical joints. Mechanical joints on drainage pipe shall be made with an elastomeric seal conforming to ASTM C 1173, ASTM D 3212 or CSA B602. Mechanical joints shall not be installed in above-ground systems, unless otherwise *approved*. Joints shall be installed in accordance with the manufacturer's instructions

705.14.2 Solvent cementing. Joint surfaces shall be clean and free from moisture. A purple primer that conforms to

ASTM F 656 shall be applied. Solvent cement not purple in color and conforming to ASTM D 2564, CSA B137.3, CSA B181.2 or CSA B182.1 shall be applied to all joint surfaces. The joint shall be made while the cement is wet and shall be in accordance with ASTM D 2855. Solvent-cement joints shall be permitted above or below ground.

705.14.3 Threaded joints. Threads shall conform to ASME B1.20.1. Schedule 80 or heavier pipe shall be permitted to be threaded with dies specifically designed for plastic pipe. *Approved* thread lubricant or tape shall be applied on the male threads only.

705.15 Vitrified clay. Joints between vitrified clay pipe or fittings shall be made with an elastomeric seal conforming to ASTM C 425, ASTM C 1173 or CSA B602.

705.16 Polyethylene plastic pipe. Joints between polyethylene plastic pipe and fittings shall be underground and shall comply with Section 705.16.1 or 705.16.2.

705.16.1 Heat-fusion joints. Joint surfaces shall be clean and free from moisture. All joint surfaces shall be cut, heated to melting temperature and joined using tools specifically designed for the operation. Joints shall be undisturbed until cool. Joints shall be made in accordance with ASTM D 2657 and the manufacturer's instructions.

705.16.2 Mechanical joints. Mechanical joints in drainage piping shall be made with an elastomeric seal conforming to ASTM C 1173, ASTM D 3212 or CSA B602. Mechanical joints shall be installed in accordance with the manufacturer's instructions.

705.17 Polyolefin plastic. Joints between polyolefin plastic pipe and fittings shall comply with Sections 705.17.1 and 705.17.2.

705.17.1 Heat-fusion joints. Heat-fusion joints for polyolefin pipe and tubing joints shall be installed with socket-type heat-fused polyolefin fittings or electrofusion polyolefin fittings. Joint surfaces shall be clean and free from moisture. The joint shall be undisturbed until cool. Joints shall be made in accordance with ASTM F 1412 or CSA B181.3.

705.17.2 Mechanical and compression sleeve joints. Mechanical and compression sleeve joints shall be installed in accordance with the manufacturer's instructions.

705.18 Polyvinylidene fluoride plastic. Joints between polyvinylidene plastic pipe and fittings shall comply with Sections 705.18.1 and 705.18.2.

705.18.1 Heat-fusion joints. Heat-fusion joints for polyvinylidene fluoride pipe and tubing joints shall be installed with socket-type heat-fused polyvinylidene fluoride fittings or electrofusion polyvinylidene fittings and couplings. Joint surfaces shall be clean and free from moisture. The joint shall be undisturbed until cool. Joints shall be made in accordance with ASTM F 1673.

705.18.2 Mechanical and compression sleeve joints. Mechanical and compression sleeve joints shall be installed in accordance with the manufacturer's instructions.

705.19 Joints between different materials. Joints between different piping materials shall be made with a mechanical joint of the compression or mechanical-sealing type conforming to ASTM C 1173, ASTM C 1460 or ASTM C 1461. Connectors and adapters shall be *approved* for the application and such joints shall have an elastomeric seal conforming to ASTM C 425, ASTM C 443, ASTM C 564, ASTM C 1440, ASTM D 1869, ASTM F 477, CSA A257.3M or CSA B602, or as required in Sections 705.19.1 through 705.19.7. Joints between glass pipe and other types of materials shall be made with adapters having a TFE seal. Joints shall be installed in accordance with the manufacturer's instructions.

705.19.1 Copper or copper-alloy tubing to cast-iron hub pipe. Joints between copper or copper-alloy tubing and cast-iron hub pipe shall be made with a brass ferrule or compression joint. The copper or copper-alloy tubing shall be soldered to the ferrule in an *approved* manner, and the ferrule shall be joined to the cast-iron hub by a caulked joint or a mechanical compression joint.

705.19.2 Copper or copper-alloy tubing to galvanized steel pipe. Joints between copper or copper-alloy tubing and galvanized steel pipe shall be made with a brass converter fitting or dielectric fitting. The copper tubing shall be soldered to the fitting in an *approved* manner, and the fitting shall be screwed to the threaded pipe.

705.19.3 Cast-iron pipe to galvanized steel or brass pipe. Joints between cast-iron and galvanized steel or brass pipe shall be made by either caulked or threaded joints or with an *approved* adapter fitting.

705.19.4 Plastic pipe or tubing to other piping material. Joints between different types of plastic pipe or between plastic pipe and other piping material shall be made with an *approved* adapter fitting. Joints between plastic pipe and cast-iron hub pipe shall be made by a caulked joint or a mechanical compression joint.

705.19.5 Lead pipe to other piping material. Joints between lead pipe and other piping material shall be made by a wiped joint to a caulking ferrule, soldering nipple, or bushing or shall be made with an *approved* adapter fitting.

705.19.6 Borosilicate glass to other materials. Joints between glass pipe and other types of materials shall be made with adapters having a TFE seal and shall be installed in accordance with the manufacturer's instructions.

705.19.7 Stainless steel drainage systems to other materials. Joints between stainless steel drainage systems and other piping materials shall be made with *approved* mechanical couplings.

705.20 Drainage slip joints. Slip joints shall comply with Section 405.8.

705.21 Caulking ferrules. Ferrules shall be of red brass and shall be in accordance with Table 705.21.

TABLE 705.21
CAULKING FERRULE SPECIFICATIONS

PIPE SIZES (inches)	INSIDE DIAMETER (inches)	LENGTH (inches)	MINIMUM WEIGHT EACH
2	$2^1/_4$	$4^1/_2$	1 pound
3	$3^1/_4$	$4^1/_2$	1 pound 12 ounces
4	$4^1/_4$	$4^1/_2$	2 pounds 8 ounces

For SI: 1 inch = 25.4 mm, 1 ounce = 28.35 g, 1 pound = 0.454 kg.

705.22 Soldering bushings. Soldering bushings shall be of red brass and shall be in accordance with Table 705.22.

TABLE 705.22
SOLDERING BUSHING SPECIFICATIONS

PIPE SIZES (inches)	MINIMUM WEIGHT EACH
$1^1/_4$	6 ounces
$1^1/_2$	8 ounces
2	14 ounces
$2^1/_2$	1 pound 6 ounces
3	2 pounds
4	3 pounds 8 ounces

For SI: 1 inch = 25.4 mm, 1 ounce = 28.35 g, 1 pound = 0.454 kg.

705.23 Stainless steel drainage systems. O-ring joints for stainless steel drainage systems shall be made with an *approved* elastomeric seal.

SECTION 706
CONNECTIONS BETWEEN DRAINAGE PIPING AND FITTINGS

706.1 Connections and changes in direction. All connections and changes in direction of the sanitary drainage system shall be made with *approved* drainage fittings. Connections between drainage piping and fixtures shall conform to Section 405.

706.2 Obstructions. The fittings shall not have ledges, shoulders or reductions capable of retarding or obstructing flow in the piping. Threaded drainage pipe fittings shall be of the recessed drainage type. This section shall not be applicable to tubular waste fittings used to convey vertical flow upstream of the trap seal liquid level of a fixture trap.

706.3 Installation of fittings. Fittings shall be installed to guide sewage and waste in the direction of flow. Change in direction shall be made by fittings installed in accordance with Table 706.3. Change in direction by combination fittings, side inlets or increasers shall be installed in accordance with Table 706.3 based on the pattern of flow created by the fitting. Double sanitary tee patterns shall not receive the discharge of back-to-back water closets and fixtures or appliances with pumping action discharge.

Exception: Back-to-back water closet connections to double sanitary tees shall be permitted where the horizontal *developed length* between the outlet of the water closet and the connection to the double sanitary tee pattern is 18 inches (457 mm) or greater.

TABLE 706.3
FITTINGS FOR CHANGE IN DIRECTION

TYPE OF FITTING PATTERN	CHANGE IN DIRECTION		
	Horizontal to vertical	Vertical to horizontal	Horizontal to horizontal
Sixteenth bend	X	X	X
Eighth bend	X	X	X
Sixth bend	X	X	X
Quarter bend	X	X[a]	X[a]
Short sweep	X	X[a,b]	X[a]
Long sweep	X	X	X
Sanitary tee	X[c]	—	—
Wye	X	X	X
Combination wye and eighth bend	X	X	X

For SI: 1 inch = 25.4 mm.

a. The fittings shall only be permitted for a 2-inch or smaller fixture drain.

b. Three inches or larger.

c. For a limitation on double sanitary tees, see Section 706.3.

706.4 Heel- or side-inlet quarter bends. Heel-inlet quarter bends shall be an acceptable means of connection, except where the quarter bend serves a water closet. A low-heel inlet shall not be used as a wet-vented connection. Side-inlet quarter bends shall be an acceptable means of connection for drainage, wet venting and *stack* venting arrangements.

SECTION 707
PROHIBITED JOINTS AND CONNECTIONS

707.1 Prohibited joints. The following types of joints and connections shall be prohibited:

1. Cement or concrete joints.

2. Mastic or hot-pour bituminous joints.

3. Joints made with fittings not *approved* for the specific installation.

4. Joints between different diameter pipes made with elastomeric rolling O-rings.

5. Solvent-cement joints between different types of plastic pipe.

6. Saddle-type fittings.

SECTION 708
CLEANOUTS

708.1 Scope. This section shall govern the size, location, installation and maintenance of drainage pipe cleanouts.

708.2 Cleanout plugs. Cleanout plugs shall be brass or plastic, or other *approved* materials. Brass cleanout plugs shall be utilized with metallic drain, waste and vent piping only, and shall conform to ASTM A 74, ASME A112.3.1 or ASME A112.36.2M. Cleanouts with plate-style *access* covers shall be fitted with corrosion-resisting fasteners. Plastic cleanout plugs shall conform to the requirements of Section 702.4. Plugs shall have raised square or countersunk square heads.

Countersunk heads shall be installed where raised heads are a trip hazard. Cleanout plugs with borosilicate glass systems shall be of borosilicate glass.

708.3 Where required. Cleanouts shall be located in accordance with Sections 708.3.1 through 708.3.6.

708.3.1 Horizontal drains within buildings. All horizontal drains shall be provided with cleanouts located not more than 100 feet (30 480 mm) apart.

708.3.2 Building sewers. Building sewers shall be provided with cleanouts located not more than 100 feet (30 480 mm) apart measured from the upstream entrance of the cleanout. For building sewers 8 inches (203 mm) and larger, manholes shall be provided and located not more than 200 feet (60 960 mm) from the junction of the *building drain* and *building sewer*, at each change in direction and at intervals of not more than 400 feet (122 m) apart. Manholes and manhole covers shall be of an *approved* type.

708.3.3 Changes of direction. Cleanouts shall be installed at each change of direction greater than 45 degrees (0.79 rad) in the *building sewer*, *building drain* and horizontal waste or soil lines. Where more than one change of direction occurs in a run of piping, only one cleanout shall be required for each 40 feet (12 192 mm) of *developed length* of the drainage piping.

708.3.4 Base of stack. A cleanout shall be provided at the base of each waste or soil *stack*.

708.3.5 Building drain and building sewer junction. There shall be a cleanout near the junction of the *building drain* and the *building sewer*. The cleanout shall be either inside or outside the building wall and shall be brought up to the finished ground level or to the basement floor level. An *approved* two-way cleanout is allowed to be used at this location to serve as a required cleanout for both the *building drain* and *building sewer*. The cleanout at the junction of the *building drain* and *building sewer* shall not be required if the cleanout on a 3-inch (76 mm) or larger diameter soil *stack* is located within a *developed length* of 10 feet (3048 mm) of the *building drain* and *building sewer* connection. The minimum size of the cleanout at the junction of the *building drain* and *building sewer* shall comply with Section 708.7.

708.3.6 Manholes. Manholes serving a *building drain* shall have secured gas-tight covers and shall be located in accordance with Section 708.3.2.

708.4 Concealed piping. Cleanouts on concealed piping or piping under a floor slab or in a crawl space of less than 24 inches (610 mm) in height or a plenum shall be extended through and terminate flush with the finished wall, floor or ground surface or shall be extended to the outside of the building. Cleanout plugs shall not be covered with cement, plaster or any other permanent finish material. Where it is necessary to conceal a cleanout or to terminate a cleanout in an area subject to vehicular traffic, the covering plate, *access* door or cleanout shall be of an *approved* type designed and installed for this purpose.

708.5 Opening direction. Every cleanout shall be installed to open to allow cleaning in the direction of the flow of the drainage pipe or at right angles thereto.

708.6 Prohibited installation. Cleanout openings shall not be utilized for the installation of new fixtures, except where *approved* and where another cleanout of equal *access* and capacity is provided.

708.7 Minimum size. Cleanouts shall be the same nominal size as the pipe they serve up to 4 inches (102 mm). For pipes larger than 4 inches (102 mm) nominal size, the size of the cleanout shall be not less than 4 inches (102 mm).

Exceptions:

1. "P" trap connections with slip joints or ground joint connections, or *stack* cleanouts that are not more than one pipe diameter smaller than the drain served, shall be permitted.

2. Cast-iron cleanout sizing shall be in accordance with referenced standards in Table 702.4, ASTM A 74 for hub and spigot fittings or ASTM A 888 or CISPI 301 for hubless fittings.

708.8 Clearances. Cleanouts on 6-inch (153 mm) and smaller pipes shall be provided with a clearance of not less than 18 inches (457 mm) for rodding. Cleanouts on 8-inch (203 mm) and larger pipes shall be provided with a clearance of not less than 36 inches (914 mm) for rodding.

708.9 Access. *Access* shall be provided to all cleanouts.

SECTION 709
FIXTURE UNITS

709.1 Values for fixtures. *Drainage fixture unit* values as given in Table 709.1 designate the relative load weight of different kinds of fixtures that shall be employed in estimating the total load carried by a soil or waste pipe, and shall be used in connection with Tables 710.1(1) and 710.1(2) of sizes for soil, waste and vent pipes for which the permissible load is given in terms of fixture units.

709.2 Fixtures not listed in Table 709.1. Fixtures not listed in Table 709.1 shall have a *drainage fixture unit* load based on the outlet size of the fixture in accordance with Table 709.2. The minimum trap size for unlisted fixtures shall be the size of the drainage outlet but not less than $1^1/_4$ inches (32 mm).

TABLE 709.2
DRAINAGE FIXTURE UNITS FOR FIXTURE DRAINS OR TRAPS

FIXTURE DRAIN OR TRAP SIZE (inches)	DRAINAGE FIXTURE UNIT VALUE
$1^1/_4$	1
$1^1/_2$	2
2	3
$2^1/_2$	4
3	5
4	6

For SI: 1 inch = 25.4 mm.

709.3 Values for continuous and semicontinuous flow. *Drainage fixture unit* values for continuous and semicontinuous flow into a drainage system shall be computed on the basis that 1 gpm (0.06 L/s) of flow is equivalent to two fixture units.

709.4 Values for indirect waste receptor. The *drainage fixture unit* load of an indirect waste receptor receiving the discharge of indirectly connected fixtures shall be the sum of the *drainage fixture unit* values of the fixtures that discharge to the receptor, but not less than the *drainage fixture unit* value given for the indirect waste receptor in Table 709.1 or 709.2.

709.4.1 Clear-water waste receptors. Where waste receptors such as floor drains, floor sinks and hub drains receive only clear-water waste from display cases, refrigerated display cases, ice bins, coolers and freezers, such receptors shall have a *drainage fixture unit* value of one-half.

SECTION 710
DRAINAGE SYSTEM SIZING

710.1 Maximum fixture unit load. The maximum number of drainage fixture units connected to a given size of *building sewer*, *building drain* or horizontal *branch* of the *building drain* shall be determined using Table 710.1(1). The maximum number of drainage fixture units connected to a given size of horizontal *branch* or vertical soil or waste *stack* shall be determined using Table 710.1(2).

TABLE 710.1(1)
BUILDING DRAINS AND SEWERS

DIAMETER OF PIPE (inches)	MAXIMUM NUMBER OF DRAINAGE FIXTURE UNITS CONNECTED TO ANY PORTION OF THE BUILDING DRAIN OR THE BUILDING SEWER, INCLUDING BRANCHES OF THE BUILDING DRAIN[a]			
	Slope per foot			
	$^1/_{16}$ inch	$^1/_8$ inch	$^1/_4$ inch	$^1/_2$ inch
$1^1/_4$	—	—	1	1
$1^1/_2$	—	—	3	3
2	—	—	21	26
$2^1/_2$	—	—	24	31
3	—	36	42	50
4	—	180	216	250
5	—	390	480	575
6	—	700	840	1,000
8	1,400	1,600	1,920	2,300
10	2,500	2,900	3,500	4,200
12	3,900	4,600	5,600	6,700
15	7,000	8,300	10,000	12,000

For SI: 1 inch = 25.4 mm, 1 inch per foot = 83.3 mm/m.

a. The minimum size of any building drain serving a water closet shall be 3 inches.

TABLE 709.1
DRAINAGE FIXTURE UNITS FOR FIXTURES AND GROUPS

FIXTURE TYPE	DRAINAGE FIXTURE UNIT VALUE AS LOAD FACTORS	MINIMUM SIZE OF TRAP (inches)
Automatic clothes washers, commercial[a,g]	3	2
Automatic clothes washers, residential[g]	2	2
Bathroom group as defined in Section 202 (1.6 gpf water closet)[f]	5	—
Bathroom group as defined in Section 202 (water closet flushing greater than 1.6 gpf)[f]	6	—
Bathtub[b] (with or without overhead shower or whirlpool attachments)	2	$1^1/_2$
Bidet	1	$1^1/_4$
Combination sink and tray	2	$1^1/_2$
Dental lavatory	1	$1^1/_4$
Dental unit or cuspidor	1	$1^1/_4$
Dishwashing machine[c], domestic	2	$1^1/_2$
Drinking fountain	$^1/_2$	$1^1/_4$
Emergency floor drain	0	2
Floor drains[h]	2[h]	2
Floor sinks	Note h	2
Kitchen sink, domestic	2	$1^1/_2$
Kitchen sink, domestic with food waste grinder and/or dishwasher	2	$1^1/_2$
Laundry tray (1 or 2 compartments)	2	$1^1/_2$
Lavatory	1	$1^1/_4$
Shower (based on the total flow rate through showerheads and body sprays) Flow rate:		
5.7 gpm or less	2	$1^1/_2$
Greater than 5.7 gpm to 12.3 gpm	3	2
Greater than 12.3 gpm to 25.8 gpm	5	3
Greater than 25.8 gpm to 55.6 gpm	6	4
Service sink	2	$1^1/_2$
Sink	2	$1^1/_2$
Urinal	4	Note d
Urinal, 1 gallon per flush or less	2[e]	Note d
Urinal, nonwater supplied	$^1/_2$	Note d
Wash sink (circular or multiple) each set of faucets	2	$1^1/_2$
Water closet, flushometer tank, public or private	4[e]	Note d
Water closet, private (1.6 gpf)	3[e]	Note d
Water closet, private (flushing greater than 1.6 gpf)	4[e]	Note d
Water closet, public (1.6 gpf)	4[e]	Note d
Water closet, public (flushing greater than 1.6 gpf)	6[e]	Note d

For SI: 1 inch = 25.4 mm, 1 gallon = 3.785 L, gpf = gallon per flushing cycle, gpm = gallon per minute.

a. For traps larger than 3 inches, use Table 709.2.

b. A showerhead over a bathtub or whirlpool bathtub attachment does not increase the drainage fixture unit value.

c. See Sections 709.2 through 709.4.1 for methods of computing unit value of fixtures not listed in this table or for rating of devices with intermittent flows.

d. Trap size shall be consistent with the fixture outlet size.

e. For the purpose of computing loads on building drains and sewers, water closets and urinals shall not be rated at a lower drainage fixture unit unless the lower values are confirmed by testing.

f. For fixtures added to a bathroom group, add the dfu value of those additional fixtures to the bathroom group fixture count.

g. See Section 406.3 for sizing requirements for fixture drain, branch drain, and drainage stack for an automatic clothes washer standpipe.

h. See Sections 709.4 and 709.4.1.

TABLE 710.1(2)
HORIZONTAL FIXTURE BRANCHES AND STACKS[a]

DIAMETER OF PIPE (inches)	MAXIMUM NUMBER OF DRAINAGE FIXTURE UNITS (dfu)			
	Total for horizontal branch	Stacks[b]		
		Total discharge into one branch interval	Total for stack of three branch Intervals or less	Total for stack greater than three branch intervals
$1^1/_2$	3	2	4	8
2	6	6	10	24
$2^1/_2$	12	9	20	42
3	20	20	48	72
4	160	90	240	500
5	360	200	540	1,100
6	620	350	960	1,900
8	1,400	600	2,200	3,600
10	2,500	1,000	3,800	5,600
12	3,900	1,500	6,000	8,400
15	7,000	Note c	Note c	Note c

For SI: 1 inch = 25.4 mm.

a. Does not include branches of the building drain. Refer to Table 710.1(1).

b. Stacks shall be sized based on the total accumulated connected load at each story or branch interval. As the total accumulated connected load decreases, stacks are permitted to be reduced in size. Stack diameters shall not be reduced to less than one-half of the diameter of the largest stack size required.

c. Sizing load based on design criteria.

710.1.1 Horizontal stack offsets. Horizontal *stack* offsets shall be sized as required for building drains in accordance with Table 710.1(1), except as required by Section 711.3.

710.1.2 Vertical stack offsets. Vertical *stack* offsets shall be sized as required for straight stacks in accordance with Table 710.1(2), except where required to be sized as a *building drain* in accordance with Section 711.1.1.

710.2 Future fixtures. Where provision is made for the future installation of fixtures, those provided for shall be considered in determining the required sizes of drain pipes.

SECTION 711
OFFSETS IN DRAINAGE PIPING IN BUILDINGS OF FIVE STORIES OR MORE

711.1 Horizontal branch connections above or below vertical stack offsets. If a horizontal *branch* connects to the *stack* within 2 feet (610 mm) above or below a vertical *stack* offset, and the offset is located more than four *branch intervals* below the top of the *stack*, the offset shall be vented in accordance with Section 906.

711.1.1 Omission of vents for vertical stack offsets. Vents for vertical offsets required by Section 711.1 shall not be required where the *stack* and its offset are sized as a *building drain* [see Table 710.1(1)].

711.2 Horizontal stack offsets. A *stack* with a horizontal offset located more than four *branch intervals* below the top

of the *stack* shall be vented in accordance with Section 907 and sized as follows:

1. The portion of the *stack* above the offset shall be sized as for a vertical *stack* based on the total number of drainage fixture units above the offset.

2. The offset shall be sized in accordance with Section 710.1.1.

3. The portion of the *stack* below the offset shall be sized as for the offset or based on the total number of drainage fixture units on the entire *stack*, whichever is larger [see Table 710.1(2), Column 5].

711.2.1 Omission of vents for horizontal stack offsets. Vents for horizontal stack offsets required by Section 711.2 shall not be required where the stack and its offset are one pipe size larger than required for a building drain [see Table 710.1(1)] and the entire stack and offset are not less in cross-sectional area than that required for a straight stack plus the area of an offset vent as provided for in Section 907.

711.3 Offsets below lowest branch. Where a vertical offset occurs in a soil or waste *stack* below the lowest horizontal *branch*, a change in diameter of the *stack* because of the offset shall not be required. If a horizontal offset occurs in a soil or waste *stack* below the lowest horizontal *branch*, the required diameter of the offset and the *stack* below it shall be determined as for a *building drain* in accordance with Table 710.1(1).

SECTION 712
SUMPS AND EJECTORS

712.1 Building subdrains. Building subdrains that cannot be discharged to the *sewer* by gravity flow shall be discharged into a tightly covered and vented sump from which the liquid shall be lifted and discharged into the building gravity drainage system by automatic pumping equipment or other *approved* method. In other than existing structures, the sump shall not receive drainage from any piping within the building capable of being discharged by gravity to the *building sewer*.

712.2 Valves required. A check valve and a full open valve located on the discharge side of the check valve shall be installed in the pump or ejector discharge piping between the pump or ejector and the gravity drainage system. *Access* shall be provided to such valves. Such valves shall be located above the sump cover required by Section 712.1 or, where the discharge pipe from the ejector is below grade, the valves shall be accessibly located outside the sump below grade in an *access* pit with a removable *access* cover.

712.3 Sump design. The sump pump, pit and discharge piping shall conform to the requirements of Sections 712.3.1 through 712.3.5.

712.3.1 Sump pump. The sump pump capacity and head shall be appropriate to anticipated use requirements.

712.3.2 Sump pit. The sump pit shall be not less than 18 inches (457 mm) in diameter and not less than 24 inches (610 mm) in depth, unless otherwise *approved*. The pit shall be accessible and located such that all drainage flows into the pit by gravity. The sump pit shall be constructed of tile, concrete, steel, plastic or other *approved* materials. The pit bottom shall be solid and provide permanent support for the pump. The sump pit shall be fitted with a gastight removable cover adequate to support anticipated loads in the area of use. The sump pit shall be vented in accordance with Chapter 9.

712.3.3 Discharge pipe and fittings. Discharge pipe and fittings serving sump pumps and ejectors shall be constructed of materials in accordance with Sections 712.3.3.1 and 712.3.3.2 and shall be *approved*.

712.3.3.1 Materials. Pipe and fitting materials shall be constructed of brass, copper, CPVC, ductile iron, PE, or PVC.

712.3.3.2 Ratings. Pipe and fittings shall be rated for the maximum system operating pressure and temperature. Pipe fitting materials shall be compatible with the pipe material. Where pipe and fittings are buried in the earth, they shall be suitable for burial.

712.3.4 Maximum effluent level. The effluent level control shall be adjusted and maintained to at all times prevent the effluent in the sump from rising to within 2 inches (51 mm) of the invert of the gravity drain inlet into the sump.

712.3.5. Pump connection to the drainage system. Pumps connected to the drainage system shall connect to a building sewer, building drain, soil stack, waste stack or horizontal branch drain. Where the discharge line connects into horizontal drainage piping, the connection shall be made through a wye fitting into the top of the drainage piping and such wye fitting shall be located not less than 10 pipe diameters from the base of any soil stack, waste stack or fixture drain.

712.4 Sewage pumps and sewage ejectors. A sewage pump or sewage ejector shall automatically discharge the contents of the sump to the building drainage system.

712.4.1 Macerating toilet systems. Macerating toilet systems shall comply with CSA B45.9 or ASME A112.3.4 and shall be installed in accordance with the manufacturer's installation instructions.

712.4.2 Capacity. A sewage pump or sewage ejector shall have the capacity and head for the application requirements. Pumps or ejectors that receive the discharge of water closets shall be capable of handling spherical solids with a diameter of up to and including 2 inches (51 mm). Other pumps or ejectors shall be capable of handling spherical solids with a diameter of up to and including 1 inch (25.4 mm). The capacity of a pump or ejector based on the diameter of the discharge pipe shall be not less than that indicated in Table 712.4.2.

Exceptions:

1. Grinder pumps or grinder ejectors that receive the discharge of water closets shall have a discharge opening of not less than $1^1/_4$ inches (32 mm).

2. Macerating toilet assemblies that serve single water closets shall have a discharge opening of not less than $^3/_4$ inch (19 mm).

TABLE 712.4.2
MINIMUM CAPACITY OF SEWAGE PUMP OR SEWAGE EJECTOR

DIAMETER OF THE DISCHARGE PIPE (inches)	CAPACITY OF PUMP OR EJECTOR (gpm)
2	21
$2^1/_2$	30
3	46

For SI: 1 inch = 25.4 mm, 1 gallon per minute = 3.785 L/m.

SECTION 713
HEALTH CARE PLUMBING

713.1 Scope. This section shall govern those aspects of health care plumbing systems that differ from plumbing systems in other structures. Health care plumbing systems shall conform to this section in addition to the other requirements of this code. The provisions of this section shall apply to the special devices and equipment installed and maintained in the following occupancies: nursing homes; homes for the aged; orphanages; infirmaries; first aid stations; psychiatric facilities; clinics; professional offices of dentists and doctors; mortuaries; educational facilities; surgery, dentistry, research and testing laboratories; establishments manufacturing pharmaceutical drugs and medicines; and other structures with similar apparatus and equipment classified as plumbing.

713.2 Bedpan washers and clinical sinks. Bedpan washers and clinical sinks shall connect to the drainage and vent system in accordance with the requirements for a water closet. Bedpan washers shall also connect to a local vent.

713.3 Indirect waste. All sterilizers, steamers and condensers shall discharge to the drainage through an indirect waste pipe by means of an *air gap*. Where a battery of not more than three sterilizers discharges to an individual receptor, the distance between the receptor and a sterilizer shall not exceed 8 feet (2438 mm). The indirect waste pipe on a bedpan steamer shall be trapped.

713.4 Vacuum system station. Ready *access* shall be provided to vacuum system station receptacles. Such receptacles shall be built into cabinets or recesses and shall be visible.

713.5 Bottle system. Vacuum (fluid suction) systems intended for collecting, removing and disposing of blood, pus or other fluids by the bottle system shall be provided with receptacles equipped with an overflow prevention device at each vacuum outlet station.

713.6 Central disposal system equipment. All central vacuum (fluid suction) systems shall provide continuous service. Systems equipped with collecting or control tanks shall provide for draining and cleaning of the tanks while the system is in operation. In hospitals, the system shall be connected to the emergency power system. The exhausts from a vacuum pump serving a vacuum (fluid suction) system shall discharge separately to open air above the roof.

713.7 Central vacuum or disposal systems. Where the waste from a central vacuum (fluid suction) system of the barometric-lag, collection-tank or bottle-disposal type is connected to the drainage system, the waste shall be directly connected to the sanitary drainage system through a trapped waste.

713.7.1 Piping. The piping of a central vacuum (fluid suction) system shall be of corrosion-resistant material with a smooth interior surface. A *branch* shall not be less than $^1/_2$ inch (12.7 mm) nominal pipe size for one outlet and shall be sized in accordance with the number of vacuum outlets. A main shall not be less than 1-inch (25 mm) nominal pipe size. The pipe sizing shall be increased in accordance with the manufacturer's instructions as stations are increased.

713.7.2 Velocity. The velocity of airflow in a central vacuum (fluid suction) system shall be less than 5,000 feet per minute (25 m/s).

713.8 Vent connections prohibited. Connections between local vents serving bedpan washers or sterilizer vents serving sterilizing apparatus and normal sanitary plumbing systems are prohibited. Only one type of apparatus shall be served by a local vent.

713.9 Local vents and stacks for bedpan washers. Bedpan washers shall be vented to open air above the roof by means of one or more local vents. The local vent for a bedpan washer shall not be less than a 2-inch-diameter (51 mm) pipe. A local vent serving a single bedpan washer is permitted to drain to the fixture served.

713.9.1 Multiple installations. Where bedpan washers are located above each other on more than one floor, a local vent *stack* is permitted to be installed to receive the local vent on the various floors. Not more than three bedpan washers shall be connected to a 2-inch (51 mm) local vent *stack*, not more than six to a 3-inch (76 mm) local vent *stack* and not more than 12 to a 4-inch (102 mm) local vent *stack*. In multiple installations, the connections between a bedpan washer local vent and a local vent *stack* shall be made with tee or tee-wye sanitary pattern drainage fittings installed in an upright position.

713.9.2 Trap required. The bottom of the local vent *stack*, except where serving only one bedpan washer, shall be drained by means of a trapped and vented waste connection to the sanitary drainage system. The trap and waste shall be the same size as the local vent *stack*.

713.9.3 Trap seal maintenance. A water supply pipe not less than $^1/_4$ inch (6.4 mm) in diameter shall be taken from the flush supply of each bedpan washer on the discharge or fixture side of the vacuum breaker, shall be trapped to form not less than a 3-inch (76 mm) water seal, and shall be connected to the local vent *stack* on each floor. The water supply shall be installed so as to provide a supply of water to the local vent *stack* for cleansing and drain trap seal maintenance each time a bedpan washer is flushed.

713.10 Sterilizer vents and stacks. Multiple installations of pressure and nonpressure sterilizers shall have the vent connections to the sterilizer vent *stack* made by means of inverted wye fittings. *Access* shall be provided to vent connections for the purpose of inspection and maintenance.

713.10.1 Drainage. The connection between sterilizer vent or exhaust openings and the sterilizer vent *stack* shall be designed and installed to drain to the funnel or basket-type waste fitting. In multiple installations, the sterilizer vent *stack* shall be drained separately to the lowest sterilizer funnel or basket-type waste fitting or receptor.

713.11 Sterilizer vent stack sizes. Sterilizer vent *stack* sizes shall comply with Sections 713.11.1 through 713.11.4.

713.11.1 Bedpan steamers. The minimum size of a sterilizer vent serving a bedpan steamer shall be $1^1/_2$ inches (38 mm) in diameter. Multiple installations shall be sized in accordance with Table 713.11.1.

TABLE 713.11.1
STACK SIZES FOR BEDPAN STEAMERS AND BOILING-TYPE STERILIZERS (Number of Connections of Various Sizes Permitted to Various-sized Sterilizer Vent Stacks)

STACK SIZE (inches)	CONNECTION SIZE		
	$1^1/_2$"		2"
$1^1/_2$[a]	1	or	0
2[a]	2	or	1
2[b]	1	and	1
3[a]	4	or	2
3[b]	2	and	2
4[a]	8	or	4
4[b]	4	and	4

For SI: 1 inch = 25.4 mm.
a. Total of each size.
b. Combination of sizes.

713.11.2 Boiling-type sterilizers. The size of a sterilizer vent *stack* shall be not less than 2 inches (51 mm) in diameter where serving a utensil sterilizer and not less than $1^1/_2$ inches (38 mm) in diameter where serving an instrument sterilizer. Combinations of boiling-type sterilizer vent connections shall be sized in accordance with Table 713.11.1.

713.11.3 Pressure sterilizers. Pressure sterilizer vent stacks shall be $2^1/_2$ inches (64 mm) minimum. Those serving combinations of pressure sterilizer exhaust connections shall be sized in accordance with Table 713.11.3.

713.11.4 Pressure instrument washer sterilizer sizes. The diameter of a sterilizer vent *stack* serving an instrument washer sterilizer shall be not less than 2 inches (51 mm). Not more than two sterilizers shall be installed on a 2-inch (51 mm) *stack*, and not more than four sterilizers shall be installed on a 3-inch (76 mm) *stack*.

TABLE 713.11.3
STACK SIZES FOR PRESSURE STERILIZERS
(Number of Connections of Various Sizes
Permitted To Various-sized Vent Stacks)

STACK SIZE (inches)	CONNECTION SIZE			
	$^3/_4$"	1"	$1^1/_4$"	$1^1/_2$"
$1^1/_2{}^a$	3 or	2 or	1	—
$1^1/_2{}^b$	2 and	1	—	—
2^a	6 or	3 or	2 or	1
2^b	3 and	2	—	—
2^b	2 and	1 and	1	—
2^b	1 and	1 and	—	1
3^a	15 or	7 or	5 or	3
3^b	1 and	1 and 5 and	2 and	2 1

For SI: 1 inch = 25.4 mm.

a. Total of each size.

b. Combination of sizes.

SECTION 714
COMPUTERIZED DRAINAGE DESIGN

714.1 Design of drainage system. The sizing, design and layout of the drainage system shall be permitted to be designed by *approved* computer design methods.

714.2 Load on drainage system. The load shall be computed from the simultaneous or sequential discharge conditions from fixtures, appurtenances and appliances or the peak usage design condition.

714.2.1 Fixture discharge profiles. The discharge profiles for flow rates versus time from fixtures and appliances shall be in accordance with the manufacturer's specifications.

714.3 Selections of drainage pipe sizes. Pipe shall be sized to prevent full-bore flow.

714.3.1 Selecting pipe wall roughness. Pipe size calculations shall be conducted with the pipe wall roughness factor (ks), in accordance with the manufacturer's specifications and as modified for aging roughness factors with deposits and corrosion.

714.3.2 Slope of horizontal drainage piping. Horizontal drainage piping shall be designed and installed at slopes in accordance with Table 704.1.

SECTION 715
BACKWATER VALVES

715.1 Sewage backflow. Where plumbing fixtures are installed on a floor with a finished floor elevation below the elevation of the manhole cover of the next upstream manhole in the *public sewer,* such fixtures shall be protected by a backwater valve installed in the *building drain,* or horizontal *branch* serving such fixtures. Plumbing fixtures installed on a floor with a finished floor elevation above the elevation of the manhole cover of the next upstream manhole in the *public sewer* shall not discharge through a backwater valve.

715.2 Material. All bearing parts of backwater valves shall be of corrosion-resistant material. Backwater valves shall comply with ASME A112.14.1, CSA B181.1 or CSA B181.2.

715.3 Seal. Backwater valves shall be so constructed as to provide a mechanical seal against backflow.

715.4 Diameter. Backwater valves, when fully opened, shall have a capacity not less than that of the pipes in which they are installed.

715.5 Location. Backwater valves shall be installed so that *access* is provided to the working parts for service and repair.

CHAPTER 8

INDIRECT/SPECIAL WASTE

SECTION 801
GENERAL

801.1 Scope. This chapter shall govern matters concerning indirect waste piping and special wastes. This chapter shall further control matters concerning food-handling establishments, sterilizers, clear-water wastes, swimming pools, methods of providing air breaks or air gaps, and neutralizing devices for corrosive wastes.

801.2 Protection. Devices, appurtenances, appliances and apparatus intended to serve some special function, such as sterilization, distillation, processing, cooling, or storage of ice or foods, and that discharge to the drainage system, shall be provided with protection against backflow, flooding, fouling, contamination and stoppage of the drain.

SECTION 802
INDIRECT WASTES

802.1 Where required. Food-handling equipment and clear-water waste shall discharge through an indirect waste pipe as specified in Sections 802.1.1 through 802.1.8. All health-care related fixtures, devices and equipment shall discharge to the drainage system through an indirect waste pipe by means of an *air gap* in accordance with this chapter and Section 713.3. Fixtures not required by this section to be indirectly connected shall be directly connected to the plumbing system in accordance with Chapter 7.

802.1.1 Food handling. Equipment and fixtures utilized for the storage, preparation and handling of food shall discharge through an indirect waste pipe by means of an *air gap*.

802.1.2 Floor drains in food storage areas. Floor drains located within walk-in refrigerators or freezers in food service and food establishments shall be indirectly connected to the sanitary drainage system by means of an *air gap*. Where a floor drain is located within an area subject to freezing, the waste line serving the floor drain shall not be trapped and shall indirectly discharge into a waste receptor located outside of the area subject to freezing.

> **Exception:** Where protected against backflow by a backwater valve, such floor drains shall be indirectly connected to the sanitary drainage system by means of an *air break* or an *air gap*.

802.1.3 Potable clear-water waste. Where devices and equipment, such as sterilizers and relief valves, discharge potable water to the building drainage system, the discharge shall be through an indirect waste pipe by means of an *air gap*.

802.1.4 Swimming pools. Where waste water from swimming pools, backwash from filters and water from pool deck drains discharge to the building drainage system, the discharge shall be through an indirect waste pipe by means of an *air gap*.

802.1.5 Nonpotable clear-water waste. Where devices and equipment such as process tanks, filters, drips and boilers discharge nonpotable water to the building drainage system, the discharge shall be through an indirect waste pipe by means of an *air break* or an *air gap*.

802.1.6 Domestic dishwashing machines. Domestic dishwashing machines shall discharge indirectly through an *air gap* or *air break* into a standpipe or waste receptor in accordance with Section 802.2, or discharge into a wye-branch fitting on the tailpiece of the kitchen sink or the dishwasher connection of a food waste grinder. The waste line of a domestic dishwashing machine discharging into a kitchen sink tailpiece or food waste grinder shall connect to a deck-mounted *air gap* or the waste line shall rise and be securely fastened to the underside of the sink rim or counter.

802.1.7 Commercial dishwashing machines. The discharge from a commercial dishwashing machine shall be through an *air gap* or *air break* into a standpipe or waste receptor in accordance with Section 802.2.

802.1.8 Food utensils, dishes, pots and pans sinks. Sinks used for the washing, rinsing or sanitizing of utensils, dishes, pots, pans or service ware used in the preparation, serving or eating of food shall discharge indirectly through an *air gap* or an *air break* to the drainage system.

802.2 Installation. Indirect waste piping shall discharge through an *air gap* or *air break* into a waste receptor. Waste receptors and standpipes shall be trapped and vented and shall connect to the building drainage system. All indirect waste piping that exceeds 30 inches (762 mm) in developed length measured horizontally, or 54 inches (1372 mm) in total developed length, shall be trapped.

> **Exception:** Where a waste receptor receives only clear-water waste and does not directly connect to a sanitary drainage system, the receptor shall not require a trap.

802.2.1 Air gap. The *air gap* between the indirect waste pipe and the *flood level rim* of the waste receptor shall be not less than twice the *effective opening* of the indirect waste pipe.

802.2.2 Air break. An *air break* shall be provided between the indirect waste pipe and the trap seal of the waste receptor or standpipe.

802.3 Waste receptors. Waste receptors shall be of an approved type. A removable strainer or basket shall cover the waste outlet of waste receptors. Waste receptors shall be installed in ventilated spaces. Waste receptors shall not be installed in bathrooms, toilet rooms, plenums, crawl spaces, attics, interstitial spaces above ceilings and below floors or in any inaccessible or unventilated space such as a closet or storeroom. Ready access shall be provided to waste receptors.

802.3.1 Size of receptors. A waste receptor shall be sized for the maximum discharge of all indirect waste pipes served by the receptor. Receptors shall be installed to prevent splashing or flooding.

802.3.2 Open hub waste receptors. Waste receptors shall be permitted in the form of a hub or pipe extending not less than 1 inch (25.4 mm) above a water-impervious floor and are not required to have a strainer.

802.4 Standpipes. Standpipes shall be individually trapped. Standpipes shall extend not less than 18 inches (457 mm) but not greater than 42 inches (1066 mm) above the trap weir. *Access* shall be provided to all standpipes and drains for rodding.

SECTION 803
SPECIAL WASTES

803.1 Waste water temperature. Steam pipes shall not connect to any part of a drainage or plumbing system and water above 140°F (60°C) shall not be discharged into any part of a drainage system. Such pipes shall discharge into an indirect waste receptor connected to the drainage system.

803.2 Neutralizing device required for corrosive wastes. Corrosive liquids, spent acids or other harmful chemicals that destroy or injure a drain, *sewer*, soil or waste pipe, or create noxious or toxic fumes or interfere with sewage treatment processes shall not be discharged into the plumbing system without being thoroughly diluted, neutralized or treated by passing through an *approved* dilution or neutralizing device. Such devices shall be automatically provided with a sufficient supply of diluting water or neutralizing medium so as to make the contents noninjurious before discharge into the drainage system. The nature of the corrosive or harmful waste and the method of its treatment or dilution shall be *approved* prior to installation.

803.3 System design. A chemical drainage and vent system shall be designed and installed in accordance with this code. Chemical drainage and vent systems shall be completely separated from the sanitary systems. Chemical waste shall not discharge to a sanitary drainage system until such waste has been treated in accordance with Section 803.2.

SECTION 804
MATERIALS, JOINTS AND CONNECTIONS

804.1 General. The materials and methods utilized for the construction and installation of indirect waste pipes and systems shall comply with the applicable provisions of Chapter 7.

CHAPTER 9

VENTS

(Note that Chapter 9 has been reorganized to present the main sections in a more logical order. For clarity of presentation, the star and double-star scheme normally used to indicate text that has been relocated was not used for this text reorganization.)

SECTION 901
GENERAL

901.1 Scope. The provisions of this chapter shall govern the materials, design, construction and installation of vent systems.

901.2 Trap seal protection. The plumbing system shall be provided with a system of vent piping that will permit the admission or emission of air so that the seal of any fixture trap shall not be subjected to a pneumatic pressure differential of more than 1 inch of water column (249 Pa).

901.2.1 Venting required. Traps and trapped fixtures shall be vented in accordance with one of the venting methods specified in this chapter.

901.3 Chemical waste vent systems. The vent system for a chemical waste system shall be independent of the sanitary vent system and shall terminate separately through the roof to the outdoors or to an air admittance valve that complies with ASSE 1049. Air admittance valves for chemical waste systems shall be constructed of materials approved in accordance with Section 702.5 and shall be tested for chemical resistance in accordance with ASTM F 1412.

901.4 Use limitations. The plumbing vent system shall not be utilized for purposes other than the venting of the plumbing system.

901.5 Tests. The vent system shall be tested in accordance with Section 312.

901.6 Engineered systems. Engineered venting systems shall conform to the provisions of Section 919.

SECTION 902
MATERIALS

902.1 Vents. The materials and methods utilized for the construction and installation of venting systems shall comply with the applicable provisions of Section 702.

902.2 Sheet copper. Sheet copper for vent pipe flashings shall conform to ASTM B 152 and shall weigh not less than 8 ounces per square foot (2.5 kg/m^2).

902.3 Sheet lead. Sheet lead for vent pipe flashings shall weigh not less than 3 pounds per square foot (15 kg/m^2) for field-constructed flashings and not less than 2^1/$_2$ pounds per square foot (12 kg/m^2) for prefabricated flashings.

SECTION 903
VENT TERMINALS

903.1 Roof extension. Open vent pipes that extend through a roof shall be terminated not less than [NUMBER] inches (mm) above the roof, except that where a roof is to be used for any purpose other than weather protection, the vent extensions shall terminate not less than 7 feet (2134 mm) above the roof.

903.2 Frost closure. Where the 97.5-percent value for outside design temperature is 0°F (-18°C) or less, every vent extension through a roof or wall shall be not less than 3 inches (76 mm) in diameter. Any increase in the size of the vent shall be made inside the structure at a point not less than 1 foot (305 mm) below the roof or inside the wall.

903.3 Flashings. The juncture of each vent pipe with the roof line shall be made water-tight by an *approved* flashing.

903.4 Prohibited use. A vent terminal shall not be used for any purpose other than a vent terminal.

903.5 Location of vent terminal. An open vent terminal from a drainage system shall not be located directly beneath any door, openable window, or other air intake opening of the building or of an adjacent building, and any such vent terminal shall not be within 10 feet (3048 mm) horizontally of such an opening unless it is 3 feet (914 mm) or more above the top of such opening.

903.6 Extension through the wall. Vent terminals extending through the wall shall terminate at a point not less than 10 feet (3048 mm) from a lot line and not less than 10 feet (3048 mm) above average ground level. Vent terminals shall not terminate under the overhang of a structure with soffit vents. Side wall vent terminals shall be protected to prevent birds or rodents from entering or blocking the vent opening.

903.7 Extension outside a structure. In climates where the 97.5-percent value for outside design temperature is less than 0°F (-18°C), vent pipes installed on the exterior of the structure shall be protected against freezing by insulation, heat or both.

SECTION 904
OUTDOOR VENT EXTENSIONS

904.1 Required vent extension. The vent system serving each *building drain* shall have not less than one vent pipe that extends to the outdoors.

904.1.1 Installation. The required vent shall be a dry vent that connects to the *building drain* or an extension of a drain that connects to the *building drain*. Such vent shall not be an island fixture vent as allowed by Section 916.

904.1.2 Size. The required vent shall be sized in accordance with Section 906.2 based on the required size of the *building drain*.

904.2 Vent stack required. A vent *stack* shall be required for every drainage *stack* that has five *branch intervals* or more.

> **Exception:** Drainage stacks installed in accordance with Section 913.

904.3 Vent termination. Vent stacks or stack vents shall terminate outdoors to the open air or to a stack-type air admittance valve in accordance with Section 918.

904.4 Vent connection at base. Vent *stacks* shall connect to the base of the drainage *stack*. The vent *stack* shall connect at or below the lowest horizontal *branch*. Where the vent *stack* connects to the *building drain*, the connection shall be located downstream of the drainage *stack* and within a distance of 10 times the diameter of the drainage *stack*.

904.5 Vent headers. *Stack vents* and vent stacks connected into a common vent header at the top of the stacks and extending to the open air at one point shall be sized in accordance with the requirements of Section 906.1. The number of fixture units shall be the sum of all fixture units on all stacks connected thereto, and the *developed length* shall be the longest vent length from the intersection at the base of the most distant *stack* to the vent terminal in the open air, as a direct extension of one *stack*.

SECTION 905
VENT CONNECTIONS AND GRADES

905.1 Connection. All individual, *branch* and circuit vents shall connect to a vent *stack*, *stack vent*, air admittance valve or extend to the open air.

905.2 Grade. All vent and *branch* vent pipes shall be so graded and connected as to drain back to the drainage pipe by gravity.

905.3 Vent connection to drainage system. Every dry vent connecting to a horizontal drain shall connect above the centerline of the horizontal drain pipe.

905.4 Vertical rise of vent. Every dry vent shall rise vertically to a point not less than 6 inches (152 mm) above the *flood level rim* of the highest trap or trapped fixture being vented.

> **Exception:** Vents for interceptors located outdoors.

905.5 Height above fixtures. A connection between a vent pipe and a vent *stack* or *stack vent* shall be made at not less than 6 inches (152 mm) above the *flood level rim* of the highest fixture served by the vent. Horizontal vent pipes forming *branch* vents, relief vents or loop vents shall be located not less than 6 inches (152 mm) above the *flood level rim* of the highest fixture served.

905.6 Vent for future fixtures. Where the drainage piping has been roughed-in for future fixtures, a rough-in connection for a vent shall be installed. The vent size shall be not less than one-half the diameter of the rough-in drain to be served. The vent rough-in shall connect to the vent system, or shall be vented by other means as provided for in this chapter. The connection shall be identified to indicate that it is a vent.

SECTION 906
VENT PIPE SIZING

906.1 Size of stack vents and vent stacks. The minimum required diameter of *stack vents* and vent stacks shall be determined from the *developed length* and the total of drainage fixture units connected thereto in accordance with Table 906.1, but in no case shall the diameter be less than one-half the diameter of the drain served or less than $1^1/_4$ inches (32 mm).

906.2 Vents other than stack vents or vent stacks. The diameter of individual vents, *branch* vents, circuit vents and relief vents shall be not less than one-half the required diameter of the drain served. The required size of the drain shall be determined in accordance with Table 710.1(2). Vent pipes shall not be less than $1^1/_4$ inches (32 mm) in diameter. Vents exceeding 40 feet (12 192 mm) in *developed length* shall be increased by one nominal pipe size for the entire *developed length* of the vent pipe. Relief vents for soil and waste stacks in buildings having more than 10 *branch intervals* shall be sized in accordance with Section 908.2.

906.3 Developed length. The *developed length* of individual, *branch,* circuit and relief vents shall be measured from the farthest point of vent connection to the drainage system to the point of connection to the vent *stack, stack vent* or termination outside of the building.

906.4 Multiple branch vents. Where multiple *branch* vents are connected to a common *branch* vent, the common *branch* vent shall be sized in accordance with this section based on the size of the common horizontal drainage *branch* that is or would be required to serve the total *drainage fixture unit (dfu)* load being vented.

906.5 Sump vents. Sump vent sizes shall be determined in accordance with Sections 906.5.1 and 906.5.2.

906.5.1 Sewage pumps and sewage ejectors other than pneumatic. Drainage piping below sewer level shall be vented in the same manner as that of a gravity system. Building sump vent sizes for sumps with sewage pumps or sewage ejectors, other than pneumatic, shall be determined in accordance with Table 906.5.1.

906.5.2 Pneumatic sewage ejectors. The air pressure relief pipe from a pneumatic sewage ejector shall be connected to an independent vent *stack* terminating as required for vent extensions through the roof. The relief pipe shall be sized to relieve air pressure inside the ejector to atmospheric pressure, but shall not be less than $1^1/_4$ inches (32 mm) in size.

TABLE 906.1
SIZE AND DEVELOPED LENGTH OF STACK VENTS AND VENT STACKS

DIAMETER OF SOIL OR WASTE STACK (inches)	TOTAL FIXTURE UNITS BEING VENTED (dfu)	MAXIMUM DEVELOPED LENGTH OF VENT (feet)[a] DIAMETER OF VENT (inches)										
		1¼	1½	2	2½	3	4	5	6	8	10	12
1¼	2	30										
1½	8	50	150	—	—	—	—	—	—	—	—	—
1½	10	30	100									
2	12		75	200								
2	20	30	50	150	—	—	—	—	—	—	—	—
2½	42	26	30	100	300							
3	10		42	150	360	1,040						
3	21	—	32	110	270	810	—	—	—	—	—	—
3	53		27	94	230	680						
3	102			86	210	620						
4	43	—	25	35	85	250	980	—	—	—	—	—
4	140			27	65	200	750					
4	320			23	55	170	640					
4	540	—	—	21	50	150	580	—	—	—	—	
5	190				28	82	320	990				
5	490				21	63	250	760				
5	940	—	—	—	18	53	210	670	—	—	—	—
5	1,400				16	49	190	590				
6	500					33	130	400	1,000			
6	1,100	—		—	—	26	100	310	780	—	—	—
6	2,000					22	84	260	660			
6	2,900					20	77	240	600			
8	1,800	—	—	—	—		31	95	240	940	—	—
8	3,400						24	73	190	729		
8	5,600						20	62	160	610		
8	7,600	—	—	—	—		18	56	140	560		—
10	4,000							31	78	310	960	
10	7,200							24	60	240	740	
10	11,000	—	—	—	—	—	—	20	51	200	630	—
10	15,000							18	46	180	571	
12	7,300								31	120	380	940
12	13,000	—	—	—	—	—	—	—	24	94	300	720
12	20,000								20	79	250	610
12	26,000								18	72	230	500
15	15,000	—	—	—	—	—	—	—		40	130	310
15	25,000									31	96	240
15	38,000									26	81	200
15	50,000	—	—	—	—	—	—	—	—	24	74	180

For SI:1 inch = 25.4 mm, 1 foot = 304.8 mm.

a.The developed length shall be measured from the vent connection to the open air.

TABLE 906.5.1
SIZE AND LENGTH OF SUMP VENTS

DISCHARGE CAPACITY OF PUMP (gpm)	MAXIMUM DEVELOPED LENGTH OF VENT (feet)[a]					
	Diameter of vent (inches)					
	1$^1/_4$	1$^1/_2$	2	2$^1/_2$	3	4
10	No limit[b]	No limit	No limit	No limit	No limit	No limit
20	270	No limit	No limit	No limit	No limit	No limit
40	72	160	No limit	No limit	No limit	No limit
60	31	75	270	No limit	No limit	No limit
80	16	41	150	380	No limit	No limit
100	10[c]	25	97	250	No limit	No limit
150	Not permitted	10[c]	44	110	370	No limit
200	Not permitted	Not permitted	20	60	210	No limit
250	Not permitted	Not permitted	10	36	132	No limit
300	Not permitted	Not permitted	10[c]	22	88	380
400	Not permitted	Not permitted	Not permitted	10[c]	44	210
500	Not permitted	Not permitted	Not permitted	Not permitted	24	130

For SI: 1 inch = 25.4 mm, 1 foot = 304.8 mm, 1 gallon per minute = 3.785 L/m.

a. Developed length plus an appropriate allowance for entrance losses and friction due to fittings, changes in direction and diameter. Suggested allowances shall be obtained from NBS Monograph 31 or other approved sources. An allowance of 50 percent of the developed length shall be assumed if a more precise value is not available.

b. Actual values greater than 500 feet.

c. Less than 10 feet.

SECTION 907
VENTS FOR STACK OFFSETS

907.1 Vent for horizontal offset of drainage stack. Horizontal offsets of drainage stacks shall be vented where five or more *branch intervals* are located above the offset. The offset shall be vented by venting the upper section of the drainage *stack* and the lower section of the drainage *stack*.

907.2 Upper section. The upper section of the drainage *stack* shall be vented as a separate *stack* with a vent *stack* connection installed in accordance with Section 904.4. The offset shall be considered the base of the *stack*.

907.3 Lower section. The lower section of the drainage *stack* shall be vented by a yoke vent connecting between the offset and the next lower horizontal *branch*. The yoke vent connection shall be permitted to be a vertical extension of the drainage *stack*. The size of the yoke vent and connection shall be a minimum of the size required for the vent *stack* of the drainage *stack*.

SECTION 908
RELIEF VENTS—STACKS OF MORE THAN 10 BRANCH INTERVALS

908.1 Where required. Soil and waste stacks in buildings having more than 10 *branch intervals* shall be provided with a relief vent at each tenth interval installed, beginning with the top floor.

908.2 Size and connection. The size of the relief vent shall be equal to the size of the vent *stack* to which it connects. The lower end of each relief vent shall connect to the soil or waste *stack* through a wye below the horizontal *branch* serving the

floor, and the upper end shall connect to the vent *stack* through a wye not less than 3 feet (914 mm) above the floor.

SECTION 909
FIXTURE VENTS

909.1 Distance of trap from vent. Each fixture trap shall have a protecting vent located so that the slope and the *developed length* in the *fixture drain* from the trap weir to the vent fitting are within the requirements set forth in Table 909.1.

Exception: The *developed length* of the *fixture drain* from the trap weir to the vent fitting for self-siphoning fixtures, such as water closets, shall not be limited.

TABLE 909.1
MAXIMUM DISTANCE OF FIXTURE TRAP FROM VENT

SIZE OF TRAP (inches)	SLOPE (inch per foot)	DISTANCE FROM TRAP (feet)
1$^1/_4$	$^1/_4$	5
1$^1/_2$	$^1/_4$	6
2	$^1/_4$	8
3	$^1/_8$	12
4	$^1/_8$	16

For SI: 1 inch = 25.4 mm, 1 foot = 304.8 mm, 1 inch per foot = 83.3 mm/m.

909.2 Venting of fixture drains. The total fall in a *fixture drain* due to pipe slope shall not exceed the diameter of the *fixture drain*, nor shall the vent connection to a *fixture drain*, except for water closets, be below the weir of the trap.

909.3 Crown vent. A vent shall not be installed within two pipe diameters of the trap weir.

SECTION 910
INDIVIDUAL VENT

910.1 Individual vent permitted. Each trap and trapped fixture is permitted to be provided with an individual vent. The individual vent shall connect to the *fixture drain* of the trap or trapped fixture being vented.

SECTION 911
COMMON VENT

911.1 Individual vent as common vent. An individual vent is permitted to vent two traps or trapped fixtures as a common vent. The traps or trapped fixtures being common vented shall be located on the same floor level.

911.2 Connection at the same level. Where the fixture drains being common vented connect at the same level, the vent connection shall be at the interconnection of the fixture drains or downstream of the interconnection.

911.3 Connection at different levels. Where the fixture drains connect at different levels, the vent shall connect as a vertical extension of the vertical drain. The vertical drain pipe connecting the two fixture drains shall be considered the vent for the lower *fixture drain*, and shall be sized in accordance with Table 911.3. The upper fixture shall not be a water closet.

TABLE 911.3
COMMON VENT SIZES

PIPE SIZE (inches)	MAXIMUM DISCHARGE FROM UPPER FIXTURE DRAIN (dfu)
$1^1/_2$	1
2	4
$2^1/_2$ to 3	6

For SI: 1 inch = 25.4 mm.

SECTION 912
WET VENTING

912.1 Horizontal wet vent permitted. Any combination of fixtures within two bathroom groups located on the same floor level is permitted to be vented by a horizontal wet vent. The wet vent shall be considered the vent for the fixtures and shall extend from the connection of the dry vent along the direction of the flow in the drain pipe to the most downstream *fixture drain* connection to the *horizontal branch drain*. Each wet-vented *fixture drain* shall connect independently to the horizontal wet vent. Only the fixtures within the bathroom groups shall connect to the wet-vented *horizontal branch drain*. Any additional fixtures shall discharge downstream of the horizontal wet vent.

912.1.1 Vertical wet vent permitted. Any combination of fixtures within two bathroom groups located on the same floor level is permitted to be vented by a vertical wet vent. The vertical wet vent shall be considered the vent for the fixtures and shall extend from the connection of the dry vent down to the lowest *fixture drain* connection. Each wet-vented fixture shall connect independently to the vertical wet vent. Water closet drains shall connect at the

same elevation. Other fixture drains shall connect above or at the same elevation as the water closet fixture drains. The dry-vent connection to the vertical wet vent shall be an individual or common vent serving one or two fixtures.

912.2 Dry vent connection. The required dry-vent connection for wet-vented systems shall comply with Sections 912.2.1 and 912.2.2.

912.2.1 Horizontal wet vent. The dry-vent connection for a horizontal wet-vent system shall be an individual vent or a common vent for any *bathroom group* fixture, except an *emergency floor drain*. Where the dry-vent connects to a water closet *fixture drain*, the drain shall connect horizontally to the horizontal wet-vent system. Not more than one wet-vented *fixture drain* shall discharge upstream of the dry-vented *fixture drain* connection.

912.2.2 Vertical wet vent. The dry-vent connection for a vertical wet-vent system shall be an individual vent or common vent for the most upstream *fixture drain*.

912.3 Size. The dry vent serving the wet vent shall be sized based on the largest required diameter of pipe within the wet-vent system served by the dry vent. The wet vent shall be of a size not less than that specified in Table 912.3, based on the fixture unit discharge to the wet vent.

TABLE 912.3
WET VENT SIZE

WET VENT PIPE SIZE (inches)	DRAINAGE FIXTURE UNIT LOAD (dfu)
$1^1/_2$	1
2	4
$2^1/_2$	6
3	12

For SI: 1 inch = 25.4 mm.

SECTION 913
WASTE STACK VENT

913.1 Waste stack vent permitted. A waste *stack* shall be considered a vent for all of the fixtures discharging to the *stack* where installed in accordance with the requirements of this section.

913.2 Stack installation. The waste *stack* shall be vertical, and both horizontal and vertical offsets shall be prohibited between the lowest *fixture drain* connection and the highest *fixture drain* connection. *Fixture drains* shall connect separately to the waste *stack*. The *stack* shall not receive the discharge of water closets or urinals.

913.3 Stack vent. A *stack vent* shall be provided for the waste *stack*. The size of the *stack vent* shall be not less than the size of the waste *stack*. Offsets shall be permitted in the *stack vent*, shall be located not less than 6 inches (152 mm) above the flood level of the highest fixture and shall be in accordance with Section 905.2. The *stack vent* shall be permitted to connect with other *stack vents* and vent stacks in accordance with Section 904.5.

913.4 Waste stack size. The waste *stack* shall be sized based on the total discharge to the *stack* and the discharge within a *branch* interval in accordance with Table 913.4. The waste *stack* shall be the same size throughout its length.

TABLE 913.4
WASTE STACK VENT SIZE

STACK SIZE (inches)	MAXIMUM NUMBER OF DRAINAGE FIXTURE UNITS (dfu)	
	Total discharge into one branch interval	Total discharge for stack
$1^1/_2$	1	2
2	2	4
$2^1/_2$	No limit	8
3	No limit	24
4	No limit	50
5	No limit	75
6	No limit	100

For SI: 1 inch = 25.4 mm.

SECTION 914
CIRCUIT VENTING

914.1 Circuit vent permitted. A maximum of eight fixtures connected to a horizontal *branch* drain shall be permitted to be circuit vented. Each *fixture drain* shall connect horizontally to the horizontal *branch* being circuit vented. The horizontal *branch* drain shall be classified as a vent from the most downstream *fixture drain* connection to the most upstream *fixture drain* connection to the horizontal *branch*.

914.1.1 Multiple circuit-vented branches. Circuit-vented horizontal *branch* drains are permitted to be connected together. Each group of a maximum of eight fixtures shall be considered a separate circuit vent and shall conform to the requirements of this section.

914.2 Vent connection. The circuit vent connection shall be located between the two most upstream fixture drains. The vent shall connect to the horizontal *branch* and shall be installed in accordance with Section 905. The circuit vent pipe shall not receive the discharge of any soil or waste.

914.3 Slope and size of horizontal branch. The slope of the vent section of the horizontal branch drain shall be not greater than one unit vertical in 12 units horizontal (8.3-percent slope). The entire length of the vent section of the horizontal branch drain shall be sized for the total drainage discharge to the branch.

914.3.1 Size of multiple circuit vent. Each separate circuit-vented horizontal *branch* that is interconnected shall be sized independently in accordance with Section 914.3. The downstream circuit-vented horizontal *branch* shall be sized for the total discharge into the *branch*, including the upstream branches and the fixtures within the *branch*.

914.4 Relief vent. A relief vent shall be provided for circuit-vented horizontal branches receiving the discharge of four or more water closets and connecting to a drainage *stack* that receives the discharge of soil or waste from upper horizontal branches.

914.4.1 Connection and installation. The relief vent shall connect to the horizontal *branch* drain between the *stack* and the most downstream *fixture drain* of the circuit vent. The relief vent shall be installed in accordance with Section 905.

914.4.2 Fixture drain or branch. The relief vent is permitted to be a *fixture drain* or fixture *branch* for fixtures located within the same *branch interval* as the circuit-vented horizontal *branch*. The maximum discharge to a relief vent shall be four fixture units.

914.5 Additional fixtures. Fixtures, other than the circuit-vented fixtures, are permitted to discharge to the horizontal *branch* drain. Such fixtures shall be located on the same floor as the circuit-vented fixtures and shall be either individually or common vented.

SECTION 915
COMBINATION WASTE AND VENT SYSTEM

915.1 Type of fixtures. A combination waste and vent system shall not serve fixtures other than floor drains, sinks, lavatories and drinking fountains. Combination waste and vent systems shall not receive the discharge from a food waste grinder or clinical sink.

915.2 Installation. The only vertical pipe of a combination waste and vent system shall be the connection between the fixture drain and the horizontal combination waste and vent pipe. The vertical distance shall not exceed 8 feet (2438 mm).

915.2.1 Slope. The slope of a horizontal combination waste and vent pipe shall not exceed one-half unit vertical in 12 units horizontal (4-percent slope) and shall not be less than that indicated in Table 704.1.

915.2.2 Connection. The combination waste and vent system shall be provided with a dry vent connected at any point within the system or the system shall connect to a horizontal drain that is vented in accordance with one of the venting methods specified in this chapter. Combination waste and vent systems connecting to building drains receiving only the discharge from a stack or stacks shall be provided with a dry vent. The vent connection to the combination waste and vent pipe shall extend vertically to a point not less than 6 inches (152 mm) above the flood level rim of the highest fixture being vented before offsetting horizontally. The horizontal length of a combination waste and vent system shall be unlimited.

915.2.3 Vent size. The vent shall be sized for the total *drainage fixture unit* load in accordance with Section 916.2.

915.2.4 Fixture branch or drain. The fixture branch or fixture drain shall connect to the combination waste and vent within a distance specified in Table 909.1. The combination waste and vent pipe shall be considered the vent for the fixture.

915.3 Size. The size of a combination waste and vent pipe shall not be less than that indicated in Table 915.3.

TABLE 915.3
SIZE OF COMBINATION WASTE AND VENT PIPE

DIAMETER PIPE (inches)	MAXIMUM NUMBER OF DRAINAGE FIXTURE UNITS (dfu)	
	Connecting to a horizontal branch or stack	Connecting to a building drain or building subdrain
2	3	4
2$^1/_2$	6	26
3	12	31
4	20	50
5	160	250
6	360	575

For SI: 1 inch = 25.4 mm.

SECTION 916
ISLAND FIXTURE VENTING

916.1 Limitation. Island fixture venting shall not be permitted for fixtures other than sinks and lavatories. Residential kitchen sinks with a dishwasher waste connection, a food waste grinder, or both, in combination with the kitchen sink waste, shall be permitted to be vented in accordance with this section.

916.2 Vent connection. The island fixture vent shall connect to the *fixture drain* as required for an individual or common vent. The vent shall rise vertically to above the drainage outlet of the fixture being vented before offsetting horizontally or vertically downward. The vent or *branch* vent for multiple island fixture vents shall extend to a point not less than 6 inches (152 mm) above the highest island fixture being vented before connecting to the outside vent terminal.

916.3 Vent installation below the fixture flood level rim. The vent located below the *flood level rim* of the fixture being vented shall be installed as required for drainage piping in accordance with Chapter 7, except for sizing. The vent shall be sized in accordance with Section 906.2. The lowest point of the island fixture vent shall connect full size to the drainage system. The connection shall be to a vertical drain pipe or to the top half of a horizontal drain pipe. Cleanouts shall be provided in the island fixture vent to permit rodding of all vent piping located below the *flood level rim* of the fixtures. Rodding in both directions shall be permitted through a cleanout.

SECTION 917
SINGLE STACK VENT SYSTEM

917.1 Where permitted. A drainage *stack* shall serve as a single stack vent system where sized and installed in accordance with Sections 917.2 through 917.9. The drainage *stack* and *branch* piping shall be the vents for the drainage system. The drainage *stack* shall have a stack vent.

917.2 Stack size. Drainage *stacks* shall be sized in accordance with Table 917.2. *Stacks* shall be uniformly sized based on the total connected *drainage fixture unit* load. The stack vent shall

be the same size as the drainage *stack*. A 3-inch (76 mm) *stack* shall serve not more than two water closets.

TABLE 917.2
SINGLE STACK SIZE

STACK SIZE (inches)	MAXIMUM CONNECTED DRAINAGE FIXTURE UNITS		
	Stacks less than 75 feet in height	Stacks 75 feet to less than 160 feet in height	Stacks 160 feet and greater in height
3	24	NP	NP
4	225	24	NP
5	480	225	24
6	1,015	480	225
8	2,320	1,015	480
10	4,500	2,320	1,015
12	8,100	4,500	2,320
15	13,600	8,100	4,500

For SI: 1 inch = 25.4 mm, 1 foot = 304.8 mm.

917.3 Branch size. Horizontal branches connecting to a single *stack vent* system shall be sized in accordance with Table 710.1(2). Not more than one water closet shall discharge into a 3-inch (76 mm) horizontal *branch* at a point within a developed length of 18 inches (457 mm) measured horizontally from the *stack*.

Where a water closet is within 18 inches (457 mm) measured horizontally from the *stack* and not more than one fixture with a drain size of not more than 1$^1/_2$ inch (38 mm) connects to a 3-inch (76 mm) horizontal branch, the branch drain connection to the *stack* shall be made with a sanitary tee.

917.4 Length of horizontal branches. The length of horizontal branches shall conform to the requirements of Sections 917.4.1 through 917.4.3.

917.4.1 Water closet connection. Water closet connections shall be not greater than 4 feet (1219 mm) in developed length measured horizontally from the *stack*.

> **Exception:** Where the connection is made with a sanitary tee, the maximum *developed length* shall be 8 feet (2438 mm).

917.4.2 Fixture connections. Fixtures other than water closets shall be located not greater than 12 feet (3657 mm) in developed length, measured horizontally from the *stack*.

917.4.3 Vertical piping in branch. The length of vertical piping in a *fixture drain* connecting to a horizontal *branch* shall not be considered in computing the fixture's distance in *developed length* measured horizontally from the *stack*.

917.5 Minimum vertical piping size from fixture. The vertical portion of piping in a *fixture drain* to a horizontal *branch* shall be 2 inches (51 mm). The minimum size of the vertical portion of piping for a water-supplied urinal or standpipe shall be 3 inches (76 mm). The maximum vertical drop shall be 4 feet (1219 mm). *Fixture drains* that are not increased in size, or have a vertical drop in excess of 4 feet (1219 mm) shall be individually vented.

917.6 Additional venting required. Additional venting shall be provided where more than one water closet discharges to a horizontal *branch* and where the distance from a fixture trap to the stack exceeds the limits in Section 917.4. Where additional venting is required, the fixture(s) shall be vented by individual vents, common vents, wet vents, circuit vents, or a combination waste and vent pipe. The dry vent extensions for the additional venting shall connect to a branch vent, vent *stack*, *stack vent*, air admittance valve, or shall terminate outdoors.

917.7 Stack offsets. Where fixture drains are not connected below a horizontal offset in a *stack*, a horizontal offset shall not be required to be vented. Where horizontal *branches* or *fixture drains* are connected below a horizontal offset in a *stack*, the offset shall be vented in accordance with Section 907. Fixture connections shall not be made to a *stack* within 2 feet (610 mm) above or below a horizontal offset.

917.8 Prohibited lower connections. *Stacks* greater than 2 *branch intervals* in height shall not receive the discharge of horizontal *branches* on the lower two floors. There shall be no connections to the *stack* between the lower two floors and a distance of not less than 10 pipe diameters downstream from the base of the single stack vented system.

917.9 Sizing building drains and sewers. The *building drain* and *building sewer* receiving the discharge of a single stack vent system shall be sized in accordance with Table 710.1(1).

SECTION 918
AIR ADMITTANCE VALVES

918.1 General. Vent systems utilizing air admittance valves shall comply with this section. Stack-type air admittance valves shall conform to ASSE 1050. Individual and branch-type air admittance valves shall conform to ASSE 1051.

918.2 Installation. The valves shall be installed in accordance with the requirements of this section and the manufacturer's installation instructions. Air admittance valves shall be installed after the DWV testing required by Section 312.2 or 312.3 has been performed.

918.3 Where permitted. Individual, *branch* and circuit vents shall be permitted to terminate with a connection to an individual or branch-type air admittance valve in accordance with Section 918.3.1. *Stack vents* and vent *stacks* shall be permitted to terminate to stack-type air admittance valves in accordance with Section 918.3.2.

918.3.1 Horizontal branches. Individual and branch-type air admittance valves shall vent only fixtures that are on the same floor level and connect to a horizontal branch drain. Where the horizontal *branch* is located more than four branch intervals from the top of the stack, the horizontal *branch* shall be provided with a relief vent that shall connect to a vent stack or stack vent, or extend outdoors to the open air. The relief vent shall connect to the horizontal *branch* drain between the stack and the most downstream fixture drain connected to the horizontal *branch* drain. The relief vent shall be sized in accordance with Section 906.2 and installed in accordance with Section

905. The relief vent shall be permitted to serve as the vent for other fixtures.

918.3.2 Stack. Stack-type air admittance valves shall be prohibited from serving as the vent terminal for vent stacks or stack vents that serve drainage *stacks* having more than six *branch intervals*.

918.4 Location. Individual and branch-type air admittance valves shall be located a minimum of 4 inches (102 mm) above the *horizontal branch drain* or *fixture drain* being vented. Stack-type air admittance valves shall be located not less than 6 inches (152 mm) above the *flood level rim* of the highest fixture being vented. The air admittance valve shall be located within the maximum *developed length* permitted for the vent. The air admittance valve shall be installed not less than 6 inches (152 mm) above insulation materials.

918.5 Access and ventilation. *Access* shall be provided to all air admittance valves. The valve shall be located within a ventilated space that allows air to enter the valve.

918.6 Size. The air admittance valve shall be rated in accordance with the standard for the size of the vent to which the valve is connected.

918.7 Vent required. Within each plumbing system, not less than one *stack vent* or vent *stack* shall extend outdoors to the open air.

918.8 Prohibited installations. Air admittance valves shall not be installed in non-neutralized special waste systems as described in Chapter 8 except where such valves are in compliance with ASSE 1049, are constructed of materials approved in accordance with Section 702.5 and are tested for chemical resistance in accordance with ASTM F 1412. Air admittance valves shall not be located in spaces utilized as supply or return air plenums. Air admittance valves without an engineered design shall not be utilized to vent sumps or tanks of any type.

SECTION 919
ENGINEERED VENT SYSTEMS

919.1 General. Engineered vent systems shall comply with this section and the design, submittal, approval, inspection and testing requirements of Section 105.4.

919.2 Individual branch fixture and individual fixture header vents. The maximum *developed length* of individual fixture vents to vent branches and vent headers shall be determined in accordance with Table 919.2 for the minimum pipe diameters at the indicated vent airflow rates.

The individual vent airflow rate shall be determined in accordance with the following:

$$Q_{h,b} = N_{n,b} \, Q_v \qquad \text{(Equation 9-1)}$$

For SI: $Q_{h,b} = N_{n,b} \, Q_v$ (0.4719 L/s)

where:

$N_{n,b}$ = Number of fixtures per header (or vent *branch*) ÷ total number of fixtures connected to vent *stack*.

$Q_{h,b}$ = Vent *branch* or vent header airflow rate (cfm).

Q_v = Total vent *stack* airflow rate (cfm).

$$Q_v \text{(gpm)} = 27.8 \, r_s^{2/3} \, (1 - r_s) \, D^{8/3}$$

$$Q_v \text{(cfm)} = 0.134 \, Q_v \text{(gpm)}$$

where:

D = Drainage *stack* diameter (inches).

Q_w = Design discharge load (gpm).

r_s = Waste water flow area to total area.

$$= \frac{Q_w}{27.8 \, D^{8/3}}$$

Individual vent airflow rates are obtained by equally distributing $Q_{h,b}$ into one-half the total number of fixtures on the *branch* or header for more than two fixtures; for an odd number of total fixtures, decrease by one; for one fixture, apply the full value of $Q_{h,b}$.

Individual vent *developed length* shall be increased by 20 percent of the distance from the vent *stack* to the fixture vent connection on the vent *branch* or header.

SECTION 920
COMPUTERIZED VENT DESIGN

920.1 Design of vent system. The sizing, design and layout of the vent system shall be permitted to be determined by *approved* computer program design methods.

920.2 System capacity. The vent system shall be based on the air capacity requirements of the drainage system under a peak load condition.

TABLE 919.2
MINIMUM DIAMETER AND MAXIMUM LENGTH OF INDIVIDUAL BRANCH FIXTURE VENTS AND INDIVIDUAL FIXTURE HEADER VENTS FOR SMOOTH PIPES

DIAMETER OF VENT PIPE (inches)	INDIVIDUAL VENT AIRFLOW RATE (cubic feet per minute)																			
	Maximum developed length of vent (feet)																			
	1	2	3	4	5	6	7	8	9	10	11	12	13	14	15	16	17	18	19	20
$1/2$	95	25	13	8	5	4	3	2	1	1	1	1	1	1	1	1	1	1	1	1
$3/4$	100	88	47	30	20	15	10	9	7	6	5	4	3	3	3	2	2	2	2	1
1	—	—	100	94	65	48	37	29	24	20	17	14	12	11	9	8	7	7	6	6
$1^1/_4$	—	—	—	—	—	—	—	100	87	73	62	53	46	40	36	32	29	26	23	21
$1^1/_2$	—	—	—	—	—	—	—	—	—	—	—	100	96	84	75	65	60	54	49	45
2	—	—	—	—	—	—	—	—	—	—	—	—	—	—	—	—	—	—	—	100

For SI: 1 inch = 25.4 mm, 1 cubic foot per minute = 0.4719 L/s, 1 foot = 304.8 mm.

CHAPTER 10

TRAPS, INTERCEPTORS AND SEPARATORS

SECTION 1001
GENERAL

1001.1 Scope. This chapter shall govern the material and installation of traps, interceptors and separators.

SECTION 1002
TRAP REQUIREMENTS

1002.1 Fixture traps. Each plumbing fixture shall be separately trapped by a liquid-seal trap, except as otherwise permitted by this code. The vertical distance from the fixture outlet to the trap weir shall not exceed 24 inches (610 mm), and the horizontal distance shall not exceed 30 inches (610 mm) measured from the centerline of the fixture outlet to the centerline of the inlet of the trap. The height of a clothes washer standpipe above a trap shall conform to Section 802.4. A fixture shall not be double trapped.

Exceptions:

1. This section shall not apply to fixtures with integral traps.

2. A combination plumbing fixture is permitted to be installed on one trap, provided that one compartment is not more than 6 inches (152 mm) deeper than the other compartment and the waste outlets are not more than 30 inches (762 mm) apart.

3. A grease interceptor intended to serve as a fixture trap in accordance with the manufacturer's installation instructions shall be permitted to serve as the trap for a single fixture or a combination sink of not more than three compartments where the vertical distance from the fixture outlet to the inlet of the interceptor does not exceed 30 inches (762 mm) and the *developed length* of the waste pipe from the most upstream fixture outlet to the inlet of the interceptor does not exceed 60 inches (1524 mm).

4. Where floor drains in multilevel parking structures are required to discharge to a combined building sewer system, the floor drains shall not be required to be individually trapped provided that they are connected to a main trap in accordance with Section 1103.1.

1002.2 Design of traps. Fixture traps shall be self-scouring. Fixture traps shall not have interior partitions, except where such traps are integral with the fixture or where such traps are constructed of an *approved* material that is resistant to corrosion and degradation. Slip joints shall be made with an *approved* elastomeric gasket and shall be installed only on the trap inlet, trap outlet and within the trap seal.

1002.3 Prohibited traps. The following types of traps are prohibited:

1. Traps that depend on moving parts to maintain the seal.

2. Bell traps.

3. Crown-vented traps.

4. Traps not integral with a fixture and that depend on interior partitions for the seal, except those traps constructed of an *approved* material that is resistant to corrosion and degradation.

5. "S" traps.

6. Drum traps.

   Exception: Drum traps used as solids interceptors and drum traps serving chemical waste systems shall not be prohibited.

1002.4 Trap seals. Each fixture trap shall have a liquid seal of not less than 2 inches (51 mm) and not more than 4 inches (102 mm), or deeper for special designs relating to accessible fixtures. Where a trap seal is subject to loss by evaporation, a trap seal primer valve shall be installed. Trap seal primer valves shall connect to the trap at a point above the level of the trap seal. A trap seal primer valve shall conform to ASSE 1018 or ASSE 1044.

1002.5 Size of fixture traps. Fixture trap size shall be sufficient to drain the fixture rapidly and not less than the size indicated in Table 709.1. A trap shall not be larger than the drainage pipe into which the trap discharges.

1002.6 Building traps. Building (house) traps shall be prohibited, except where local conditions necessitate such traps. Building traps shall be provided with a cleanout and a relief vent or fresh air intake on the inlet side of the trap. The size of the relief vent or fresh air intake shall not be less than one-half the diameter of the drain to which the relief vent or air intake connects. Such relief vent or fresh air intake shall be carried above grade and shall be terminated in a screened outlet located outside the building.

1002.7 Trap setting and protection. Traps shall be set level with respect to the trap seal and, where necessary, shall be protected from freezing.

1002.8 Recess for trap connection. A recess provided for connection of the underground trap, such as one serving a bathtub in slab-type construction, shall have sides and a bottom of corrosion-resistant, insect- and verminproof construction.

1002.9 Acid-resisting traps. Where a vitrified clay or other brittleware, acid-resisting trap is installed underground, such trap shall be embedded in concrete extending 6 inches (152 mm) beyond the bottom and sides of the trap.

1002.10 Plumbing in mental health centers. In mental health centers, pipes and traps shall not be exposed.

SECTION 1003
INTERCEPTORS AND SEPARATORS

1003.1 Where required. Interceptors and separators shall be provided to prevent the discharge of oil, grease, sand and other substances harmful or hazardous to the public sewer, the private sewage system or the sewage treatment plant or processes.

1003.2 Approval. The size, type and location of each interceptor and of each separator shall be designed and installed in accordance with the manufacturer's instructions and the requirements of this section based on the anticipated conditions of use. Wastes that do not require treatment or separation shall not be discharged into any interceptor or separator.

1003.3 Grease interceptors. Grease interceptors shall comply with the requirements of Sections 1003.3.1 through 1003.3.5.

1003.3.1 Grease interceptors and automatic grease removal devices required. A grease interceptor or automatic grease removal device shall be required to receive the drainage from fixtures and equipment with grease-laden waste located in food preparation areas, such as in restaurants, hotel kitchens, hospitals, school kitchens, bars, factory cafeterias and clubs. Fixtures and equipment shall include pot sinks, prerinse sinks; soup kettles or similar devices; wok stations; floor drains or sinks into which kettles are drained; automatic hood wash units and dishwashers without prerinse sinks. Grease interceptors and automatic grease removal devices shall receive waste only from fixtures and equipment that allow fats, oils or grease to be discharged. Where lack of space or other constraints prevent the installation or replacement of a grease interceptor, one or more grease interceptors shall be permitted to be installed on or above the floor and upstream of an existing grease interceptor.

1003.3.2 Food waste grinders. Where food waste grinders connect to grease interceptors, a solids interceptor shall separate the discharge before connecting to the grease interceptor. Solids interceptors and grease interceptors shall be sized and rated for the discharge of the food waste grinder. Emulsifiers, chemicals, enzymes and bacteria shall not discharge into the food waste grinder.

1003.3.3 Grease interceptors and automatic grease removal devices not required. A grease interceptor or an automatic grease removal device shall not be required for individual dwelling units or any private living quarters.

1003.3.4 Hydromechanical grease interceptors and automatic grease removal devices. *Hydromechanical grease interceptors* and automatic grease removal devices shall be sized in accordance with ASME A112.14.3 Appendix A, ASME 112.14.4, CSA B481.3 or PDI G101. Hydromechanical grease interceptors and automatic grease removal devices shall be designed and tested in accordance with ASME A112.14.3 Appendix A, ASME 112.14.4, CSA B481.1, PDI G101 or PDI G102. Hydro-

mechanical grease interceptors and automatic grease removal devices shall be installed in accordance with the manufacturer's instructions. Where manufacturer's instructions are not provided, hydromechanical grease interceptors and grease removal devices shall be installed in compliance with ASME A112.14.3, ASME 112.14.4, CSA B481.3 or PDI G101. This section shall not apply to *gravity grease interceptors*.

1003.3.4.1 Grease interceptor capacity. Grease interceptors shall have the grease retention capacity indicated in Table 1003.3.4.1 for the flow-through rates indicated.

TABLE 1003.3.4.1
CAPACITY OF GREASE INTERCEPTORS[a]

TOTAL FLOW-THROUGH RATING (gpm)	GREASE RETENTION CAPACITY (pounds)
4	8
6	12
7	14
9	18
10	20
12	24
14	28
15	30
18	36
20	40
25	50
35	70
50	100
75	150
100	200

For SI: 1 gallon per minute = 3.785 L/m, 1 pound = 0.454 kg.

a. For total flow-through ratings greater than 100 (gpm), double the flow-through rating to determine the grease retention capacity (pounds).

1003.3.4.2 Rate of flow controls. Grease interceptors shall be equipped with devices to control the rate of water flow so that the water flow does not exceed the rated flow. The flow-control device shall be vented and terminate not less than 6 inches (152 mm) above the flood rim level or be installed in accordance with the manufacturer's instructions.

1003.3.5 Automatic grease removal devices. Where automatic grease removal devices are installed, such devices shall be located downstream of each fixture or multiple fixtures in accordance with the manufacturer's instructions. The automatic grease removal device shall be sized to pretreat the measured or calculated flows for all connected fixtures or equipment. Ready *access* shall be provided for inspection and maintenance.

1003.4 Oil separators required. At repair garages, car-washing facilities, at factories where oily and flammable liquid wastes are produced and in hydraulic elevator pits, separators shall be installed into which all oil-bearing, grease-bearing or flammable wastes shall be discharged before emptying into the building drainage system or other point of disposal.

Exception: An oil separator is not required in hydraulic elevator pits where an *approved* alarm system is installed.

1003.4.1 Separation of liquids. A mixture of treated or untreated light and heavy liquids with various specific gravities shall be separated in an *approved* receptacle.

1003.4.2 Oil separator design. Oil separators shall be listed and labeled, or designed in accordance with Sections 1003.4.2.1 and 1003.4.2.2.

1003.4.2.1 General design requirements. Oil separators shall have a depth of not less than 2 feet (610 mm) below the invert of the discharge drain. The outlet opening of the separator shall have not less than an 18-inch (457 mm) water seal.

1003.4.2.2 Garages and service stations. Where automobiles are serviced, greased, repaired or washed or where gasoline is dispensed, oil separators shall have a capacity of not less than 6 cubic feet (0.168 m^3) for the first 100 square feet (9.3 m^2) of area to be drained, plus 1 cubic foot (0.28 m^3) for each additional 100 square feet (9.3 m^2) of area to be drained into the separator. Parking garages in which servicing, repairing or washing is not conducted, and in which gasoline is not dispensed, shall not require a separator. Areas of commercial garages utilized only for storage of automobiles are not required to be drained through a separator.

1003.5 Sand interceptors in commercial establishments. Sand and similar interceptors for heavy solids shall be designed and located so as to be provided with ready *access* for cleaning, and shall have a water seal of not less than 6 inches (152 mm).

1003.6 Laundries. Laundry facilities not installed within an individual dwelling unit or intended for individual family use shall be equipped with an interceptor with a wire basket or similar device, removable for cleaning, that prevents passage into the drainage system of solids $^1/_2$ inch (12.7 mm) or larger in size, string, rags, buttons or other materials detrimental to the public sewage system.

1003.7 Bottling establishments. Bottling plants shall discharge process wastes into an interceptor that will provide for the separation of broken glass or other solids before discharging waste into the drainage system.

1003.8 Slaughterhouses. Slaughtering room and dressing room drains shall be equipped with *approved* separators. The separator shall prevent the discharge into the drainage system of feathers, entrails and other materials that cause clogging.

1003.9 Venting of interceptors and separators. Interceptors and separators shall be designed so as not to become air bound where tight covers are utilized. Each interceptor or separator shall be vented where subject to a loss of trap seal.

1003.10 Access and maintenance of interceptors and separators. *Access* shall be provided to each interceptor and separator for service and maintenance. Interceptors and separators shall be maintained by periodic removal of accumulated grease, scum, oil, or other floating substances and solids deposited in the interceptor or separator.

SECTION 1004
MATERIALS, JOINTS AND CONNECTIONS

1004.1 General. The materials and methods utilized for the construction and installation of traps, interceptors and separators shall comply with this chapter and the applicable provisions of Chapters 4 and 7. The fittings shall not have ledges, shoulders or reductions capable of retarding or obstructing flow of the piping.

CHAPTER 11

STORM DRAINAGE

SECTION 1101
GENERAL

1101.1 Scope. The provisions of this chapter shall govern the materials, design, construction and installation of storm drainage.

1101.2 Where required. All roofs, paved areas, yards, courts and courtyards shall drain into a separate storm *sewer* system, or a combined *sewer* system, or to an *approved* place of disposal. For one- and two-family dwellings, and where *approved*, storm water is permitted to discharge onto flat areas, such as streets or lawns, provided that the storm water flows away from the building.

1101.3 Prohibited drainage. Storm water shall not be drained into sewers intended for sewage only.

1101.4 Tests. The conductors and the building *storm drain* shall be tested in accordance with Section 312.

1101.5 Change in size. The size of a drainage pipe shall not be reduced in the direction of flow.

1101.6 Fittings and connections. All connections and changes in direction of the storm drainage system shall be made with *approved* drainage-type fittings in accordance with Table 706.3. The fittings shall not obstruct or retard flow in the system.

1101.7 Roof design. Roofs shall be designed for the maximum possible depth of water that will pond thereon as determined by the relative levels of roof deck and overflow weirs, scuppers, edges or serviceable drains in combination with the deflected structural elements. In determining the maximum possible depth of water, all primary roof drainage means shall be assumed to be blocked.

1101.8 Cleanouts required. Cleanouts shall be installed in the storm drainage system and shall comply with the provisions of this code for sanitary drainage pipe cleanouts.

Exception: Subsurface drainage system.

1101.9 Backwater valves. Storm drainage systems shall be provided with backwater valves as required for sanitary drainage systems in accordance with Section 715.

SECTION 1102
MATERIALS

1102.1 General. The materials and methods utilized for the construction and installation of storm drainage systems shall comply with this section and the applicable provisions of Chapter 7.

1102.2 Inside storm drainage conductors. Inside storm drainage conductors installed above ground shall conform to one of the standards listed in Table 702.1.

1102.3 Underground building storm drain pipe. Underground building *storm drain* pipe shall conform to one of the standards listed in Table 702.2.

1102.4 Building storm sewer pipe. Building storm *sewer* pipe shall conform to one of the standards listed in Table 1102.4.

TABLE 1102.4
BUILDING STORM SEWER PIPE

MATERIAL	STANDARD
Acrylonitrile butadiene styrene (ABS) plastic pipe	ASTM D 2661; ASTM D 2751; ASTM F 628; CSA B181.1; CSA B182.1
Asbestos-cement pipe	ASTM C 428
Cast-iron pipe	ASTM A 74; ASTM A 888; CISPI 301
Concrete pipe	ASTM C 14; ASTM C 76; CSA A257.1M; CSA A257.2M
Copper or copper-alloy tubing (Type K, L, M or DWV)	ASTM B 75; ASTM B 88; ASTM B 251; ASTM B 306
Polyethylene (PE) plastic pipe	ASTM F 2306/F 2306M
Polyvinyl chloride (PVC) plastic pipe (Type DWV, SDR26, SDR35, SDR41, PS50 or PS100)	ASTM D 2665; ASTM D 3034; ASTM F 891; CSA B182.4; CSA B181.2; CSA B182.2
Vitrified clay pipe	ASTM C 4; ASTM C 700
Stainless steel drainage systems, Type 316L	ASME A112.3.1

1102.5 Subsoil drain pipe. Subsoil drains shall be open-jointed, horizontally split or perforated pipe conforming to one of the standards listed in Table 1102.5.

TABLE 1102.5
SUBSOIL DRAIN PIPE

MATERIAL	STANDARD
Asbestos-cement pipe	ASTM C 508
Cast-iron pipe	ASTM A 74; ASTM A 888; CISPI 301
Polyethylene (PE) plastic pipe	ASTM F 405; CSA B182.1; CSA B182.6; CSA B182.8
Polyvinyl chloride (PVC) Plastic pipe (type sewer pipe, PS25, PS50 or PS100)	ASTM D 2729; ASTM F 891; CSA B182.2; CSA B182.4
Stainless steel drainage systems, Type 316L	ASME A 112.3.1
Vitrified clay pipe	ASTM C 4; ASTM C 700

1102.6 Roof Drains. Roof drains shall conform to ASME A112.6.4 or ASME A112.3.1.

1102.7 Fittings. Pipe fittings shall be *approved* for installation with the piping material installed, and shall conform to the respective pipe standards or one of the standards listed in Table 1102.7. The fittings shall not have ledges, shoulders or reductions capable of retarding or obstructing flow in the piping. Threaded drainage pipe fittings shall be of the recessed drainage type.

TABLE 1102.7
PIPE FITTING

MATERIAL	STANDARD
Acrylonitrile butadiene styrene (ABS) plastic	ASTM D 2661; ASTM D 3311; CSA B181.1
Cast-iron	ASME B16.4; ASME B16.12; ASTM A 888; CISPI 301; ASTM A 74
Coextruded composite ABS and drain DR-PS in PS35, PS50, PS100, PS140, PS200	ASTM D 2751
Coextruded composite ABS DWV Schedule 40 IPS pipe (solid or cellular core)	ASTM D 2661; ASTM D 3311; ASTM F 628
Coextruded composite PVC DWV Schedule 40 IPS-DR, PS140, PS200 (solid or cellular core)	ASTM D 2665; ASTM D 3311; ASTM F 891
Coextruded composite PVC sewer and drain DR-PS in PS35, PS50, PS100, PS140, PS200	ASTM D 3034
Copper or copper alloy	ASME B16.15; ASME B16.18; ASME B16.22; ASME B16.23; ASME B16.26; ASME B16.29
Gray iron and ductile iron	AWWA C110/A21.10
Malleable iron	ASME B16.3
Plastic, general	ASTM F 409
Polyethylene (PE) plastic pipe	ASTM F 2306/F 2306M
Polyvinyl chloride (PVC) plastic	ASTM D 2665; ASTM D 3311; ASTM F 1866
Steel	ASME B16.9; ASME B16.11; ASME B16.28
Stainless steel drainage systems, Type 316L	ASME A112.3.1

SECTION 1103 TRAPS

1103.1 Main trap. Leaders and storm drains connected to a combined *sewer* shall be trapped. Individual storm water traps shall be installed on the storm water drain *branch* serving each conductor, or a single trap shall be installed in the main *storm drain* just before its connection with the combined *building sewer* or the *public sewer*.

1103.2 Material. Storm water traps shall be of the same material as the piping system to which they are attached.

1103.3 Size. Traps for individual conductors shall be the same size as the horizontal drain to which they are connected.

1103.4 Cleanout. An accessible cleanout shall be installed on the building side of the trap.

SECTION 1104 CONDUCTORS AND CONNECTIONS

1104.1 Prohibited use. Conductor pipes shall not be used as soil, waste or vent pipes, and soil, waste or vent pipes shall not be used as conductors.

1104.2 Combining storm with sanitary drainage. The sanitary and storm drainage systems of a structure shall be entirely separate except where combined *sewer* systems are utilized. Where a combined *sewer* is utilized, the building *storm drain* shall be connected in the same horizontal plane through a single-wye fitting to the combined *sewer* not less than 10 feet (3048 mm) downstream from any soil *stack*.

1104.3 Floor drains. Floor drains shall not be connected to a *storm drain*.

SECTION 1105 ROOF DRAINS

1105.1 General. Roof drains shall be installed in accordance with the manufacturer's instructions. The inside opening for the roof drain shall not be obstructed by the roofing membrane material.

1105.2 Roof drain flashings. The connection between roofs and roof drains which pass through the roof and into the interior of the building shall be made water-tight by the use of *approved* flashing material.

SECTION 1106 SIZE OF CONDUCTORS, LEADERS AND STORM DRAINS

1106.1 General. The size of the vertical conductors and leaders, building storm drains, building storm sewers, and any horizontal branches of such drains or sewers shall be based on the 100-year hourly rainfall rate indicated in Figure 1106.1 or on other rainfall rates determined from *approved* local weather data.

For SI: 1 inch = 25.4 mm.
Source: National Weather Service, National Oceanic and Atmospheric Administration, Washington D.C.

FIGURE 1106.1
100-YEAR, 1-HOUR RAINFALL (INCHES) EASTERN UNITED STATES

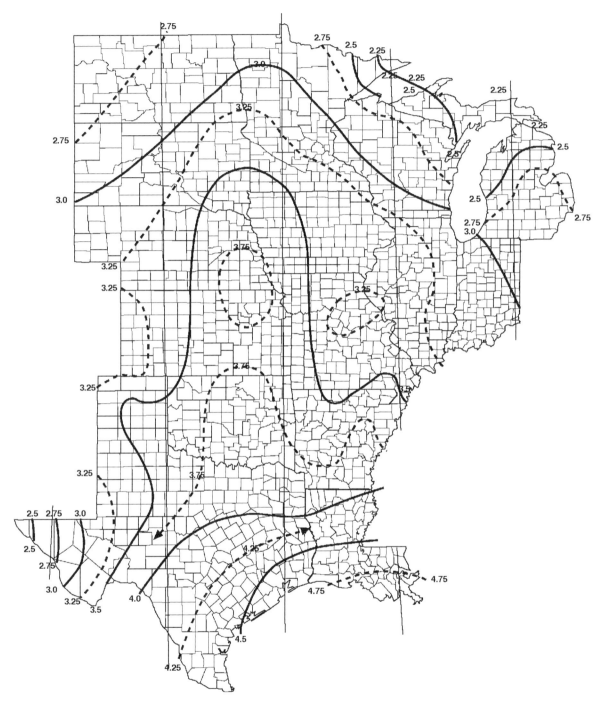

For SI: 1 inch = 25.4 mm.

Source: National Weather Service, National Oceanic and Atmospheric Administration, Washington D.C.

FIGURE 1106.1—continued
100-YEAR, 1-HOUR RAINFALL (INCHES) CENTRAL UNITED STATES

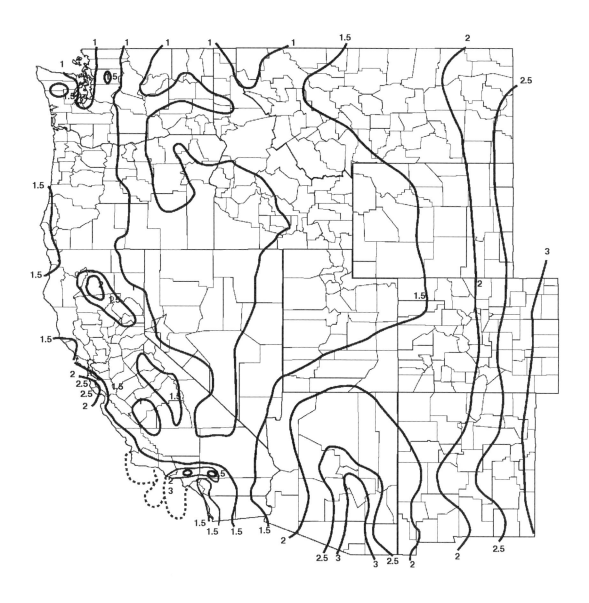

For SI: 1 inch = 25.4 mm.
Source: National Weather Service, National Oceanic and Atmospheric Administration, Washington D.C.

FIGURE 1106.1—continued
100-YEAR, 1-HOUR RAINFALL (INCHES) WESTERN UNITED STATES

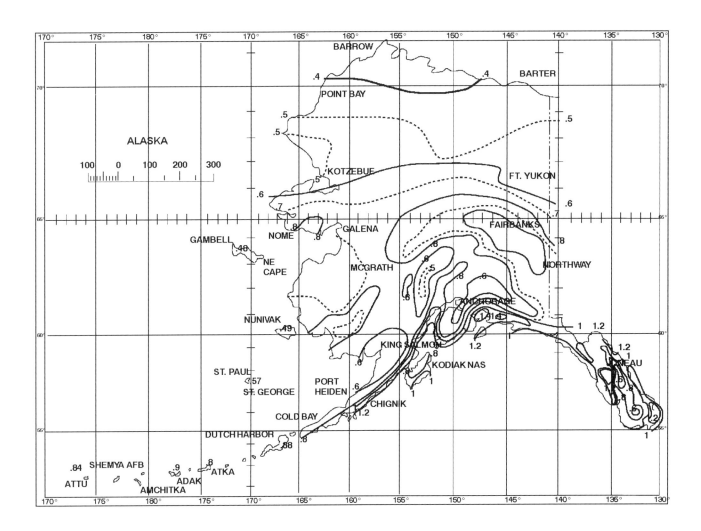

For SI: 1 inch = 25.4 mm.
Source: National Weather Service, National Oceanic and Atmospheric Administration, Washington D.C.

FIGURE 1106.1—continued
100-YEAR, 1-HOUR RAINFALL (INCHES) ALASKA

FIGURE 1106.1—continued
100-YEAR, 1-HOUR RAINFALL (INCHES) HAWAII

For SI: 1 inch = 25.4 mm.
Source: National Weather Service, National Oceanic and Atmospheric Administration, Washington D.C.

1106.2 Vertical conductors and leaders. Vertical conductors and leaders shall be sized for the maximum projected roof area, in accordance with Table 1106.2(1) and Table 1106.2(2).

1106.3 Building storm drains and sewers. The size of the building *storm drain*, building storm *sewer* and their horizontal branches having a slope of one-half unit or less vertical in 12 units horizontal (4-percent slope) shall be based on the maximum projected roof area in accordance with Table 1106.3. The slope of horizontal branches shall be not less than one-eighth unit vertical in 12 units horizontal (1-percent slope) unless otherwise *approved*.

1106.4 Vertical walls. In sizing roof drains and storm drainage piping, one-half of the area of any vertical wall that diverts rainwater to the roof shall be added to the projected roof area for inclusion in calculating the required size of vertical conductors, leaders and horizontal storm drainage piping.

1106.5 Parapet wall scupper location. Parapet wall roof drainage scupper and overflow scupper location shall comply with the requirements of Section 1503.4 of the *International Building Code.*

1106.6 Size of roof gutters. The size of semicircular gutters shall be based on the maximum projected roof area in accordance with Table 1106.6.

TABLE 1106.2(1)
SIZE OF CIRCULAR VERTICAL CONDUCTORS AND LEADERS

| DIAMETER OF LEADER (inches)[a] | HORIZONTALLY PROJECTED ROOF AREA (square feet) | | | | | | | | | | | |
| | Rainfall rate (inches per hour) | | | | | | | | | | | |
	1	2	3	4	5	6	7	8	9	10	11	12
2	2,880	1,440	960	720	575	480	410	360	320	290	260	240
3	8,800	4,400	2,930	2,200	1,760	1,470	1,260	1,100	980	880	800	730
4	18,400	9,200	6,130	4,600	3,680	3,070	2,630	2,300	2,045	1,840	1,675	1,530
5	34,600	17,300	11,530	8,650	6,920	5,765	4,945	4,325	3,845	3,460	3,145	2,880
6	54,000	27,000	17,995	13,500	10,800	9,000	7,715	6,750	6,000	5,400	4,910	4,500
8	116,000	58,000	38,660	29,000	23,200	19,315	16,570	14,500	12,890	11,600	10,545	9,600

For SI: 1 inch = 25.4 mm, 1 square foot = 0.0929 m².

a. Sizes indicated are the diameter of circular piping. This table is applicable to piping of other shapes, provided the cross-sectional shape fully encloses a circle of the diameter indicated in this table. For rectangular leaders, see Table 1106.2(2). Interpolation is permitted for pipe sizes that fall between those listed in this table.

TABLE 1106.2(2)
SIZE OF RECTANGULAR VERTICAL CONDUCTORS AND LEADERS

| DIMENSIONS OF COMMON LEADER SIZES width x length (inches)[a, b] | HORIZONTALLY PROJECTED ROOF AREA (square feet) | | | | | | | | | | | |
| | Rainfall rate (inches per hour) | | | | | | | | | | | |
	1	2	3	4	5	6	7	8	9	10	11	12
$1^3/_4 \times 2^1/_2$	3,410	1,700	1,130	850	680	560	480	420	370	340	310	280
2×3	5,540	2,770	1,840	1,380	1,100	920	790	690	610	550	500	460
$2^3/_4 \times 4^1/_4$	12,830	6,410	4,270	3,200	2,560	2,130	1,830	1,600	1,420	1,280	1,160	1,060
3×4	13,210	6,600	4,400	3,300	2,640	2,200	1,880	1,650	1,460	1,320	1,200	1,100
$3^1/_2 \times 4$	15,900	7,950	5,300	3,970	3,180	2,650	2,270	1,980	1,760	1,590	1,440	1,320
$3^1/_2 \times 5$	21,310	10,650	7,100	5,320	4,260	3,550	3,040	2,660	2,360	2,130	1,930	1,770
$3^3/_4 \times 4^3/_4$	21,960	10,980	7,320	5,490	4,390	3,660	3,130	2,740	2,440	2,190	1,990	1,830
$3^3/_4 \times 5^1/_4$	25,520	12,760	8,500	6,380	5,100	4,250	3,640	3,190	2,830	2,550	2,320	2,120
$3^1/_2 \times 6$	27,790	13,890	9,260	6,940	5,550	4,630	3,970	3,470	3,080	2,770	2,520	2,310
4×6	32,980	16,490	10,990	8,240	6,590	5,490	4,710	4,120	3,660	3,290	2,990	2,740
$5^1/_2 \times 5^1/_2$	44,300	22,150	14,760	11,070	8,860	7,380	6,320	5,530	4,920	4,430	4,020	3,690
$7^1/_2 \times 7^1/_2$	100,500	50,250	33,500	25,120	20,100	16,750	14,350	12,560	11,160	10,050	9,130	8,370

For SI: 1 inch = 25.4 mm, 1 square foot = 0.0929 m².

a. Sizes indicated are nominal width × length of the opening for rectangular piping.

b. For shapes not included in this table, Equation 11-1 shall be used to determine the equivalent circular diameter, D_e, of rectangular piping for use in interpolation using the data from Table 1106.2(1).

$D_e = [\text{width} \times \text{length}]^{1/2}$ **(Equation 11-1)**

where:

D_e = equivalent circular diameter and D_e, width and length are in inches.

TABLE 1106.3
SIZE OF HORIZONTAL STORM DRAINGE PIPING

SIZE OF HORIZONTAL PIPING (inches)	HORIZONTALLY PROJECTED ROOF AREA (square feet)					
	Rainfall rate (inches per hour)					
	1	2	3	4	5	6
$^1/_8$ unit vertical in 12 units horizontal (1-percent slope)						
3	3,288	1,644	1,096	822	657	548
4	7,520	3,760	2,506	1,800	1,504	1,253
5	13,360	6,680	4,453	3,340	2,672	2,227
6	21,400	10,700	7,133	5,350	4,280	3,566
8	46,000	23,000	15,330	11,500	9,200	7,600
10	82,800	41,400	27,600	20,700	16,580	13,800
12	133,200	66,600	44,400	33,300	26,650	22,200
15	218,000	109,000	72,800	59,500	47,600	39,650
$^1/_4$ unit vertical in 12 units horizontal (2-percent slope)						
3	4,640	2,320	1,546	1,160	928	773
4	10,600	5,300	3,533	2,650	2,120	1,766
5	18,880	9,440	6,293	4,720	3,776	3,146
6	30,200	15,100	10,066	7,550	6,040	5,033
8	65,200	32,600	21,733	16,300	13,040	10,866
10	116,800	58,400	38,950	29,200	23,350	19,450
12	188,000	94,000	62,600	47,000	37,600	31,350
15	336,000	168,000	112,000	84,000	67,250	56,000
$^1/_2$ unit vertical in 12 units horizontal (4-percent slope)						
3	6,576	3,288	2,295	1,644	1,310	1,096
4	15,040	7,520	5,010	3,760	3,010	2,500
5	26,720	13,360	8,900	6,680	5,320	4,450
6	42,800	21,400	13,700	10,700	8,580	7,140
8	92,000	46,000	30,650	23,000	18,400	15,320
10	171,600	85,800	55,200	41,400	33,150	27,600
12	266,400	133,200	88,800	66,600	53,200	44,400
15	476,000	238,000	158,800	119,000	95,300	79,250

For SI: 1 inch = 25.4 mm, 1 square foot = 0.0929 m².

TABLE 1106.6
SIZE OF SEMICIRCULAR ROOF GUTTERS

DIAMETER OF GUTTERS (inches)	HORIZONTALLY PROJECTED ROOF AREA (square feet)					
	Rainfall rate (inches per hour)					
	1	2	3	4	5	6
$^1/_{16}$ unit vertical in 12 units horizontal (0.5-percent slope)						
3	680	340	226	170	136	113
4	1,440	720	480	360	288	240
5	2,500	1,250	834	625	500	416
6	3,840	1,920	1,280	960	768	640
7	5,520	2,760	1,840	1,380	1,100	918
8	7,960	3,980	2,655	1,990	1,590	1,325
10	14,400	7,200	4,800	3,600	2,880	2,400
$^1/_8$ unit vertical 12 units horizontal (1-percent slope)						
3	960	480	320	240	192	160
4	2,040	1,020	681	510	408	340
5	3,520	1,760	1,172	880	704	587
6	5,440	2,720	1,815	1,360	1,085	905
7	7,800	3,900	2,600	1,950	1,560	1,300
8	11,200	5,600	3,740	2,800	2,240	1,870
10	20,400	10,200	6,800	5,100	4,080	3,400
$^1/_4$ unit vertical in 12 units horizontal (2-percent slope)						
3	1,360	680	454	340	272	226
4	2,880	1,440	960	720	576	480
5	5,000	2,500	1,668	1,250	1,000	834
6	7,680	3,840	2,560	1,920	1,536	1,280
7	11,040	5,520	3,860	2,760	2,205	1,840
8	15,920	7,960	5,310	3,980	3,180	2,655
10	28,800	14,400	9,600	7,200	5,750	4,800
$^1/_2$ unit vertical in 12 units horizontal (4-percent slope)						
3	1,920	960	640	480	384	320
4	4,080	2,040	1,360	1,020	816	680
5	7,080	3,540	2,360	1,770	1,415	1,180
6	11,080	5,540	3,695	2,770	2,220	1,850
7	15,600	7,800	5,200	3,900	3,120	2,600
8	22,400	11,200	7,460	5,600	4,480	3,730
10	40,000	20,000	13,330	10,000	8,000	6,660

For SI: 1 inch = 25.4 mm, 1 square foot = 0.0929 m².

SECTION 1107
SIPHONIC ROOF DRAINAGE SYSTEMS

1107.1 General. Siphonic roof drains and drainage systems shall be designed in accordance with ASME A112.6.9 and ASPE 45.

SECTION 1108
SECONDARY (EMERGENCY) ROOF DRAINS

1108.1 Secondary (emergency overflow) drains or scuppers. Where roof drains are required, secondary (emergency overflow) roof drains or scuppers shall be provided where the roof perimeter construction extends above the roof in such a manner that water will be entrapped if the primary drains allow buildup for any reason.

1108.2 Separate systems required. Secondary roof drain systems shall have the end point of discharge separate from the primary system. Discharge shall be above grade, in a location that would normally be observed by the building occupants or maintenance personnel.

1108.3 Sizing of secondary drains. Secondary (emergency) roof drain systems shall be sized in accordance with Section 1106 based on the rainfall rate for which the primary system is sized in Tables 1106.2(1), 1106.2(2), 1106.3 and 1106.6. Scuppers shall be sized to prevent the depth of ponding water from exceeding that for which the roof was designed as determined by Section 1101.7. Scuppers shall have an opening dimension of not less than 4 inches (102 mm). The flow through the primary system shall not be considered when sizing the secondary roof drain system.

SECTION 1109
COMBINED SANITARY AND STORM SYSTEM

1109.1 Size of combined drains and sewers. The size of a combination sanitary and *storm drain* or se*wer s*hall be computed in accordance with the method in Section 1106.3. The fixture units shall be converted into an equivalent projected roof or paved area. Where the total fixture load on the combined drain is less than or equal to 256 fixture units, the equivalent drainage area in horizontal projection shall be taken as 4,000 square feet (372 m²). Where the total fixture load exceeds 256 fixture units, each additional fixture unit shall be considered the equivalent of 15.6 square feet (1.5 m²) of drainage area. These values are based on a rainfall rate of 1 inch (25 mm) per hour.

SECTION 1110
VALUES FOR CONTINUOUS FLOW

1110.1 Equivalent roof area. Where there is a continuous or semicontinuous discharge into the building *storm drain* or building storm *sewer*, such as from a pump, ejector, air conditioning plant or similar device, each gallon per minute (L/m) of such discharge shall be computed as being equivalent to 96 square feet (9 m²) of roof area, based on a rainfall rate of 1 inch (25.4 mm) per hour.

SECTION 1111
CONTROLLED FLOW ROOF DRAIN SYSTEMS

1111.1 General. The roof of a structure shall be designed for the storage of water where the storm drainage system is engineered for controlled flow. The controlled flow roof drain system shall be an engineered system in accordance with this section and the design, submittal, approval, inspection and testing requirements of Section 105.4. The controlled flow system shall be designed based on the required rainfall rate in accordance with Section 1106.1.

1111.2 Control devices. The control devices shall be installed so that the rate of discharge of water per minute shall not exceed the values for continuous flow as indicated in Section 1109.1.

1111.3 Installation. Runoff control shall be by control devices. Control devices shall be protected by strainers.

1111.4 Minimum number of roof drains. Not less than two roof drains shall be installed in roof areas 10,000 square feet (929 m²) or less and not less than four roof drains shall be installed in roofs over 10,000 square feet (929 m²) in area.

SECTION 1112
SUBSOIL DRAINS

1112.1 Subsoil drains. Subsoil drains shall be open-jointed, horizontally split or perforated pipe conforming to one of the standards listed in Table 1102.5. Such drains shall not be less than 4 inches (102 mm) in diameter. Where the building is subject to backwater, the subsoil drain shall be protected by an accessibly located backwater valve. Subsoil drains shall discharge to a trapped area drain, sump, dry well or *approved* location above ground. The subsoil sump shall not be required to have either a gas-tight cover or a vent. The sump and pumping system shall comply with Section 1114.1.

SECTION 1113
BUILDING SUBDRAINS

1113.1 Building subdrains. Building subdrains located below the *public sewer* level shall discharge into a sump or receiving tank, the contents of which shall be automatically lifted and discharged into the drainage system as required for building sumps. The sump and pumping equipment shall comply with Section 1114.1.

SECTION 1114
SUMPS AND PUMPING SYSTEMS

1114.1 Pumping system. The sump pump, pit and discharge piping shall conform to Sections 1114.1.1 through 1114.1.4.

1114.1.1 Pump capacity and head. The sump pump shall be of a capacity and head appropriate to anticipated use requirements.

1114.1.2 Sump pit. The sump pit shall not be less than 18 inches (457 mm) in diameter and not less than 24 inches (610 mm) in depth, unless otherwise *approved*. The pit shall be accessible and located such that all drainage flows into the pit by gravity. The sump pit shall be constructed of tile, steel, plastic, cast-iron, concrete or other *approved* material, with a removable cover adequate to support anticipated loads in the area of use. The pit floor shall be solid and provide permanent support for the pump.

1114.1.3 Electrical. Electrical service outlets, when required, shall meet the requirements of NFPA 70.

1114.1.4 Piping. Discharge piping shall meet the requirements of Section 1102.2, 1102.3 or 1102.4 and shall include a gate valve and a full flow check valve. Pipe and fittings shall be the same size as, or larger than, pump discharge tapping.

> **Exception:** In one- and two-family dwellings, only a check valve shall be required, located on the discharge piping from the pump or ejector.

CHAPTER 12

SPECIAL PIPING AND STORAGE SYSTEMS

SECTION 1201
GENERAL

1201.1 Scope. The provisions of this chapter shall govern the design and installation of piping and storage systems for non-flammable medical gas systems and nonmedical oxygen systems. All maintenance and operations of such systems shall be in accordance with the *International Fire Code*.

SECTION 1202
MEDICAL GASES

[F] 1202.1 Nonflammable medical gases. Nonflammable medical gas systems, inhalation anesthetic systems and vacuum piping systems shall be designed and installed in accordance with NFPA 99C.

Exceptions:

1. This section shall not apply to portable systems or cylinder storage.

2. Vacuum system exhaust terminations shall comply with the *International Mechanical Code*.

SECTION 1203
OXYGEN SYSTEMS

[F] 1203.1 Design and installation. Nonmedical oxygen systems shall be designed and installed in accordance with NFPA 50 and NFPA 51.

CHAPTER 13

GRAY WATER RECYCLING SYSTEMS

SECTION 1301
GENERAL

1301.1 Scope. The provisions of Chapter 13 shall govern the materials, design, construction and installation of gray water systems for flushing of water closets and urinals and for subsurface landscape irrigation. See Figures 1301.1(1) and 1301.1(2).

1301.2 Installation. In addition to the provisions of Section 1301, systems for flushing of water closets and urinals shall comply with Section 1302 and systems for subsurface landscape irrigation shall comply with Section 1303. Except as provided for in this chapter, all systems shall comply with the provisions of the other chapters of this code.

1301.3 Materials. Above-ground drain, waste and vent piping for gray water systems shall conform to one of the standards listed in Table 702.1. Gray water underground building drainage and vent pipe shall conform to one of the standards listed in Table 702.2.

1301.4 Tests. Drain, waste and vent piping for gray water systems shall be tested in accordance with Section 312.

1301.5 Inspections. Gray water systems shall be inspected in accordance with Section 107.

1301.6 Potable water connections. Only connections in accordance with Section 1302.3 shall be made between a gray water recycling system and a potable water system.

1301.7 Waste water connections. Gray water recycling systems shall receive only the waste discharge of bathtubs, showers, lavatories, clothes washers or laundry trays.

1301.8 Collection reservoir. Gray water shall be collected in an approved reservoir constructed of durable, nonabsorbent and corrosion-resistant materials. The reservoir shall be a closed and gas-tight vessel. Access openings shall be provided to allow inspection and cleaning of the reservoir interior.

1301.9 Filtration. Gray water entering the reservoir shall pass through an approved filter such as a media, sand or diatomaceous earth filter.

> **1301.9.1 Required valve.** A full-open valve shall be installed downstream of the last fixture connection to the gray water discharge pipe before entering the required filter.

1301.10 Overflow. The collection reservoir shall be equipped with an overflow pipe having the same or larger diameter as the influent pipe for the gray water. The overflow pipe shall be trapped and shall be indirectly connected to the sanitary drainage system.

1301.11 Drain. A drain shall be located at the lowest point of the collection reservoir and shall be indirectly connected to the sanitary drainage system. The drain shall be the same diameter as the overflow pipe required in Section 1301.10.

1301.12 Vent required. The reservoir shall be provided with a vent sized in accordance with Chapter 9 and based on the diameter of the reservoir influent pipe.

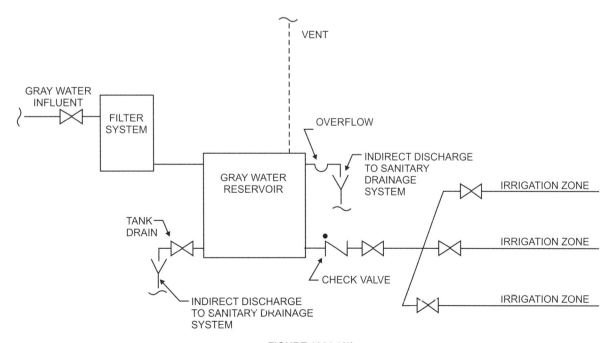

FIGURE 1301.1(1)
GRAY WATER RECYCLING SYSTEM FOR SUBSURFACE LANDSCAPE IRRIGATION

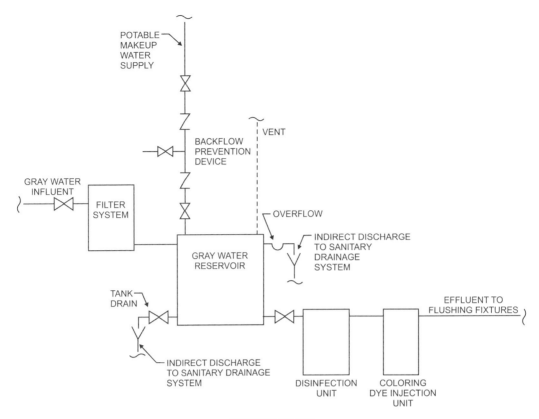

FIGURE 1301.1(2)
GRAY WATER RECYCLING SYSTEM FOR FLUSHING WATER CLOSETS AND URINALS

SECTION 1302
SYSTEMS FOR FLUSHING WATER CLOSETS AND URINALS

1302.1 Collection reservoir. The holding capacity of the reservoir shall be a minimum of twice the volume of water required to meet the daily flushing requirements of the fixtures supplied with gray water, but not less than 50 gallons (189 L). The reservoir shall be sized to limit the retention time of gray water to a maximum of 72 hours.

1302.2 Disinfection. Gray water shall be disinfected by an approved method that employs one or more disinfectants such as chlorine, iodine or ozone that are recommended for use with the pipes, fittings and equipment by the manufacturer of the pipes, fittings and equipment.

1302.3 Makeup water. Potable water shall be supplied as a source of makeup water for the gray water system. The potable water supply shall be protected against backflow in accordance with Section 608. There shall be a full-open valve located on the makeup water supply line to the collection reservoir.

1302.4 Coloring. The gray water shall be dyed blue or green with a food grade vegetable dye before such water is supplied to the fixtures.

1302.5 Materials. Distribution piping shall conform to one of the standards listed in Table 605.4.

1302.6 Identification. Distribution piping and reservoirs shall be identified as containing nonpotable water. Piping identification shall be in accordance with Section 608.8.

SECTION 1303
SUBSURFACE LANDSCAPE IRRIGATION SYSTEMS

1303.1 Collection reservoir. Reservoirs shall be sized to limit the retention time of gray water to a maximum of 24 hours.

1303.1.1 Identification. The reservoir shall be identified as containing nonpotable water.

1303.2 Valves required. A check valve and a full-open valve located on the discharge side of the check valve shall be installed on the effluent pipe of the collection reservoir.

1303.3 Makeup water. Makeup water shall not be required for subsurface landscape irrigation systems. Where makeup water is provided, the installation shall be in accordance with Section 1302.3.

1303.4 Disinfection. Disinfection shall not be required for gray water used for subsurface landscape irrigation systems.

1303.5 Coloring. Gray water used for subsurface landscape irrigation systems shall not be required to be dyed.

1303.6 Estimating gray water discharge. The system shall be sized in accordance with the gallons-per-day-per-occupant number based on the type of fixtures connected to the gray water system. The discharge shall be calculated by the following equation:

$$C = A \ x \ B \qquad \textbf{(Equation 13-1)}$$

where:

A = Number of occupants:

Residential–Number of occupants shall be determined by the actual number of occupants, but not less than two occupants for one bedroom and one occupant for each additional bedroom.

Commercial–Number of occupants shall be determined by the *International Building Code®*.

B = Estimated flow demands for each occupant:

Residential–25 gallons per day (94.6 lpd) per occupant for showers, bathtubs and lavatories and 15 gallons per day (56.7 lpd) per occupant for clothes washers or laundry trays.

Commercial–Based on type of fixture or water use records minus the discharge of fixtures other than those discharging gray water.

C = Estimated gray water discharge based on the total number of occupants.

1303.7 Percolation tests. The permeability of the soil in the proposed absorption system shall be determined by percolation tests or permeability evaluation.

1303.7.1 Percolation tests and procedures. At least three percolation tests in each system area shall be conducted. The holes shall be spaced uniformly in relation to the bottom depth of the proposed absorption system. More percolation tests shall be made where necessary, depending on system design.

1303.7.1.1 Percolation test hole. The test hole shall be dug or bored. The test hole shall have vertical sides and a horizontal dimension of 4 inches to 8 inches (102 mm to 203 mm). The bottom and sides of the hole shall be scratched with a sharp-pointed instrument to expose the natural soil. All loose material shall be removed from the hole and the bottom shall be covered with 2 inches (51 mm) of gravel or coarse sand.

1303.7.1.2 Test procedure, sandy soils. The hole shall be filled with clear water to a minimum of 12 inches (305 mm) above the bottom of the hole for tests in sandy soils. The time for this amount of water to seep away shall be determined, and this procedure shall be repeated if the water from the second filling of the hole seeps away in 10 minutes or less. The test shall proceed as follows: Water shall be added to a point not more than 6 inches (152 mm) above the gravel or coarse sand. Thereupon, from a fixed reference point, water levels shall be measured at 10-minute intervals for a period of 1 hour. Where 6 inches (152 mm) of water seeps away in less than 10 minutes, a shorter interval between measurements shall be used, but in no case shall the water depth exceed 6 inches (152 mm). Where 6 inches (152 mm) of water seeps away in less than 2 minutes, the test shall be

stopped and a rate of less than 3 minutes per inch (7.2 s/mm) shall be reported. The final water level drop shall be used to calculate the percolation rate. Soils not meeting the above requirements shall be tested in accordance with Section 1303.7.1.3.

1303.7.1.3 Test procedure, other soils. The hole shall be filled with clear water, and a minimum water depth of 12 inches (305 mm) shall be maintained above the bottom of the hole for a 4-hour period by refilling whenever necessary or by use of an automatic siphon. Water remaining in the hole after 4 hours shall not be removed. Thereafter, the soil shall be allowed to swell not less than 16 hours or more than 30 hours. Immediately after the soil swelling period, the measurements for determining the percolation rate shall be made as follows: any soil sloughed into the hole shall be removed and the water level shall be adjusted to 6 inches (152 mm) above the gravel or coarse sand. Thereupon, from a fixed reference point, the water level shall be measured at 30-minute intervals for a period of 4 hours, unless two successive water level drops do not vary by more than $^{1}/_{16}$ inch (1.59 mm). At least three water level drops shall be observed and recorded. The hole shall be filled with clear water to a point not more than 6 inches (152 mm) above the gravel or coarse sand whenever it becomes nearly empty. Adjustments of the water level shall not be made during the three measurement periods except to the limits of the last measured water level drop. When the first 6 inches (152 mm) of water seeps away in less than 30 minutes, the time interval between measurements shall be 10 minutes and the test run for 1 hour. The water depth shall not exceed 5 inches (127 mm) at any time during the measurement period. The drop that occurs during the final measurement period shall be used in calculating the percolation rate.

1303.7.1.4 Mechanical test equipment. Mechanical percolation test equipment shall be of an approved type.

1303.7.2 Permeability evaluation. Soil shall be evaluated for estimated percolation based on structure and texture in accordance with accepted soil evaluation practices. Borings shall be made in accordance with Section 1303.7.1 for evaluating the soil.

1303.8 Subsurface landscape irrigation site location. The surface grade of all soil absorption systems shall be located at a point lower than the surface grade of any water well or reservoir on the same or adjoining lot. Where this is not possible, the site shall be located so surface water drainage from the site is not directed toward a well or reservoir. The soil absorption system shall be located with a minimum horizontal distance between various elements as indicated in Table 1303.8. Private sewage disposal systems in compacted areas, such as parking lots and driveways, are prohibited. Surface water shall be diverted away from any soil absorption site on the same or neighboring lots.

TABLE 1303.8
LOCATION OF GRAY WATER SYSTEM

ELEMENT	MINIMUM HORIZONTAL DISTANCE	
	HOLDING TANK (feet)	IRRIGATION DISPOSAL FIELD (feet)
Buildings	5	2
Lot line adjoining private property	5	5
Water wells	50	100
Streams and lakes	50	50
Seepage pits	5	5
Septic tanks	0	5
Water service	5	5
Public water main	10	10

For SI: 1 foot = 304.8 mm.

1303.9 Installation. Absorption systems shall be installed in accordance with Sections 1303.9.1 through 1303.9.5 to provide landscape irrigation without surfacing of gray water.

1303.9.1 Absorption area. The total absorption area required shall be computed from the estimated daily gray water discharge and the design-loading rate based on the percolation rate for the site. The required absorption area equals the estimated gray water discharge divided by the design-loading rate from Table 1303.9.1.

TABLE 1303.9.1
DESIGN LOADING RATE

PERCOLATION RATE (minutes per inch)	DESIGN LOADING FACTOR (gallons per square foot per day)
0 to less than 10	1.2
10 to less than 30	0.8
30 to less than 45	0.72
45 to 60	0.4

For SI: 1 minute per inch = min/25.4 mm,
1 gallon per square foot = 40.7 L/m^2.

1303.9.2 Seepage trench excavations. Seepage trench excavations shall be not less than 1 foot (304 mm) in width and not greater than 5 feet (1524 mm) in width. Trench excavations shall be spaced not less than 2 feet (610 mm) apart. The soil absorption area of a seepage trench shall be computed by using the bottom of the trench area (width) multiplied by the length of pipe. Individual seepage trenches shall be not greater than 100 feet (30 480 mm) in developed length.

1303.9.3 Seepage bed excavations. Seepage bed excavations shall be not less than 5 feet (1524 mm) in width and have more than one distribution pipe. The absorption area of a seepage bed shall be computed by using the bottom of the trench area. Distribution piping in a seepage bed shall be uniformly spaced not greater than 5 feet (1524 mm) and not less than 3 feet (914 mm) apart, and greater than 3 feet (914 mm) and not less than 1 foot (305 mm) from the sidewall or headwall.

1303.9.4 Excavation and construction. The bottom of a trench or bed excavation shall be level. Seepage trenches or beds shall not be excavated where the soil is so wet that such material rolled between the hands forms a soil wire. All smeared or compacted soil surfaces in the sidewalls or bottom of seepage trench or bed excavations shall be scarified to the depth of smearing or compaction and the loose material removed. Where rain falls on an open excavation, the soil shall be left until sufficiently dry so a soil wire will not form when soil from the excavation bottom is rolled between the hands. The bottom area shall then be scarified and loose material removed.

1303.9.5 Aggregate and backfill. Not less than 6 inches in depth of aggregate ranging in size from $^1/_2$ to $2^1/_2$ inches (12.7 mm to 64 mm) shall be laid into the trench below the distribution piping elevation. The aggregate shall be evenly distributed not less than 2 inches (51 mm) in depth over the top of the distribution pipe. The aggregate shall be covered with approved synthetic materials or 9 inches (229 mm) of uncompacted marsh hay or straw. Building paper shall not be used to cover the aggregate. Not less than 9 inches (229 mm) of soil backfill shall be provided above the covering.

1303.10 Distribution piping. Distribution piping shall be not less than 3 inches (76 mm) in diameter. Materials shall comply with Table 1303.10. The top of the distribution pipe shall be not less than 8 inches (203 mm) below the original surface. The slope of the distribution pipes shall be not less than 2 inches (51 mm) and not greater than 4 inches (102 mm) per 100 feet (30 480 mm).

TABLE 1303.10
DISTRIBUTION PIPE

MATERIAL	STANDARD
Polyethylene (PE) plastic pipe	ASTM F 405
Polyvinyl chloride (PVC) plastic pipe	ASTM D 2729
Polyvinyl chloride (PVC) plastic pipe with a 3.5 inch O.D. and solid cellular core or composite wall.	ASTM F 1488

1303.11 Joints. Joints in distribution pipe shall be made in accordance with Section 705 of this code.

CHAPTER 14

REFERENCED STANDARDS

This chapter lists the standards that are referenced in various sections of this document. The standards are listed herein by the promulgating agency of the standard, the standard identification, the effective date and title, and the section or sections of this document that reference the standard. The application of the referenced standards shall be as specified in Section 102.8.

ANSI

American National Standards Institute
25 West 43rd Street, Fourth Floor
New York, NY 10036

Standard reference number	Title	Referenced in code section number
A118.10—99	Specifications for Load Bearing, Bonded, Waterproof Membranes for Thin Set Ceramic Tile and Dimension Stone Installation	417.5.2.5, 417.5.2.6
Z4.3—95	Minimum Requirements for Nonsewered Waste-disposal Systems	311.1
Z21.22—99 (R2003)	Relief Valves for Hot Water Supply Systems with Addenda Z21.22a—2000 (R2003) and Z21.22b—2001 (R2003)	504.2, 504.4, 504.4.1
Z124.1.2—2005	Plastic Bathtub and Shower Units	407.1, 417.1
Z124.3—95	Plastic Lavatories	416.1, 416.2, 417.1
Z124.4—96	Plastic Water Closet Bowls and Tanks	420.1
Z124.6—97	Plastic Sinks	415.1, 418.1
Z124.9—94	Plastic Urinal Fixtures	419.1

AHRI

Air-Conditioning, Heating, & Refrigeration Institute
4100 North Fairfax Drive, Suite 200
Arlington, VA 22203

Standard reference number	Title	Referenced in code section number
1010—02	Self-contained, Mechanically Refrigerated Drinking-Water Coolers	410.1

ASME

American Society of Mechanical Engineers
Three Park Avenue
New York, NY 10016-5990

Standard reference number	Title	Referenced in code section number
A112.1.2—2004	Air Gaps in Plumbing Systems	Table 608.1, 608.13.1
A112.1.3—2000 (Reaffirmed 2005)	Air Gap Fittings for Use with Plumbing Fixtures, Appliances and Appurtenances	Table 608.1, 608.13.1, 1102.6
A112.3.1—2007	Stainless Steel Drainage Systems for Sanitary, DWV, Storm and Vacuum Applications Above and Below Ground	412.1, Table 702.1, Table 702.2, Table 702.3, Table 702.4, 708.2, Table 1102.4, Table 1102.5, Table 1102.7
A112.3.4—2000 (Reaffirmed 2004)	Macerating Toilet Systems and Related Components	712.4.1
A112.4.1—1993 (R2002)	Water Heater Relief Valve Drain Tubes	504.6
A112.4.2—2003(R2008)	Water Closet Personal Hygiene Devices	424.9
A112.4.3—1999 (Reaffirmed 2004)	Plastic Fittings for Connecting Water Closets to the Sanitary Drainage System	405.4
A112.6.1M—1997 (R2002)	Floor-affixed Supports for Off-the-floor Plumbing Fixtures for Public Use	405.4.3
A112.6.2—2000 (Reaffirmed 2004)	Framing-affixed Supports for Off-the-floor Water Closets with Concealed Tanks	405.4.3
A112.6.3—2001 (Reaffirmed 2007)	Floor and Trench Drains	412.1

ASME—continued

A112.6.4—2003 (R2008)	Roof, Deck, and Balcony Drains	1102.6
A112.6.7—2001 (Reaffirmed 2007)	Enameled and Epoxy-coated Cast-iron and PVC Plastic Sanitary Floor Sinks	427.1
A112.6.9—05	Siphonic Roof Drains	1107.1
A112.14.1—2003	Backwater Valves	715.2
A112.14.3—2000	Grease Interceptors	1003.3.4
A112.14.4—2001 (Reaffirmed 2007)	Grease Removal Devices	1003.3.4
A112.18.1—2005/ CSA B125.1—2005	Plumbing Supply Fittings	424.1, 424.2, 424.3, 424.4, 424.6, 607.4, 608.2
A112.18.2—2005/ CSA B125.1—2010	Plumbing Waste Fittings with 2007 and 2008 Supplements	424.1.2
A112.18.3—2002 (Reaffirmed 2008)	Performance Requirements for Backflow Protection Devices and Systems in Plumbing Fixture Fittings	424.2, 424.6
A112.18.6/ CSA B125.6—2010	Flexible Water Connectors	605.6
A112.18.7—1999 (Reaffirmed 2004)	Deck mounted Bath/Shower Transfer Valves with Integral Backflow Protection	424.8
A112.19.1M/ CSA B45.2—2008	Enameled Cast Iron and Enameled Steel Plumbing Fixtures	407.1, 410.1, 415.1, 416.1, 418.1
A112.19.2—2008/ CSA B45.1—08	Ceramic Plumbing Fixtures	401.2, 405.9, 407.1, 408.1, 410.1, 415.1, 416.1, 417.1, 418.1, 419.1, 420.1
A112.19.3M—2008/ CSA B45.4—08	Stainless Steel Plumbing Fixtures	405.9, 407.1, 415.1, 416.1, 418.1, 420.1
A112.19.5/ CSA B45.15—2009	Trim for Water-closet Bowls, Tanks and Urinals	425.4
A112.19.7M/ CSA B45.10—2009	Hydromassage Bathtub Appliances	421.1, 421.4
A112.19.9M—1991 (R2002)	Nonvitreous Ceramic Plumbing Fixtures with 2002 Supplement	407.1, 408.1, 410.1, 419.1
A112.19.12—2006	Wall Mounted and Pedestal Mounted, Adjustable, Elevating, Tilting and Pivoting Lavatory, Sink and Shampoo Bowl Carrier Systems and Drain Systems	416.4, 418.3
A112.19.15—2005	Bathtub/Whirlpool Bathtubs with Pressure Sealed Doors	407.4, 421.6
A112.19.19—2006	Vitreous China Nonwater Urinals	419.1
A112.36.2M—1991 (R2008)	Cleanouts	708.2
B1.20.1—1983(R2006)	Pipe Threads, General Purpose (inch)	605.10.3, 605.12.3, 605.14.4, 605.16.3, 605.18.1, 605.22.3, 705.2.3, 705.4.3, 705.9.4, 705.12.1, 705.14.3
B 16.3—2006	Malleable Iron Threaded Fittings Classes 150 and 300	Table 605.5, Table 702.4, Table 1102.7
B 16.4—2006	Gray Iron Threaded Fittings Classes 125 and 250	Table 605.5, Table 702.4, Table 1102.7
B 16.9—2007	Factory-made Wrought Steel Buttwelding Fittings	Table 605.5, Table 702.4, Table 1102.7
B 16.11—2005	Forged Fittings, Socket-welding and Threaded	Table 605.5, Table 702.4, Table 1102.7
B16.12—1998 (Reaffirmed 2006)	Cast-iron Threaded Drainage Fittings	Table 605.5, Table 702.4, Table 1102.7
B16.15—2006	Cast Bronze Threaded Fittings	Table 605.5, Table 702.4, Table 1102.7
B 16.18—2001 (Reaffirmed 2005)	Cast Copper Alloy Solder Joint Pressure Fittings	Table 605.5, Table 702.4, Table 1102.7
B16.22—2001 (Reaffirmed 2005)	Wrought Copper and Copper Alloy Solder Joint Pressure Fittings	Table 605.5, Table 702.4, Table 1102.7
B16.23—2002 (Reaffirmed 2006)	Cast Copper Alloy Solder Joint Drainage Fittings DWV	Table 605.5, Table 702.4, Table 1102.7
B 16.26—2006	Cast Copper Alloy Fittings for Flared Copper Tubes	Table 605.5, Table 702.4, Table 1102.7
B 16.28—1994	Wrought Steel Buttwelding Short Radius Elbows and Returns	Table 605.5, Table 702.4, Table 1102.7
B 16.29—2007	Wrought Copper and Wrought Copper Alloy Solder Joint Drainage Fittings (DWV)	Table 605.5, Table 702.4, Table 1102.7

ASPE

American Society of Plumbing Engineers
8614 Catalpa Avenue, Suite 1007
Chicago, IL 60656-1116

Standard reference number	Title	Referenced in code section number
45—2007	Siphonic Roof Drainage Systems	1107.1

ASSE

American Society of Sanitary Engineering
901 Canterbury Road, Suite A
Westlake, OH 44145

Standard reference number	Title	Referenced in code section number
1001—08	Performance Requirements for Atmospheric Type Vacuum Breakers	425.2, Table 608.1, 608.13.6, 608.16.4.1
1002—08	Performance Requirements for Antisiphon Fill Valves (Ballcocks) for Gravity Water Closet Flush Tanks	425.3.1, Table 608.1
1003—09	Performance Requirements for Water Pressure Reducing Valves	604.8
1004—08	Performance Requirements for Backflow Prevention Requirements for Commercial Dishwashing Machines	409.1
1005—99	Performance Requirements for Water Heater Drain Valves	501.3
1008—06	Performance Requirements for Plumbing Aspects of Food Waste Disposer Units	413.1
1010—04	Performance Requirements for Water Hammer Arresters	604.9
1011—04	Performance Requirements for Hose Connection Vacuum Breakers	Table 608.1, 608.13.6
1012—09	Performance Requirements for Backflow Preventers with Intermediate Atmospheric Vent	Table 608.1, 608.13.3, 608.16.2
1013—09	Performance Requirements for Reduced Pressure Principle Backflow Preventers and Reduced Pressure Principle Fire Protection Backflow Preventers	Table 608.1, 608.13.2, 608.16.2
1015—09	Performance Requirements for Double Check Backflow Prevention Assemblies and Double Check Fire Protection Backflow Prevention Assemblies	Table 608.1, 608.13.7
1016—2010	Performance Requirements for Individual Thermostatic, Pressure Balancing and Combination Control Valves for Individual Fixture Fittings	424.3, 424.4, 607.4
1017—2010	Performance Requirements for Temperature Actuated Mixing Valves for Hot Water Distribution Systems	501.2, 613.1
1018—2010	Performance Requirements for Trap Seal Primer Valves; Potable Water Supplied	1002.4
1019—2010	Performance Requirements for Vacuum Breaker Wall Hydrants, Freeze Resistant, Automatic Draining Type	Table 608.1, 608.13.6
1020—04	Performance Requirements for Pressure Vacuum Breaker Assembly	Table 608.1, 608.13.5
1022—03	Performance Requirements for Backflow Preventer for Beverage Dispensing Equipment	Table 608.1, 608.16.1, 608.16.10
1024—04	Performance Requirements for Dual Check Valve Type Backflow Preventers (for Residential Supply Service or Individual Outlets)	605.3.1, Table 608.1
1035—08	Performance Requirements for Laboratory Faucet Backflow Preventers	Table 608.1, 608.13.6
1037—90	Performance Requirements for Pressurized Flushing Devices for Plumbing Fixtures	425.2
1044—2010	Performance Requirements for Trap Seal Primer Devices Drainage Types and Electronic Design Types	1002.4
1047—2009	Performance Requirements for Reduced Pressure Detector Fire Protection Backflow Prevention Assemblies	Table 608.1, 608.13.2
1048—2009	Performance Requirements for Double Check Detector Fire Protection Backflow Prevention Assemblies	Table 608.1, 608.13.7
1049—2009	Performance Requirements for Individual and Branch Type Air Admittance Valves for Chemical Waste Systems	901.3, 918.8
1050—2009	Performance Requirements for Stack Air Admittance Valves for Sanitary Drainage Systems	918.1
1051—2009	Performance Requirements for Individual and Branch Type Air Admittance Valves for Sanitary Drainage Systems-fixture and Branch Devices	917.1
1052—04	Performance Requirements for Hose Connection Backflow Preventers	Table 608.1, 608.13.6
1055—2009	Performance Requirements for Chemical Dispensing Systems	608.13.9
1056—2010	Performance Requirements for Spill Resistant Vacuum Breaker	Table 608.1, 608.13.5, 608.13.8
1060—2006	Performance Requirements for Outdoor Enclosures for Fluid Conveying Components	608.14.1
1061—2010	Performance Requirements for Removable and Nonremovable Push Fit Fittings	Table 605.5
1062—2006	Performance Requirements for Temperature Actuated, Flow Reduction Valves to Individual Supply Fittings	424.7
1066—97	Performance Requirements for Individual Pressure Balancing In-line Valves for Individual Fixture Fittings	604.11
1069—05	Performance Requirements for Automatic Temperature Control Mixing Valves	424.4
1070—04	Performance Requirements for Water-temperature Limiting Devices	408.3, 416.5, 424.5, 607.1.2
1079—2005	Performance Requirements for Dielectric Pipe Unions	605.24.1, 605.24.3
5013—2009	Performance Requirements for Testing Reduced Pressure Principle Backflow Prevention Assembly (RPA) and Reduced Pressure Fire Protection Backflow Preventers (RFP)	312.10.2
5015—2009	Performance Requirements for Testing Double Check Valve Backflow Prevention Assemblies (DC) and Double Check Fire Protection Backflow Prevention Assemblies (DCF)	312.10.2
5020—2009	Performance Requirements for Testing Pressure Vacuum Breaker Assemblies (PVBA)	312.10.2

ASSE—continued

5047—98	Performance Requirements for Testing Reduced Pressure Detector Fire Protection Backflow Prevention Assemblies (RPDA)	312.10.2
5048—2009	Performance Requirements for Testing Double Check Valve Detector Assembly (DCDA)	312.10.2
5052—98	Performance Requirements for Testing Hose Connection Backflow Preventers	312.10.2
5056—98	Performance Requirements for Testing Spill Resistant Vacuum Breaker (SRVB)	312.10.2

ASTM

ASTM International
100 Barr Harbor Drive
West Conshohocken, PA 19428-2959

Standard reference number	Title	Referenced in code section number
A 53/A 53M—07	Specification for Pipe, Steel, Black and Hot-dipped, Zinc-coated Welded and Seamless	Table 605.3, Table 605.4, Table 702.1
A 74—09	Specification for Cast-iron Soil Pipe and Fittings	Table 702.1, Table 702.2, Table 702.3, Table 702.4, 708.2, 708.7, Table 1102.4, Table 1102.5, Table 1102.7
A 312/A 312M—08a	Specification for Seamless, Welded, And Heavily Cold Worked Austenitic Stainless Steel Pipes	Table 605.3, Table 605.4, Table 605.5, 605.23.2
A 733—03	Specification for Welded and Seamless Carbon Steel and Austenitic Stainless Steel Pipe Nipples	Table 605.8
A 778—01	Specification for Welded Unannealed Austenitic Stainless Steel Tubular Products	Table 605.3, Table 605.4, Table 605.5
A 888—09	Specification for Hubless Cast-iron Soil Pipe and Fittings for Sanitary and Storm Drain, Waste, and Vent Piping Application	Table 702.1, Table 702.2, Table 702.3, Table 702.4, 708.7, Table 1102.4, Table 1102.5, Table 1102.7
B 32—08	Specification for Solder Metal	605.14.3, 605.15.4, 705.9.3, 705.10.3
B 42—02e01	Specification for Seamless Copper Pipe, Standard Sizes	Table 605.3, Table 605.4, Table 702.1
B 43—98(2004)	Specification for Seamless Red Brass Pipe, Standard Sizes	Table 605.3, Table 605.4, Table 702.1
B 75—02	Specification for Seamless Copper Tube	Table 605.3, Table 605.4, Table 702.1, Table 702.2, Table 702.3, Table 1102.4
B 88—03	Specification for Seamless Copper Water Tube	Table 605.3, Table 605.4, Table 702.1, Table 702.2, Table 702.3, Table 1102.4
B 152/B 152M—06a	Specification for Copper Sheet, Strip Plate and Rolled Bar	402.3, , 417.5.2.4, 425.3.3, 902.2
B 251—02e01	Specification for General Requirements for Wrought Seamless Copper and Copper-alloy Tube	Table 605.3, Table 605.4, Table 702.1, Table 702.2, Table 702.3, Table 1102.4
B 302—07	Specification for Threadless Copper Pipe, Standard Sizes	Table 605.3, Table 605.4, Table 702.1
B 306—09	Specification for Copper Drainage Tube (DWV)	Table 702.1, Table 702.2, Table 1102.4
B 447—07	Specification for Welded Copper Tube	Table 605.3, Table 605.4
B 687—99(2005)e01	Specification for Brass, Copper and Chromium-plated Pipe Nipples	Table 605.8
B 813—00(2009)	Specification for Liquid and Paste Fluxes for Soldering of Copper and Copper Alloy Tube	605.14.3, 605.15.4, 705.9.3, 705.10.3
B 828—02	Practice for Making Capillary Joints by Soldering of Copper and Copper Alloy Tube and Fittings	605.14.3, 605.15.4, 705.9.3, 705.10.3
C 4—04e01	Specification for Clay Drain Tile and Perforated Clay Drain Tile	Table 702.3, Table 1102.4, Table 1102.5
C 14—07	Specification for Nonreinforced Concrete Sewer, Storm Drain and Culvert Pipe	Table 702.3, Table 1102.4
C 76—08a	Specification for Reinforced Concrete Culvert, Storm Drain and Sewer Pipe	Table 702.3, Table 1102.4
C 296—00(2004)	Specification for Asbestos-cement Pressure Pipe	Table 605.3
C 425—04	Specification for Compression Joints for Vitrified Clay Pipe and Fittings	705.15, 705.19
C 428—05(2006)	Specification for Asbestos-cement Nonpressure Sewer Pipe	Table 702.2, Table 702.3, Table 702.4, Table 1102.4
C 443—05a	Specification for Joints for Concrete Pipe and Manholes, Using Rubber Gaskets	705.6, 705.19
C 508—00(2004)	Specification for Asbestos-cement Underdrain Pipe	Table 1102.5
C 564—08	Specification for Rubber Gaskets for Cast-iron Soil Pipe and Fittings	705.5.2, 705.5.3, 705.19
C 700—07a	Specification for Vitrified Clay Pipe, Extra Strength, Standard Strength, and Perforated	Table 702.3, Table 702.4, Table 1102.4, Table 1102.5
C 1053—00(2005)	Specification for Borosilicate Glass Pipe and Fittings for Drain, Waste, and Vent (DWV) Applications	Table 702.1, Table 702.4

ASTM—continued

ASTM—continued

D 4551—96(2008) e1	Specification for Poly (Vinyl Chloride) (PVC) Plastic Flexible Concealed Water-containment Membrane	417.5.2.1
F 405—05	Specification for Corrugated Polyethylene (PE) Pipe and Fittings	Table 1102.5, Table 1303.10
F 409—02 (2008)	Specification for Thermoplastic Accessible and Replaceable Plastic Tube and Tubular Fittings	424.1.2, Table 1102.7
F 437—06	Specification for Threaded Chlorinated Poly (Vinyl Chloride) (CPVC) Plastic Pipe Fittings, Schedule 80	Table 605.5
F 438—04	Specification for Socket-type Chlorinated Poly (Vinyl Chloride) (CPVC) Plastic Pipe Fittings, Schedule 40	Table 605.5
F 439—06	Standard Specification for Chlorinated Poly (Vinyl Chloride) (CPVC) Plastic Pipe Fittings, Schedule 80	Table 605.5
F 441/F 441M—02 (2008)	Specification for Chlorinated Poly (Vinyl Chloride) (CPVC) Plastic Pipe, Schedules 40 and 80	Table 605.3, Table 605.4
F 442/F 442M—99(2005) e1	Specification for Chlorinated Poly (Vinyl Chloride) (CPVC) Plastic Pipe (SDR-PR)	Table 605.3, Table 605.4
F 477—08	Specification for Elastomeric Seals (Gaskets) for Joining Plastic Pipe	605.24, 705.19
F 493—04	Specification for Solvent Cements for Chlorinated Poly (Vinyl Chloride) (CPVC) Plastic Pipe and Fittings	605.16.2
F 628—08	Specification for Acrylonitrile-Butadiene-Styrene (ABS) Schedule 40 Plastic Drain, Waste, and Vent Pipe with a Cellular Core	Table 702.1, Table 702.2, Table 702.3, Table 702.4, 705.2.2, 705.7.2, Table 1102.4, Table 1102.7
F 656—08	Specification for Primers for Use in Solvent Cement Joints of Poly (Vinyl Chloride) (PVC) Plastic Pipe and Fittings	605.22.2, 705.8.2, 705.14.2
F 714—08	Specification for Polyethylene (PE) Plastic Pipe (SDR-PR) Based on Outside Diameter	Table 702.3
F 876—08b	Specification for Cross-linked Polyethylene (PEX) Tubing	Table 605.3, Table 605.4
F 877—07	Specification for Cross-linked Polyethylene (PEX) Plastic Hot and Cold Water Distribution Systems	Table 605.3, Table 605.4, Table 605.5
F 891—07	Specification for Coextruded Poly (Vinyl Chloride) (PVC) Plastic Pipe with a Cellular Core	Table 702.1 Table 702.2, Table 702.3, Table 1102.4, Table 1102.5, Table 1102.7
F 1055—98(2006)	Standard Specification for Electrofusion Type Polyethylene Fittings for Outside Diameter Controlled Polyethylene Pipe and Tubing	Table 605.5
F 1281—07	Specification for Cross-linked Polyethylene/Aluminum/ Cross-linked Polyethylene (PEX-AL-PEX) Pressure Pipe	Table 605.3, Table 605.4, Table 605.5, 605.21.1
F 1282—06	Specification for Polyethylene/Aluminum/Polyethylene (PE-AL-PE) Composite Pressure Pipe	Table 605.3, Table 605.4, Table 605.5, 605.21.1
F 1412—09	Specification for Polyolefin Pipe and Fittings for Corrosive Waste Drainage	Table 702.1, Table 702.2, Table 702.4, 705.17.1
F 1488—03	Specification for Coextruded Composite Pipe	Table 702.1, Table 702.2, Table 702.3, Table 1303.10
F 1673—04	Polyvinylidene Fluoride (PVDF) Corrosive Waste Drainage Systems	Table 702.1, Table 702.2, Table 702.3, Table 702.4, 705.18.1
F 1807—08	Specification for Metal Insert Fittings Utilizing a Copper Crimp Ring for SDR9 Cross-linked Polyethylene (PEX) Tubing	Table 605.5
F 1866—07	Specification for Poly (Vinyl Chloride) (PVC) Plastic Schedule 40 Drainage and DWV Fabricated Fittings	Table 702.4, Table 1102.7
F 1960—09	Specification for Cold Expansion Fittings with PEX Reinforcing Rings for use with Cross-linked Polyethylene (PEX) Tubing	Table 605.5
F 1974—08	Specification for Metal Insert Fittings for Polyethylene/Aluminum/Polyethylene and Cross-linked Polyethylene/Aluminum/Cross-linked Polyethylene Composite Pressure Pipe	Table 605.5, 605.21.1
F 1986—01(2006)	Specification for Multilayer Pipe, Type 2, Compression Fittings and Compression Joints for Hot and Cold Drinking Water Systems	Table 605.3, Table 605.4, Table 605.5
F 2080—08	Specifications for Cold-expansion Fittings with Metal Compression-sleeves for Cross-linked Polyethylene (PEX) Pipe	Table 605.5
F 2098—08	Standard specification for Stainless Steel Clamps for Securing SDR9 Cross-linked Polyethylene (PEX) Tubing to Metal Insert Fittings	Table 605.5
F 2159—05	Specification for Plastic Insert Fittings Utilizing a Copper Crimp Ring for SDR9 Cross-linked Polyethylene (PEX) Tubing	Table 605.5
F 2262—05	Specification for Cross-linked Polyethylene/Aluminum/Cross-linked Polyethylene Tubing OD Controlled SDR9	Table 605.3, Table 605.4
F 2306/F 2306M—08	12" to 60" Annular Corrugated Profile-wall Polyethylene (PE) Pipe and Fittings for Gravity Flow Storm Sewer and Subsurface Drainage Applications	Table 1102.4, Table 1102.7
F 2389—07e1	Specification for Pressure-rated Polypropylene (PP) Piping Systems	Table 605.3, Table 605.4, Table 605.5, 605.20.1
F 2434—08	Standard Specification for Metal Insert Fittings Utilizing a Copper Crimp Ring for SDR9 Cross-linked Polyethylene (PEX) Tubing and SDR9 Cross-linked Polyethylene/Aluminum/ Cross-linked Polyethylene (PEX AL-PEX) Tubing	Table 605.5

ASTM—continued

F 2735—09	Standard Specification for SDR9 Cross-linked Polyethylene (PEX) and Raised Temperature (PE-RT) Tubing	Table 605.5
F 2769—09	Polyethylene or Raised Temperature (PE-RT) Plastic Hot and Cold Water Tubing and Distribution Systems	Table 605.3, Table 605.4

AWS

American Welding Society
550 N.W. LeJeune Road
Miami, FL 33126

Standard reference number	Title	Referenced in code section number
A5.8—04	Specifications for Filler Metals for Brazing and Braze Welding	605.12.1, 605.14.1, 605.15.1, 705.4.1, 705.9.1, 705.10.1

AWWA

American Water Works Association
6666 West Quincy Avenue
Denver, CO 80235

Standard reference number	Title	Referenced in code section number
C104—98	Standard for Cement-mortar Lining for Ductile-iron Pipe and Fittings for Water	605.3, 605.5
C110/A21.10—03	Standard for Ductile-iron and Gray-iron Fittings, 3 Inches through 48 Inches, for Water	Table 605.5, Table 702.4, Table 1102.7
C111—00	Standard for Rubber-gasket Joints for Ductile-iron Pressure Pipe and Fittings	605.13
C115/A21.15—99	Standard for Flanged Ductile-iron Pipe with Ductile-iron or Gray-iron Threaded Flanges	Table 605.3, Table 605.4
C151/A21.51—02	Standard for Ductile-iron Pipe, Centrifugally Cast for Water	Table 605.3, Table 605.4
C153—00/A21.53—00	Standard for Ductile-iron Compact Fittings for Water Service	Table 605.5
C510—00	Double Check Valve Backflow Prevention Assembly	Table 608.1, 608.13.7
C511—00	Reduced-pressure Principle Backflow Prevention Assembly	Table 608.1, 608.13.2, 608.16.2
C651—99	Disinfecting Water Mains	610.1
C652—02	Disinfection of Water-storage Facilities	610.1
C901—08	Polyethylene (PE) Pressure Pipe and Tubing $^1/_2$ inch (13 mm) Through 3 inch (76 mm) for Water Service	Table 605.3
C904—08	Cross-linked Polyethylene (PEX) Pressure Pipe $^1/_2$ inch (13 mm) Through 3 inch (76 mm) for Water Service	Table 605.3

CISPI

Cast Iron Soil Pipe Institute
5959 Shallowford Road, Suite 419
Chattanooga, TN 37421

Standard reference number	Title	Referenced in code section number
301—04a	Specification for Hubless Cast-iron Soil Pipe and Fittings for Sanitary and Storm Drain, Waste and Vent Piping Applications	Table 702.1, Table 702.2, Table 702.3, Table 702.4, 708.7, Table 1102.4, Table 1102.5, Table 1102.7
310—04	Specification for Coupling for Use in Connection with Hubless Cast-iron Soil Pipe and Fittings for Sanitary and Storm Drain, Waste and Vent Piping Applications	705.5.3

CSA

Canadian Standards Association
5060 Spectrum Way
Mississauga, Ontario, Canada L4W 5N6

Standard reference number	Title	Referenced in code section number
A257.1M—92	Circular Concrete Culvert, Storm Drain, Sewer Pipe and Fittings	Table 702.3, Table 1102.4
A257.2M—92	Reinforced Circular Concrete Culvert, Storm Drain, Sewer Pipe and Fittings	Table 702.3, Table 1102.4

CSA—continued

Standard	Title	Sections
A257.3M—92	Joints for Circular Concrete Sewer and Culvert Pipe, Manhole Sections and Fittings Using Rubber Gaskets	705.6, 705.19
ASME A112.18.1/ CSA B125.1—05	Plumbing Supply Fittings	424.1, 424.2, 424.3, 424.4, 424.6, 607.4, 608.2
ASME A112.18.2/ CSA B125.2—05	Plumbing Waste Fittings	424.1.2
ASME A112.9.2/B45.1—08	Ceramic Plumbing Fixtures	401.2, 405.9, 407.1, 408.1, 410.1, 415.1, 416.1, 417.1, 418.1, 419.1, 420.1
ASME A112.19.1/ CSA B45.2—08	Enameled Cast-iron and Enameled Steel Plumbing Fixtures	407.1, 410.1, 415.1, 416.1, 418.1
ASME A112.19.3/ CSA B45.4—08	Stainless-steel Plumbing Fixtures	405.9, 407.1, 415.1, 416.1, 418.1, 420.1
ASME A112.19.5/ CSA B45.15—09	Trim for Water Closet Bowls and Tanks	425.4
ASME A112.19.7/ CSA B45.10—09	Hydromassage Bathtub Appliances	421.1, 421.4
B45.5—02(R2008)	Plastic Plumbing Fixtures	407.1, 416.2, 417.1, 419.1, 420.1
B45.9—99(R2008)	Macerating Systems and Related Components	712.4.1
B64.1.1—01	Vacuum Breakers, Atmospheric Type (AVB)	425.2, Table 608.1, 608.13.6, 608.16.4.1
B64.1.2—07	Pressure Vacuum Breakers, (PVB)	Table 608.1, 608.13.5
B64.1.3—07	Spill Resistant Pressure Vacuum Breaks (SRPVB)	608.13.8
B64.2—01	Vacuum Breakers, Hose Connection Type (HCVB)	Table 608.1, 608.13.6
B64.2.1—07	Vacuum Breakers, Hose Connection (HCVB) with Manual Draining Feature	Table 608.1, 608.13.6
B64.2.1.1—07	Hose Connection Dual Check Vacuum Breakers, (HCDVB)	Table 608.1, 608.13.6
B64.2.2—01	Vacuum Breakers, Hose Connection Type (HCVB) with Automatic Draining Feature	Table 608.1, 608.13.6
B64.3—01	Backflow Preventers, Dual Check Valve Type with Atmospheric Port (DCAP)	Table 608.1, 608.13.3, 608.16.2
B64.4—01	Backflow Preventers, Reduced Pressure Principle Type (RP)	Table 608.1, 608.13.2, 608.16.2
B64.4.1—07	Reduced Pressure Principle for Fire Sprinklers (RPF)	Table 608.1, 608.13.2
B64.5—07	Double Check Backflow Preventers (DCVA)	Table 608.1, 608.13.7
B64.5.1—07	Double Check Valve Backflow Preventer for Fire Systems (DCVAF)	Table 608.1 608.13.7
B64.6—07	Dual Check Backflow PreventersValve (DuC)	605.3.1, Table 608.1
B64.7—07	Laboratory Faucet Vacuum Breakers (LFVB)	Table 608.1, 608.13.6
B64.10—07	Manual for the Selection and Installation of Backflow Prevention Devices	312.10.2
B64.10—01	Manual for the Selection, Installation, Maintenance and Field Testing of Backflow Prevention Devices	312.10.2
B79—08	Commercial and Residential Drains, and Cleanouts	412.1
B125.3—2005	Plumbing Fittings	408.3, 416.5, 424.4, 424.5, 425.2, 425.3.1, Table 608.1
B137.1—05	Polyethylene (PE) Pipe, Tubing and Fittings for Cold Water Pressure Services	Table 605.3, Table 605.5
B137.2—05	Polyvinylchloride PVC Injection-moulded Gasketed Fittings for Pressure Applications	Table 605.5
B137.3—05	Rigid Poly (Vinyl Chloride) (PVC) Pipe for Pressure Applications	Table 605.3, Table 605.4, Table 605.5, 605.22.2, 705.8.2, 705.14.2
B137.5—05	Cross-linked Polyethylene (PEX) Tubing Systems for Pressure Applications	Table 605.3, Table 605.4, Table 605.5
B137.6—05	CPVC Pipe, Tubing and Fittings for Hot and Cold Water Distribution Systems	Table 605.3, Table 605.4
B137.9—02	Polyethylene/Aluminum/Polyethylene Composite Pressure Pipe Systems	Table 605.3, Table 605.5, 605.21.1
B137.10M—02	Cross-linked Polyethylene/Aluminum/Polyethylene Composite Pressure Pipe Systems	Table 605.3, Table 605.4, Table 605.5, 605.21.1
B137.11—05	Polypropylene (PP-R) Pipe and Fittings for Pressure Applications	Table 605.3, Table 605.4, Table 605.5
B181.1—06	Acrylonitrile-butadiene-styrene ABS Drain, Waste and Vent Pipe and Pipe Fittings	Table 702.1, Table 702.2, Table 702.3, Table 702.4, 705.2.2, 705.7.2, 715.2, Table 1102.4, Table 1102.7
B 181.2—06	Polyvinylchloride PVC and chlorinated polyvinylchloride (CPVC) Drain, Waste, and Vent Pipe and Pipe Fittings	Table 702.1 Table 702.2, 705.8.2, 705.14.2, 715.2
B181.3—02	Polyolefin Laboratory Drainage Systems	Table 702.1, Table 702.2, Table 702.3, Table 702.4, 705.17.1
B182.1—06	Plastic Drain and Sewer Pipe and Pipe Fittings	705.8.2, 705.14.2, Table 1102.4
B182.2—06	PSM Type Polyvinylchloride PVC Sewer Pipe and Fittings	Table 702.3, Table 1102.4, Table 1102.5
B182.4—02	Profile PVC Sewer Pipe and Fittings	Table 702.3, Table 1102.4, Table 1102.5
B182.4—06	Profile Polyvinylchloride PVC Sewer Pipe and Fittings	Table 702.3, Table 1102.4, Table 1102.5
B182.6—06	Profile Polyethylene Sewer Pipe and	Table 1102.5
B182.8—06	Profile Polyethylene (PE) Storm Sewer and Drainage Pipe and Fittings	Table 1102.5
B356—00(2005)	Water Pressure Reducing Valves for Domestic Water Systems	604.8
B483.1—07	Drinking Water Treatment Units	611.1, 611.2
B481.3	Sizing, Selection, Location and Installation of Grease Interceptors	1003.3.4

| B481.1—07 | Testing and Rating of Grease interceptors Using Lard | 1003.3.4 |
| B602—02 | Mechanical Couplings for Drain, Waste and Vent Pipe and Sewer Pipe | 705.2.1, 705.5.3, 705.6, 705.7.1, 705.14.1, 705.15, 705.16.2, 705.19 |

ICC

International Code Council, Inc.
500 New Jersey Ave, NW
6th Floor
Washington, DC 20001

Standard reference number	Title	Referenced in code section number
IBC—12	International Building Code®	201.3, 307.1, 307.2, 307.3, 308.2, 309.1, 309.2, 310.1, 310.3, 315.1, 403.1, Table 403.1, 403.4, 404.1, 407.3, 417.6, 502.4, 606.5.2, 1106.5
IECC—12	International Energy Conservation Code®	313.1
IFC—12	International Fire Code®	201.3, 1201.1
IFGC—12	International Fuel Gas Code®	101.2, 201.3, 502.1
IMC—12	International Mechanical Code®	201.3, 307.6, 310.1, 422.9, 502.1, 612.1, 1202.1
IPSDC—12	International Private Sewage Disposal Code®	701.2
IRC—12	International Residential Code®	101.2

ISEA

International Safety Equipment Association
1901 N. Moore Street, Suite 808
Arlington, VA 22209

Standard reference number	Title	Referenced in code section number
Z358.1—98	Emergency Eyewash and Shower Equipment	411.1

NFPA

National Fire Protection Association
1 Batterymarch Park
Quincy, MA 02169-7471

Standard reference number	Title	Referenced in code section number
50—01	Bulk Oxygen Systems at Consumer Sites	1203.1
51—07	Design and Installation of Oxygen-fuel Gas Systems for Welding, Cutting and Allied Processes	1203.1
70—11	National Electric Code	502.1, 504.3, 1114.1.3
99C—05	Gas and Vacuum Systems	1202.1

NSF

NSF International
789 Dixboro Road
Ann Arbor, MI 48105

Standard reference number	Title	Referenced in code section number
3—2008	Commercial Warewashing Equipment	409.1
14—2008e	Plastic Piping System Components and Related Materials	303.3, 611.3
18—2007	Manual Food and Beverage Dispensing Equipment	426.1
42—2007ae	Drinking Water Treatment Units-Aesthetic Effects	611.1, 611.3
44—2007	Residential Cation Exchange Water Softeners	611.1, 611.3
53—2007a	Drinking Water Treatment Units-Health Effects	611.1, 611.3
58—2007	Reverse Osmosis Drinking Water Treatment Systems	611.2, 611.3
61—2008	Drinking Water System Components-Health Effects	410.1, 424.1, 605.3, 605.4, 605.5, 605.7, 611.3
62—2007	Drinking Water Distillation Systems	611.1

PDI

Plumbing and Drainage Institute
800 Turnpike Street, Suite 300
North Andover, MA 01845

Standard reference number	Title	Referenced in code section number
G101(2003)	Testing and Rating Procedure for Grease Interceptors with Appendix of Sizing and Installation Data . .1003.3.4	
G102	Testing and Certification for Grease Interceptors with Fog Sensing and Alarm Devices 1003.3.4	

UL

Underwriters Laboratories, Inc.
333 Pfingsten Road
Northbrook, IL 60062-2096

Standard reference number	Title	Referenced in code section number
508—99	Industrial Control Equipment - with revision through April 2010 . . . . .314.2.3	

APPENDIX A

PLUMBING PERMIT FEE SCHEDULE

The provisions contained in this appendix are not mandatory unless specifically referenced in the adopting ordinance.

Permit Issuance

1. For issuing each permit. . . . . $ _____

2. For issuing each supplemental permit . . . . _____

Unit Fee Schedule

1. For each plumbing fixture or trap or set of fixtures on one trap (including water, drainage piping and backflow protection thereof). . . . . _____

2. For each building sewer and each trailer park sewer . . . . _____

3. Rainwater systems—per drain (inside building) . . . . _____

4. For each cesspool (where permitted) . . . . _____

5. For each private sewage disposal system . . . . _____

6. For each water heater and/or vent. . . . . _____

7. For each industrial waste pretreatment interceptor including its trap and vent, excepting kitchen-type grease interceptors functioning as fixture traps . . . . _____

8. For installation, alteration or repair of water-piping and/or water-treating equipment, each . . . . _____

9. For repair or alteration of drainage or vent piping, each fixture . . . . _____

10. For each lawn sprinkler system on any one meter including backflow protection devices therefor . . . . _____

11. For atmospheric-type vacuum breakers not included in Item 2:

    1 to 5. . . . . _____

    over 5, each. . . . . _____

12. For each backflow protective device other than atmospheric-type vacuum breakers:

    2 inches (51 mm) and smaller . . . . _____

    Over 2 inches (51 mm). . . . . _____

Other Inspections and Fees

1. Inspections outside of normal business hours (minimum charge two hours). _____ per hour

2. Reinspection fee assessed under provisions of Section 107.4.3. . . . . _____each

3. Inspections for which no fee is specifically indicated (minimum charge one-half hour) . . . . _____ per hour

4. Additional plan review required by changes, additions or revisions to approved plans (minimum charge one-half hour). . . . . _____ per hour

APPENDIX B

RATES OF RAINFALL FOR VARIOUS CITIES

This appendix is informative and is not part of the code.

Rainfall rates, in inches per hour, are based on a storm of 1-hour duration and a 100-year return period. The rainfall rates shown in the appendix are derived from Figure 1106.1.

Alabama:
Birmingham 3.8
Huntsville 3.6
Mobile 4.6
Montgomery 4.2

Alaska:
Fairbanks 1.0
Juneau 0.6

Arizona:
Flagstaff 2.4
Nogales 3.1
Phoenix 2.5
Yuma 1.6

Arkansas:
Fort Smith 3.6
Little Rock 3.7
Texarkana 3.8

California:
Barstow 1.4
Crescent City 1.5
Fresno 1.1
Los Angeles 2.1
Needles 1.6
Placerville 1.5
San Fernando 2.3
San Francisco 1.5
Yreka 1.4

Colorado:
Craig 1.5
Denver 2.4
Durango 1.8
Grand Junction 1.7
Lamar 3.0
Pueblo 2.5

Connecticut:
Hartford 2.7
New Haven 2.8
Putnam 2.6

Delaware:
Georgetown 3.0
Wilmington 3.1

District of Columbia:
Washington 3.2

Florida:
Jacksonville 4.3
Key West 4.3
Miami 4.7
Pensacola 4.6
Tampa 4.5

Georgia:
Atlanta 3.7
Dalton 3.4
Macon 3.9
Savannah 4.3
Thomasville 4.3

Hawaii:
Hilo 6.2
Honolulu 3.0
Wailuku 3.0

Idaho:
Boise 0.9
Lewiston 1.1
Pocatello 1.2

Illinois:
Cairo 3.3
Chicago 3.0
Peoria 3.3
Rockford 3.2
Springfield 3.3

Indiana:
Evansville 3.2
Fort Wayne 2.9
Indianapolis 3.1

Iowa:
Davenport 3.3
Des Moines 3.4
Dubuque 3.3
Sioux City 3.6

Kansas:
Atwood 3.3
Dodge City 3.3
Topeka 3.7
Wichita 3.7

Kentucky:
Ashland 3.0
Lexington 3.1

Louisville 3.2
Middlesboro 3.2
Paducah 3.3

Louisiana:
Alexandria 4.2
Lake Providence 4.0
New Orleans 4.8
Shreveport 3.9

Maine:
Bangor 2.2
Houlton 2.1
Portland 2.4

Maryland:
Baltimore 3.2
Hagerstown 2.8
Oakland 2.7
Salisbury 3.1

Massachusetts:
Boston 2.5
Pittsfield 2.8
Worcester 2.7

Michigan:
Alpena 2.5
Detroit 2.7
Grand Rapids 2.6
Lansing 2.8
Marquette 2.4
Sault Ste. Marie 2.2

Minnesota:
Duluth 2.8
Grand Marais 2.3
Minneapolis 3.1
Moorhead 3.2
Worthington 3.5

Mississippi:
Biloxi 4.7
Columbus 3.9
Corinth 3.6
Natchez 4.4
Vicksburg 4.1

Missouri:
Columbia 3.2
Kansas City 3.6

Springfield 3.4
St. Louis 3.2

Montana:
Ekalaka 2.5
Havre 1.6
Helena 1.5
Kalispell 1.2
Missoula 1.3

Nebraska:
North Platte 3.3
Omaha 3.8
Scottsbluff 3.1
Valentine 3.2

Nevada:
Elko 1.0
Ely 1.1
Las Vegas 1.4
Reno 1.1

New Hampshire:
Berlin 2.5
Concord 2.5
Keene 2.4

New Jersey:
Atlantic City 2.9
Newark 3.1
Trenton 3.1

New Mexico:
Albuquerque 2.0
Hobbs 3.0
Raton 2.5
Roswell 2.6
Silver City 1.9

New York:
Albany 2.5
Binghamton 2.3
Buffalo 2.3
Kingston 2.7
New York 3.0
Rochester 2.2

North Carolina:
Asheville 4.1
Charlotte 3.7
Greensboro 3.4
Wilmington 4.2

North Dakota:
Bismarck 2.8
Devils Lake 2.9
Fargo 3.1
Williston. 2.6

Ohio:
Cincinnati. 2.9
Cleveland. 2.6
Columbus. 2.8
Toledo 2.8

Oklahoma:
Altus. 3.7
Boise City 3.3
Durant 3.8
Oklahoma City. 3.8

Oregon:
Baker 0.9
Coos Bay 1.5
Eugene 1.3
Portland 1.2

Pennsylvania:
Erie. 2.6
Harrisburg 2.8
Philadelphia 3.1
Pittsburgh. 2.6
Scranton 2.7
Rhode Island:
Block Island 2.75
Providence 2.6

South Carolina:
Charleston 4.3
Columbia 4.0
Greenville. 4.1

South Dakota:
Buffalo 2.8
Huron 3.3
Pierre 3.1
Rapid City 2.9
Yankton 3.6

Tennessee:
Chattanooga 3.5
Knoxville 3.2
Memphis 3.7
Nashville 3.3

Texas:
Abilene. 3.6
Amarillo. 3.5
Brownsville 4.5

Dallas 4.0
Del Rio 4.0
El Paso. 2.3
Houston. 4.6
Lubbock 3.3
Odessa. 3.2
Pecos 3.0
San Antonio. 4.2

Utah:
Brigham City. 1.2
Roosevelt. 1.3
Salt Lake City 1.3
St. George 1.7

Vermont:
Barre 2.3
Bratteboro 2.7
Burlington 2.1
Rutland 2.5

Virginia:
Bristol 2.7
Charlottesville. 2.8
Lynchburg. 3.2
Norfolk 3.4
Richmond 3.3

Washington:
Omak. 1.1
Port Angeles 1.1
Seattle 1.4
Spokane. 1.0
Yakima 1.1

West Virginia:
Charleston 2.8
Morgantown 2.7

Wisconsin:
Ashland 2.5
Eau Claire 2.9
Green Bay 2.6
La Crosse. 3.1
Madison. 3.0
Milwaukee. 3.0

Wyoming:
Cheyenne. 2.2
Fort Bridger. 1.3
Lander 1.5
New Castle 2.5
Sheridan 1.7
Yellowstone Park 1.4

APPENDIX C

VACUUM DRAINAGE SYSTEM

The provisions contained in this appendix are not mandatory unless specifically referenced in the adopting ordinance.

SECTION C101
VACUUM DRAINAGE SYSTEM

C101.1 Scope. This appendix provides general guidelines for the requirements for vacuum drainage systems.

C101.2 General requirements.

C101.2.1 System design. Vacuum drainage systems shall be designed in accordance with manufacturer's recommendations. The system layout, including piping layout, tank assemblies, vacuum pump assembly and other components/designs necessary for proper function of the system shall be per manufacturer's recommendations. Plans, specifications and other data for such systems shall be submitted to the local administrative authority for review and approval prior to installation.

C101.2.2 Fixtures. Gravity-type fixtures used in vacuum drainage systems shall comply with Chapter 4 of this code.

C101.2.3 Drainage fixture units. Fixture units for gravity drainage systems which discharge into or receive discharge from vacuum drainage systems shall be based on values in Chapter 7 of this code.

C101.2.4 Water supply fixture units. Water supply fixture units shall be based on values in Chapter 6 of this code with the addition that the fixture unit of a vacuum-type water closet shall be "1."

C101.2.5 Traps and cleanouts. Gravity-type fixtures shall be provided with traps and cleanouts in accordance with Chapters 7 and 10 of this code.

C101.2.6 Materials. Vacuum drainage pipe, fitting and valve materials shall be as recommended by the vacuum drainage system manufacturer and as permitted by this code.

C101.3 Testing and demonstrations. After completion of the entire system installation, the system shall be subjected to a vacuum test of 19 inches (483 mm) of mercury and shall be operated to function as required by the administrative authority and the manufacturer. Recorded proof of all tests shall be submitted to the administrative authority.

C101.4 Written instructions. Written instructions for the operations, maintenance, safety and emergency procedures shall be provided by the building owner as verified by the administrative authority.

APPENDIX D

DEGREE DAY AND DESIGN TEMPERATURES

This appendix is informative and is not part of the code.

TABLE D101
DEGREE DAY AND DESIGN TEMPERATURES[a] FOR CITIES IN THE UNITED STATES

STATE	STATION[b]	HEATING DEGREE DAYS (yearly total)	DESIGN TEMPERATURES			DEGREES NORTH LATITUDE[c]
			Winter	Summer		
			97^1/$_2$%	Dry bulb 2^1/$_2$%	Wet bulb 2^1/$_2$%	
AL	Birmingham	2,551	21	94	77	33°30′
	Huntsville	3,070	16	96	77	34°40′
	Mobile	1,560	29	93	79	30°40′
	Montgomery	2,291	25	95	79	32°20′
AK	Anchorage	10,864	-18	68	59	61°10′
	Fairbanks	14,279	-47	78	62	64°50′
	Juneau	9,075	1	70	59	58°20′
	Nome	14,171	-27	62	56	64°30′
AZ	Flagstaff	7,152	4	82	60	35°10′
	Phoenix	1,765	34	107	75	33°30′
	Tuscon	1,800	32	102	71	33°10′
	Yuma	974	39	109	78	32°40′
AR	Fort Smith	3,292	17	98	79	35°20′
	Little Rock	3,219	20	96	79	34°40′
	Texarkana	2,533	23	96	79	33°30′
CA	Fresno	2,611	30	100	71	36°50′
	Long Beach	1,803	43	80	69	33°50′
	Los Angeles	2,061	43	80	69	34°00′
	Los Angeles[d]	1,349	40	89	71	34°00′
	Oakland	2,870	36	80	64	37°40′
	Sacramento	2,502	32	98	71	38°30′
	San Diego	1,458	44	80	70	32°40′
	San Francisco	3,015	38	77	64	37°40′
	San Francisco[d]	3,001	40	71	62	37°50′
CO	Alamosa	8,529	-16	82	61	37°30′
	Colorado Springs	6,423	2	88	62	38°50′
	Denver	6,283	1	91	63	39°50′
	Grand Junction	5,641	7	94	63	39°10′
	Pueblo	5,462	0	95	66	38°20′
CT	Bridgeport	5,617	9	84	74	41°10′
	Hartford	6,235	7	88	75	41°50′
	New Haven	5,897	7	84	75	41°20′
DE	Wilmington	4,930	14	89	76	39°40′
DC	Washington	4,224	17	91	77	38°50′
FL	Daytona	879	35	90	79	29°10′
	Fort Myers	442	44	92	79	26°40′
	Jacksonville	1,239	32	94	79	30°30′
	Key West	108	57	90	79	24°30′
	Miami	214	47	90	79	25°50′
	Orlando	766	38	93	78	28°30′
	Pensacola	1,463	29	93	79	30°30′
	Tallahassee	1,485	30	92	78	30°20′
	Tampa	683	40	91	79	28°00′
	West Palm Beach	253	45	91	79	26°40′

(continued)

TABLE D101—continued
DEGREE DAY AND DESIGN TEMPERATURES[a] FOR CITIES IN THE UNITED STATES

| STATE | STATION[b] | HEATING DEGREE DAYS (yearly total) | DESIGN TEMPERATURES | | | DEGREES NORTH LATITUDE[c] |
| | | | Winter | Summer | | |
			$97^1/_2$%	Dry bulb $2^1/_2$%	Wet bulb $2^1/_2$%	
GA	Athens	2,929	22	92	77	34°00'
	Atlanta	2,961	22	92	76	33°40'
	Augusta	2,397	23	95	79	33°20'
	Columbus	2,383	24	93	78	32°30'
	Macon	2,136	25	93	78	32°40'
	Rome	3,326	22	93	78	34°20'
	Savannah	1,819	27	93	79	32°10'
HI	Hilo	0	62	83	74	19°40'
	Honolulu	0	63	86	75	21°20'
ID	Boise	5,809	10	94	66	43°30'
	Lewiston	5,542	6	93	66	46°20'
	Pocatello	7,033	-1	91	63	43°00'
IL	Chicago (Midway)	6,155	0	91	75	41°50'
	Chicago (O'Hare)	6,639	-4	89	76	42°00'
	Chicago[d]	5,882	2	91	77	41°50'
	Moline	6,408	-4	91	77	41°30'
	Peoria	6,025	-4	89	76	40°40'
	Rockford	6,830	-4	89	76	42°10'
	Springfield	5,429	2	92	77	39°50'
IN	Evansville	4,435	9	93	78	38°00'
	Fort Wayne	6,205	1	89	75	41°00'
	Indianapolis	5,699	2	90	76	39°40'
	South Bend	6,439	1	89	75	41°40'
IA	Burlington	6,114	-3	91	77	40°50'
	Des Moines	6,588	-5	91	77	41°30'
	Dubuque	7,376	-7	88	75	42°20'
	Sioux City	6,951	-7	92	77	42°20'
	Waterloo	7,320	-10	89	77	42°30'
KS	Dodge City	4,986	5	97	73	37°50'
	Goodland	6,141	0	96	70	39°20'
	Topeka	5,182	4	96	78	39°00'
	Wichita	4,620	7	98	76	37°40'
KY	Covington	5,265	6	90	75	39°00'
	Lexington	4,683	8	91	76	38°00'
	Louisville	4,660	10	93	77	38°10'
LA	Alexandria	1,921	27	94	79	31°20'
	Baton Rouge	1,560	29	93	80	30°30'
	Lake Charles	1,459	31	93	79	30°10'
	New Orleans	1,385	33	92	80	30°00'
	Shreveport	2,184	25	96	79	32°30'
ME	Caribou	9,767	-13	81	69	46°50'
	Portland	7,511	-1	84	72	43°40'
MD	Baltimore	4,654	13	91	77	39°10'
	Baltimore[d]	4,111	17	89	78	39°20'
	Frederick	5,087	12	91	77	39°20'

(continued)

TABLE D101—continued
DEGREE DAY AND DESIGN TEMPERATURES[a] FOR CITIES IN THE UNITED STATES

STATE	STATION[b]	HEATING DEGREE DAYS (yearly total)	DESIGN TEMPERATURES			DEGREES NORTH LATITUDE[c]
			Winter	Summer		
			97¹/₂%	Dry bulb 2¹/₂%	Wet bulb 2¹/₂%	
MA	Boston	5,634	9	88	74	42°20′
	Pittsfield	7,578	-3	84	72	42°30′
	Worcester	6,969	4	84	72	42°20′
MI	Alpena	8,506	-6	85	72	45°00′
	Detroit (City)	6,232	6	88	74	42°20′
	Escanaba[d]	8,481	-7	83	71	45°40′
	Flint	7,377	1	87	74	43°00′
	Grand Rapids	6,894	5	88	74	42°50′
	Lansing	6,909	1	87	74	42°50′
	Marquette[d]	8,393	-8	81	70	46°30′
	Muskegon	6,696	6	84	73	43°10′
	Sault Ste. Marie	9,048	-8	81	70	46°30′
MN	Duluth	10,000	-16	82	70	46°50′
	Minneapolis	8,382	-12	89	5	44°50′
	Rochester	8,295	-12	87	75	44°00′
MS	Jackson	2,239	25	95	78	32°20′
	Meridian	2,289	23	95	79	32°20′
	Vicksburg[d]	2,041	26	95	80	32°20′
MO	Columbia	5,046	4	94	77	39°00′
	Kansas City	4,711	6	96	77	39°10′
	St. Joseph	5,484	2	93	79	39°50′
	St. Louis	4,900	6	94	77	38°50′
	St. Louis[d]	4,484	8	94	77	38°40′
	Springfield	4,900	9	93	77	37°10′
MT	Billings	7,049	-10	91	66	45°50′
	Great Falls	7,750	-15	88	62	47°30′
	Helena	8,129	-16	88	62	46°40′
	Missoula	8,125	-6	88	63	46°50′
NE	Grand Island	6,530	-3	94	74	41°00′
	Lincoln[d]	5,864	-2	95	77	40°50′
	Norfolk	6,979	-4	93	77	42°00′
	North Platte	6,684	-4	94	72	41°10′
	Omaha	6,612	-3	91	77	41°20′
	Scottsbluff	6,673	-3	92	68	41°50′
NV	Elko	7,433	-2	92	62	40°50′
	Ely	7,733	-4	87	59	39°10′
	Las Vegas	2,709	28	106	70	36°10′
	Reno	6,332	10	92	62	39°30′
	Winnemucca	6,761	3	94	62	40°50′
NH	Concord	7,383	-3	87	73	43°10′
NJ	Atlantic City	4,812	13	89	77	39°30′
	Newark	4,589	14	91	76	40°40′
	Trenton[d]	4,980	14	88	76	40°10′
NM	Albuquerque	4,348	16	94	65	35°00′
	Raton	6,228	1	89	64	36°50′
	Roswell	3,793	18	98	70	33°20′
	Silver City	3,705	10	94	64	32°40′

(continued)

TABLE D101—continued
DEGREE DAY AND DESIGN TEMPERATURES[a] FOR CITIES IN THE UNITED STATES

STATE	STATION[b]	HEATING DEGREE DAYS (yearly total)	DESIGN TEMPERATURES			DEGREES NORTH LATITUDE[c]
			Winter	Summer		
			97¹/₂%	Dry bulb 2¹/₂%	Wet bulb 2¹/₂%	
NY	Albany	6,875	-1	88	74	42°50′
	Albany[d]	6,201	1	88	74	42°50′
	Binghamton	7,286	1	83	72	42°10′
	Buffalo	7,062	6	85	73	43°00′
	NY (Central Park)[d]	4,871	15	89	75	40°50′
	NY (Kennedy)	5,219	15	87	75	40°40′
	NY(LaGuardia)	4,811	15	89	75	40°50′
	Rochester	6,748	5	88	73	43°10′
	Schenectady[d]	6,650	1	87	74	42°50′
	Syracuse	6,756	2	87	73	43°10′
NC	Charlotte	3,181	22	93	76	35°10′
	Greensboro	3,805	18	91	76	36°10′
	Raleigh	3,393	20	92	77	35°50′
	Winston-Salem	3,595	20	91	75	36°10′
ND	Bismarck	8,851	-19	91	71	46°50′
	Devils Lake[d]	9,901	-21	88	71	48°10′
	Fargo	9,226	-18	89	74	46°50′
	Williston	9,243	-21	88	70	48°10′
OH	Akron-Canton	6,037	6	86	73	41°00′
	Cincinnati[d]	4,410	6	90	75	39°10′
	Cleveland	6,351	5	88	74	41°20′
	Columbus	5,660	5	90	75	40°00′
	Dayton	5,622	4	89	75	39°50′
	Mansfield	6,403	5	87	74	40°50′
	Sandusky[d]	5,796	6	91	74	41°30′
	Toledo	6,494	1	88	75	41°40′
	Youngstown	6,417	4	86	73	41°20′
OK	Oklahoma City	3,725	13	97	77	35°20′
	Tulsa	3,860	13	98	78	36°10′
OR	Eugene	4,726	22	89	67	44°10′
	Medford	5,008	23	94	68	42°20′
	Portland	4,635	23	85	67	45°40′
	Portland[d]	4,109	24	86	67	45°30′
	Salem	4,754	23	88	68	45°00′
PA	Allentown	5,810	9	88	75	40°40′
	Erie	6,451	9	85	74	42°10′
	Harrisburg	5,251	11	91	76	40°10′
	Philadelphia	5,144	14	90	76	39°50′
	Pittsburgh	5,987	5	86	73	40°30′
	Pittsburgh[d]	5,053	7	88	73	40°30′
	Reading[d]	4,945	13	89	75	40°20′
	Scranton	6,254	5	87	73	41°20′
	Williamsport	5,934	7	89	74	41°10′
RI	Providence	5,954	9	86	74	41°40′
SC	Charleston	2,033	27	91	80	32°50′
	Charleston[d]	1,794	28	92	80	32°50′
	Columbia	2,484	24	95	78	34°00′

(continued)

TABLE D101—continued
DEGREE DAY AND DESIGN TEMPERATURES[a] FOR CITIES IN THE UNITED STATES

STATE	STATION[b]	HEATING DEGREE DAYS (yearly total)	DESIGN TEMPERATURES			DEGREES NORTH LATITUDE[c]
			Winter	Summer		
			97$^1/_2$%	Dry bulb 2$^1/_2$%	Wet bulb 2$^1/_2$%	
SD	Huron	8,223	-14	93	75	44°30'
	Rapid City	7,345	-7	92	69	44°00'
	Sioux Falls	7,839	-11	91	75	43°40'
TN	Bristol	4,143	14	89	75	36°30'
	Chattanooga	3,254	18	93	77	35°00'
	Knoxville	3,494	19	92	76	35°50'
	Memphis	3,232	18	95	79	35°00'
	Nashville	3,578	14	94	77	36°10'
TX	Abilene	2,624	20	99	74	32°30'
	Austin	1,711	28	98	77	30°20'
	Dallas	2,363	22	100	78	32°50'
	El Paso	2,700	24	98	68	31°50'
	Houston	1,396	32	94	79	29°40'
	Midland	2,591	21	98	72	32°00'
	San Angelo	2,255	22	99	74	31°20'
	San Antonio	1,546	30	97	76	29°30'
	Waco	2,030	26	99	78	31°40'
	Wichita Falls	2,832	18	101	76	34°00'
UT	Salt Lake City	6,052	8	95	65	40°50'
VT	Burlington	8,269	-7	85	72	44°30'
VA	Lynchburg	4,166	16	90	76	37°20'
	Norfolk	3,421	22	91	78	36°50'
	Richmond	3,865	17	92	78	37°30'
	Roanoke	4,150	16	91	74	37°20'
WA	Olympia	5,236	22	83	66	47°00'
	Seattle-Tacoma	5,145	26	80	64	47°30'
	Seattle[d]	4,424	27	82	67	47°40'
	Spokane	6,655	2	90	64	47°40'
WV	Charleston	4,476	11	90	75	38°20'
	Elkins	5,675	6	84	72	38°50'
	Huntington	4,446	10	91	77	38°20'
	Parkersburg[d]	4,754	11	90	76	39°20'
WI	Green Bay	8,029	-9	85	74	44°30'
	La Crosse	7,589	-9	88	75	43°50'
	Madison	7,863	-7	88	75	43°10'
	Milwaukee	7,635	-4	87	74	43°00'
WY	Casper	7,410	-5	90	61	42°50'
	Cheyenne	7,381	-1	86	62	41°10'
	Lander	7,870	-11	88	63	42°50'
	Sheridan	7,680	-8	91	65	44°50'

a. All data were extracted from the 1985 ASHRAE Handbook, Fundamentals Volume.

b. Design data developed from airport temperature observations unless noted.

c. Latitude is given to the nearest 10 minutes. For example, the latitude for Miami, Florida, is given as 25°50', or 25 degrees 50 minutes.

d. Design data developed from office locations within an urban area, not from airport temperature observations.

APPENDIX E

SIZING OF WATER PIPING SYSTEM

The provisions contained in this appendix are not mandatory unless specifically referenced in the adopting ordinance.

SECTION E101
GENERAL

E101.1 Scope.

E101.1.1 This appendix outlines two procedures for sizing a water piping system (see Sections E103.3 and E201.1). The design procedures are based on the minimum static pressure available from the supply source, the head changes in the system caused by friction and elevation, and the rates of flow necessary for operation of various fixtures.

E101.1.2 Because of the variable conditions encountered in hydraulic design, it is impractical to specify definite and detailed rules for sizing of the water piping system. Accordingly, other sizing or design methods conforming to good engineering practice standards are acceptable alternatives to those presented herein.

SECTION E102
INFORMATION REQUIRED

E102.1 Preliminary. Obtain the necessary information regarding the minimum daily static service pressure in the area where the building is to be located. If the building supply is to be metered, obtain information regarding friction loss relative to the rate of flow for meters in the range of sizes likely to be used. Friction loss data can be obtained from most manufacturers of water meters.

E102.2 Demand load.

E102.2.1 Estimate the supply demand of the building main and the principal branches and risers of the system by totaling the corresponding demand from the applicable part of Table E103.3(3).

E102.2.2 Estimate continuous supply demands in gallons per minute (L/m) for lawn sprinklers, air conditioners, etc., and add the sum to the total demand for fixtures. The result is the estimated supply demand for the building supply.

SECTION E103
SELECTION OF PIPE SIZE

E103.1 General. Decide from Table 604.3 what is the desirable minimum residual pressure that should be maintained at the highest fixture in the supply system. If the highest group of fixtures contains flushometer valves, the pressure for the group should be not less than 15 pounds per square inch (psi) (103.4 kPa) flowing. For flush tank supplies, the available pressure should be not less than 8 psi (55.2 kPa) flowing,

except blowout action fixtures must be not less than 25 psi (172.4 kPa) flowing.

E103.2 Pipe sizing.

E103.2.1 Pipe sizes can be selected according to the following procedure or by other design methods conforming to acceptable engineering practice and *approved* by the administrative authority. The sizes selected must not be less than the minimum required by this code.

E103.2.2 Water pipe sizing procedures are based on a system of pressure requirements and losses, the sum of which must not exceed the minimum pressure available at the supply source. These pressures are as follows:

1. Pressure required at fixture to produce required flow. See Sections 604.3 and 604.5.

2. Static pressure loss or gain (due to head) is computed at 0.433 psi per foot (9.8 kPa/m) of elevation change.

   Example: Assume that the highest fixture supply outlet is 20 feet (6096 mm) above or below the supply source. This produces a static pressure differential of 20 feet by 0.433 psi/foot (2096 mm by 9.8 kPa/m) and an 8.66 psi (59.8 kPa) loss.

3. Loss through water meter. The friction or pressure loss can be obtained from meter manufacturers.

4. Loss through taps in water main.

5. Losses through special devices such as filters, softeners, backflow prevention devices and pressure regulators. These values must be obtained from the manufacturers.

6. Loss through valves and fittings. Losses for these items are calculated by converting to equivalent length of piping and adding to the total pipe length.

7. Loss due to pipe friction can be calculated when the pipe size, the pipe length and the flow through the pipe are known. With these three items, the friction loss can be determined. For piping flow charts not included, use manufacturers' tables and velocity recommendations.

   Note: For the purposes of all examples, the following metric conversions are applicable:

   1 cubic foot per minute = 0.4719 L/s

   1 square foot = 0.0929 m²

   1 degree = 0.0175 rad

   1 pound per square inch = 6.895 kPa

   1 inch = 25.4 mm

1 foot = 304.8 mm

1 gallon per minute = 3.785 L/m

E103.3 Segmented loss method. The size of water service mains, *branch* mains and risers by the segmented loss method, must be determined according to water supply demand [gpm (L/m)], available water pressure [psi (kPa)] and friction loss caused by the water meter and *developed length* of pipe [feet (m)], including equivalent length of fittings. This design procedure is based on the following parameters:

- Calculates the friction loss through each length of the pipe.
- Based on a system of pressure losses, the sum of which must not exceed the minimum pressure available at the street main or other source of supply.
- Pipe sizing based on estimated peak demand, total pressure losses caused by difference in elevation, equipment, *developed length* and pressure required at most remote fixture, loss through taps in water main, losses through fittings, filters, backflow prevention devices, valves and pipe friction.

Because of the variable conditions encountered in hydraulic design, it is impractical to specify definite and detailed rules for sizing of the water piping system. Current sizing methods do not address the differences in the probability of use and flow characteristics of fixtures between types of occupancies. Creating an exact model of predicting the demand for a building is impossible and final studies assessing the impact of water conservation on demand are not yet complete. The following steps are necessary for the segmented loss method.

1. **Preliminary.** Obtain the necessary information regarding the minimum daily static service pressure in the area where the building is to be located. If the building supply is to be metered, obtain information regarding friction loss relative to the rate of flow for meters in the range of sizes to be used. Friction loss data can be obtained from manufacturers of water meters. It is essential that enough pressure be available to overcome all system losses caused by friction and elevation so that plumbing fixtures operate properly. Section 604.6 requires the water distribution system to be designed for the minimum pressure available taking into consideration pressure fluctuations. The lowest pressure must be selected to guarantee a continuous, adequate supply of water. The lowest pressure in the public main usually occurs in the summer because of lawn sprinkling and supplying water for air-conditioning cooling towers. Future demands placed on the public main as a result of large growth or expansion should also be considered. The available pressure will decrease as additional loads are placed on the public system.

2. **Demand load.** Estimate the supply demand of the building main and the principal branches and risers of the system by totaling the corresponding demand from the applicable part of Table E103.3(3). When estimating peak demand sizing methods typically use water supply fixture units (w.s.f.u.) [see Table E103.3(2)]. This numerical factor measures the load-producing effect of a single plumbing fixture of a given kind. The

use of such fixture units can be applied to a single basic probability curve (or table), found in the various sizing methods [Table E103.3(3)]. The fixture units are then converted into gallons per minute (L/m) flow rate for estimating demand.

2.1. Estimate continuous supply demand in gallons per minute (L/m) for lawn sprinklers, air conditioners, etc., and add the sum to the total demand for fixtures. The result is the estimated supply demand for the building supply. Fixture units cannot be applied to constant use fixtures such as hose bibbs, lawn sprinklers and air conditioners. These types of fixtures must be assigned the gallon per minute (L/m) value.

3. **Selection of pipe size.** This water pipe sizing procedure is based on a system of pressure requirements and losses, the sum of which must not exceed the minimum pressure available at the supply source. These pressures are as follows:

3.1. Pressure required at the fixture to produce required flow. See Section 604.3 and Section 604.5.

3.2. Static pressure loss or gain (because of head) is computed at 0.433 psi per foot (9.8 kPa/m) of elevation change.

3.3. Loss through a water meter. The friction or pressure loss can be obtained from the manufacturer.

3.4. Loss through taps in water main [see Table E103.3(4)].

3.5. Losses through special devices such as filters, softeners, backflow prevention devices and pressure regulators. These values must be obtained from the manufacturers.

3.6. Loss through valves and fittings [see Tables E103.3(5) and E103.3(6)]. Losses for these items are calculated by converting to equivalent length of piping and adding to the total pipe length.

3.7. Loss caused by pipe friction can be calculated when the pipe size, the pipe length and the flow through the pipe are known. With these three items, the friction loss can be determined using Figures E103.3(2) through E103.3(7). When using charts, use pipe inside diameters. For piping flow charts not included, use manufacturers' tables and velocity recommendations. Before attempting to size any water supply system, it is necessary to gather preliminary information which includes available pressure, piping material, select design velocity, elevation differences and *developed length* to most remote fixture. The water supply system is divided into sections at major changes in elevation or where branches lead to fixture groups. The peak demand must be determined in each part of the hot and cold water supply system which includes the corresponding water supply fixture unit and conversion to gallons per minute (L/m) flow rate to be expected through each section. Sizing methods require the determination of the "most hydraulically remote" fixture to compute the pressure loss caused by pipe and fittings. The hydraulically remote fixture represents the most downstream fixture along the circuit of piping requiring the most available pressure to operate properly. Consideration must be given to all pressure demands and losses, such as friction caused by pipe, fittings and equipment, elevation and the residual pressure required by Table 604.3. The two most common and frequent complaints about the water supply system operation are lack of adequate pressure and noise.

Problem: What size Type L copper water pipe, service and distribution will be required to serve a two-story factory building having on each floor, back-to-back, two toilet rooms each equipped with hot and cold water? The highest fixture is 21 feet (6401 mm) above the street main, which is tapped with a 2-inch (51 mm) corporation cock at which point the minimum pressure is 55 psi (379.2 kPa). In the building basement, a 2-inch (51 mm) meter with a maximum pressure drop of 11 psi (75.8 kPa) and 3-inch (76 mm) reduced pressure principle backflow preventer with a maximum pressure drop of 9 psi (621 kPa) are to be installed. The system is shown by Figure E103.3(1). To be determined are the pipe sizes for the service main and the cold and hot water distribution pipes.

Solution: A tabular arrangement such as shown in Table E103.3(1) should first be constructed. The steps to be followed are indicated by the tabular arrangement itself as they are in sequence, columns 1 through 10 and lines A through L.

Step 1

Columns 1 and 2: Divide the system into sections breaking at major changes in elevation or where branches lead to fixture groups. After point B [see Figure E103.3(1)], separate consideration will be given to the hot and cold water piping. Enter the sections to be considered in the service and cold water piping in Column 1 of the tabular arrangement. Column 1 of Table E103.3(1) provides a line-by-line recommended tabular arrangement for use in solving pipe sizing.

The objective in designing the water supply system is to ensure an adequate water supply and pressure to all fixtures and equipment. Column 2 provides the pounds per square inch (psi) to be considered separately from the minimum pressure available at the main. Losses to take into consideration are the following: the differences in elevations between the water supply source and the highest water supply outlet, meter pressure losses, the tap in main loss, special fixture devices such as water softeners and backflow prevention devices and the pressure required at the most remote fixture outlet. The difference in elevation can result in an increase or decrease in available pressure at the main. Where the water supply outlet is located above the source, this results in a loss in the available pressure and is subtracted from the pressure at the water source. Where the highest water supply outlet is located below the water supply source, there will be an increase in pressure that is added to the available pressure of the water source.

Column 3: According to Table E103.3(3), determine the gpm (L/m) of flow to be expected in each section of the system. These flows range from 28.6 to 108 gpm. Load values for fixtures must be determined as water supply fixture units and then converted to a gallon-per-minute (gpm) rating to determine peak demand. When calculating peak demands, the water supply fixture units are added and then converted to the gallon-per-minute rating. For continuous flow fixtures such as hose bibbs and lawn sprinkler systems, add the gallon-per-minute demand to the intermittent demand of fixtures. For example, a total of 120 water supply fixture units is converted to a demand of 48 gallons per minute. Two hose bibbs × 5 gpm demand = 10 gpm. Total gpm rating = 48.0 gpm + 10 gpm = 58.0 gpm demand.

Step 2

Line A: Enter the minimum pressure available at the main source of supply in Column 2. This is 55 psi (379.2 kPa). The local water authorities generally keep records of pressures at different times of day and year. The available pressure can also be checked from nearby buildings or from fire department hydrant checks.

Line B: Determine from Table 604.3 the highest pressure required for the fixtures on the system, which is 15 psi (103.4 kPa), to operate a flushometer valve. The most remote fixture outlet is necessary to compute the pressure loss caused by pipe and fittings, and represents the most downstream fixture along the circuit of piping requiring the available pressure to operate properly as indicated by Table 604.3.

Line C: Determine the pressure loss for the meter size given or assumed. The total water flow from the main through the service as determined in Step 1 will serve to aid in the meter selected. There are three common types of water meters; the pressure losses are determined by the American Water Works Association Standards for displacement type, compound type and turbine type. The maximum pressure loss of such devices takes into consideration the meter size, safe operating capacity (gpm) and maximum rates for continuous operations (gpm). Typically, equipment imparts greater pressure losses than piping.

Line D: Select from Table E103.3(4) and enter the pressure loss for the tap size given or assumed. The loss of pressure through taps and tees in pounds per square inch (psi) are based on the total gallon-per-minute flow rate and size of the tap.

Line E: Determine the difference in elevation between the main and source of supply and the highest fixture on the system. Multiply this figure, expressed in feet, by 0.43 psi (2.9 kPa). Enter the resulting psi loss on Line E. The difference in elevation between the water supply source and the highest water supply outlet has a significant impact on the sizing of the water supply system. The difference in elevation usually results in a loss in the available pressure because the water supply outlet is generally located above the water supply source. The loss is caused by the pressure required to lift the water to the outlet. The pressure loss is subtracted from the pressure at the water source. Where the highest water supply outlet is located below the water source, there will be an increase in pressure which is added to the available pressure of the water source.

Lines F, G and H: The pressure losses through filters, backflow prevention devices or other special fixtures must be obtained from the manufacturer or estimated and entered on these lines. Equipment such as backflow prevention devices, check valves, water softeners, instantaneous or tankless water heaters, filters and strainers can impart a much greater pressure loss than the piping. The pressure losses can range from 8 psi to 30 psi.

Step 3

Line I: The sum of the pressure requirements and losses that affect the overall system (Lines B through H) is entered on this line. Summarizing the steps, all of the system losses are subtracted from the minimum water pressure. The remainder is the pressure available for friction, defined as the energy available to push the water through the pipes to each fixture. This force can be used as an average pressure loss, as long as the pressure available for friction is not exceeded. Saving a certain amount for available water supply pressures as an area incurs growth, or because of aging of the pipe or equipment added to the system is recommended.

Step 4

Line J: Subtract Line I from Line A. This gives the pressure that remains available from overcoming friction losses in the system. This figure is a guide to the pipe size that is chosen for each section, incorporating the total friction losses to the most remote outlet (measured length is called developed length).

> **Exception:** When the main is above the highest fixture, the resulting psi must be considered a pressure gain (static head gain) and omitted from the sums of Lines B through H and added to Line J.

The maximum friction head loss that can be tolerated in the system during peak demand is the difference between the static pressure at the highest and most remote outlet at no-flow conditions and the minimum flow pressure required at that outlet. If the losses are within the required limits, then every run of pipe will also be within the required friction head loss. Static pressure loss is the most remote outlet in feet × 0.433 = loss in psi caused by elevation differences.

Step 5

Column 4: Enter the length of each section from the main to the most remote outlet (at Point E). Divide the water supply system into sections breaking at major changes in elevation or where branches lead to fixture groups.

Step 6

E103.3.3. Selection of pipe size, Step 6 Column 5: When selecting a trial pipe size, the length from the water service or meter to the most remote fixture outlet must be measured to determine the developed length. However, in systems having a flushometer valve or temperature controlled shower at the top most floors the developed length would be from the water meter to the most remote flushometer valve on the system. A rule of thumb is that size will become progressively smaller as the system extends farther from the main source of supply. Trial pipe size may be arrived at by the following formula:

Line J: (Pressure available to overcome pipe friction) × 100/equivalent length of run total *developed length* to most remote fixture × percentage factor of 1.5 (note: a percentage factor is used only as an estimate for friction losses imposed for fittings for initial trial pipe size) = psi (average pressure drops per 100 feet of pipe).

For trial pipe size see Figure E 103.3(3) (Type L copper) based on 2.77 psi and a 108 gpm = $2^1/_2$ inches. To determine the equivalent length of run to the most remote outlet, the *developed length* is determined and added to the friction losses for fittings and valves. The developed lengths of the designated pipe sections are as follows:

A - B	54 ft
B - C	8 ft
C - D	13 ft
D - E	150 ft
Total developed length = 225 ft	

The equivalent length of the friction loss in fittings and valves must be added to the *developed length* (most remote outlet). Where the size of fittings and valves is not known, the added friction loss should be approximated. A general rule that has been used is to add 50 percent of the *developed length* to allow for fittings and valves. For example, the equivalent length of run equals the *developed length* of run (225 ft × 1.5 = 338 ft). The total equivalent length of run for determining a trial pipe size is 338 feet.

> **Example:** 9.36 (pressure available to overcome pipe friction) × 100/ 338 (equivalent length of run = 225 × 1.5) = 2.77 psi (average pressure drop per 100 feet of pipe).

Step 7

Column 6: Select from Table E103.3(6) the equivalent lengths for the trial pipe size of fittings and valves on each pipe section. Enter the sum for each section in Column 6. (The number of fittings to be used in this example must be an estimate.) The equivalent length of piping is the developed length plus the equivalent lengths of pipe corresponding to friction head losses for fittings and valves. Where the size of fittings and valves is not known, the added friction head losses must be approximated. An estimate for this example is found in Table E.1.

TABLE E.1

COLD WATER PIPE SECTION	FITTINGS/VALVES	PRESSURE LOSS EXPRESSED AS EQUIVALENT LENGTH OF TUBE (feet)	HOT WATER PIPE SECTION	FITTINGS/ VALVES	PRESSURE LOSS EXPRESSED AS EQUIVALENT OF TUBE (feet)
A-B	3-2$^1/_2$″ Gate valves	3	A-B	3-2$^1/_2$″ Gate valves	3
	1-2$^1/_2$″ Side branch tee	12		1-2$^1/_2$″ Side branch tee	12
B-C	1-2$^1/_2$″ Straight run tee	0.5	B-C	1-2″ Straight run tee	7
				1-2″ 90-degree ell	0.5
C-F	1-2$^1/_2$″ Side branch tee	12	C-F	1-1$^1/_2$″ Side branch tee	7
C-D	1-2$^1/_2$″ 90-degree ell	7	C-D	1-$^1/_2$″ 90-degree ell	4
D-E	1-2$^1/_2$″ Side branch tee	12	D-E	1-1$^1/_2$″ Side branch tee	7

For SI: 1 foot = 304.8 mm, i inch = 25.4 mm.

Step 8

Column 7: Add the figures from Column 4 and Column 6, and enter in Column 7. Express the sum in hundreds of feet.

Step 9

Column 8: Select from Figure E103.3(3) the friction loss per 100 feet (30 480 mm) of pipe for the gallon-per-minute flow in a section (Column 3) and trial pipe size (Column 5). Maximum friction head loss per 100 feet is determined on the basis of total pressure available for friction head loss and the longest equivalent length of run. The selection is based on the gallon-per-minute demand, the uniform friction head loss, and the maximum design velocity. Where the size indicated by hydraulic table indicates a velocity in excess of the selected velocity, a size must be selected which produces the required velocity.

Step 10

Column 9: Multiply the figures in Columns 7 and 8 for each section and enter in Column 9.

Total friction loss is determined by multiplying the friction loss per 100 feet (30 480 mm) for each pipe section in the total *developed length* by the pressure loss in fittings expressed as equivalent length in feet. Note: Section C-F should be considered in the total pipe friction losses only if greater loss occurs in Section C-F than in pipe section D-E. Section C-F is not considered in the total *developed length*. Total friction loss in equivalent length is determined in Table E.2.

Step 11

Line K: Enter the sum of the values in Column 9. The value is the total friction loss in equivalent length for each designated pipe section.

Step 12

Line L: Subtract Line J from Line K and enter in Column 10.

The result should always be a positive or plus figure. If it is not, repeat the operation using Columns 5, 6, 8 and 9 until a balance or near balance is obtained. If the difference between Lines J and K is a high positive number, it is an indication that the pipe sizes are too large and should be reduced, thus saving materials. In such a case, the operations using Columns 5, 6, 8 and 9 should again be repeated.

The total friction losses are determined and subtracted from the pressure available to overcome pipe friction for trial pipe size. This number is critical as it provides a guide to whether the pipe size selected is too large and the process should be repeated to obtain an economically designed system.

Answer: The final figures entered in Column 5 become the design pipe size for the respective sections. Repeating this operation a second time using the same sketch but considering the demand for hot water, it is possible to size the hot water distribution piping. This has been worked up as a part of the overall problem in the tabular arrangement used for sizing the service and water distribution piping. Note that consideration must be given to the pressure losses from the street main to the water heater (Section A-B) in determining the hot water pipe sizes.

TABLE E.2

PIPE SECTIONS	FRICTION LOSS EQUIVALENT LENGTH (feet)	
	Cold Water	Hot Water
A-B	0.69 × 3.2 = 2.21	0.69 × 3.2 = 2.21
B-C	0.085 × 3.1 = 0.26	0.16 × 1.4 = 0.22
C-D	0.20 × 1.9 = 0.38	0.17 × 3.2 = 0.54
D-E	1.62 × 1.9 = 3.08	1.57 × 3.2 = 5.02
Total pipe friction losses (Line K)	5.93	7.99

For SI: 1 foot = 304.8 mm, 1 gpm = 3.785 L/m.

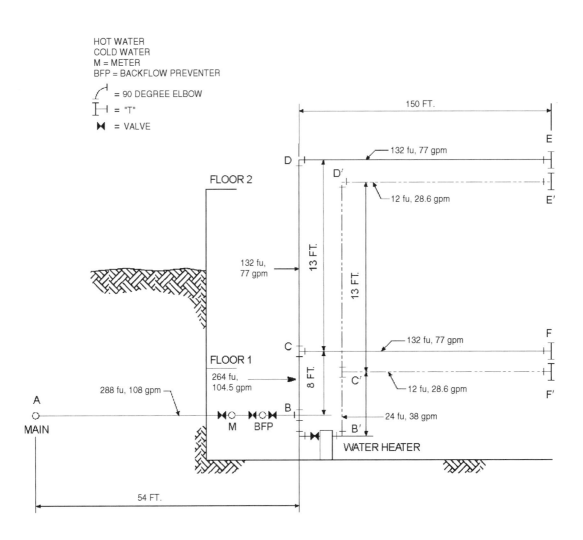

For SI: foot = 304.8 mm, 1 gpm = 3.785 L/m.

FIGURE E103.3(1)
EXAMPLE-SIZING

TABLE E103.3(1)
RECOMMENDED TABULAR ARRANGEMENT FOR USE IN SOLVING PIPE SIZING PROBLEMS

COLUMN		1	2	3	4	5	6	7	8	9	10
Line		Description	Lb per square inch (psi)	Gal. per min through section	Length of section (feet)	Trial pipe size (inches)	Equivalent length of fittings and valves (feet)	Total equivalent length col. 4 and col. 6 (100 feet)	Friction loss per 100 feet of trial size pipe (psi)	Friction loss in equivalent length col. 8 x col. 7 (psi)	Excess pressure over friction losses (psi)
A	Service and cold water distribution piping[a]	Minimum pressure available at main . . 55.00									
B		Highest pressure required at a fixture (Table 604.3) 15.00									
C		Meter loss 2″ meter 11.00									
D		Tap in main loss 2″ tap (Table E103A). . 1.61									
E		Static head loss 21 × 43 psi 9.03									
F		Special fixture loss backflow preventer 9.00									
G		Special fixture loss—Filter 0.00									
H		Special fixture loss—Other 0.00									
I		Total overall losses and requirements (Sum of Lines B through H) 45.64									
J		Pressure available to overcome pipe friction (Line A minus Lines B to H) 9.36									
	DESIGNATION Pipe section (from diagram) Cold water Distribution piping	FU 264									
		AB 288	108.0	54	2¹/₂	15.00	0.69	3.2	2.21	—
		BC 264	104.5	8	2¹/₂	0.5	0.85	3.1	0.26	—
		CD 132	77.0	13	2¹/₂	7.00	0.20	1.9	0.38	—
		CF[b] 132	77.0	150	2¹/₂	12.00	1.62	1.9	3.08	—
		DE[b] 132	77.0	150	2¹/₂	12.00	1.62	1.9	3.08	—
K	Total pipe friction losses (cold)		—	—	—	—	—	—	5.93	—
L	Difference (Line J minus Line K)		—	—	—	—	—	—	—	3.43
	Pipe section (from diagram) Diagram Hot water Distribution Piping	A′B′ 288	108.0	54	2¹/₂	12.00	0.69	3.3	2.21	—
		B′C′ 24	38.0	8	2	7.5	0.16	1.4	0.22	—
		C′D′ 12	28.6	13	1¹/₂	4.0	0.17	3.2	0.54	—
		C′F′[b] 12	28.6	150	1¹/₂	7.00	1.57	3.2	5.02	—
		D′E′[b] 12	28.6	150	1¹/₂	7.00	1.57	3.2	5.02	—
K	Total pipe friction losses (hot)		—	—	—	—	—	—	7.99	—
L	Difference (Line J minus Line K)		—	—	—	—	—	—	—	1.37

For SI: 1 inch = 25.4 mm, 1 foot = 304.8 mm, 1 psi = 6.895 kPa, 1 gpm = 3.785 L/m.

a. To be considered as pressure gain for fixtures below main (to consider separately, omit from "I" and add to "J").

b. To consider separately, in K use C-F only if greater loss than above.

TABLE E103.3(2)
LOAD VALUES ASSIGNED TO FIXTURES[a]

FIXTURE	OCCUPANCY	TYPE OF SUPPLY CONTROL	LOAD VALUES, IN WATER SUPPLY FIXTURE UNITS (wsfu)		
			Cold	Hot	Total
Bathroom group	Private	Flush tank	2.7	1.5	3.6
Bathroom group	Private	Flushometer valve	6.0	3.0	8.0
Bathtub	Private	Faucet	1.0	1.0	1.4
Bathtub	Public	Faucet	3.0	3.0	4.0
Bidet	Private	Faucet	1.5	1.5	2.0
Combination fixture	Private	Faucet	2.25	2.25	3.0
Dishwashing machine	Private	Automatic	—	1.4	1.4
Drinking fountain	Offices, etc.	$^3/_8''$ valve	0.25	—	0.25
Kitchen sink	Private	Faucet	1.0	1.0	1.4
Kitchen sink	Hotel, restaurant	Faucet	3.0	3.0	4.0
Laundry trays (1 to 3)	Private	Faucet	1.0	1.0	1.4
Lavatory	Private	Faucet	0.5	0.5	0.7
Lavatory	Public	Faucet	1.5	1.5	2.0
Service sink	Offices, etc.	Faucet	2.25	2.25	3.0
Shower head	Public	Mixing valve	3.0	3.0	4.0
Shower head	Private	Mixing valve	1.0	1.0	1.4
Urinal	Public	1″ flushometer valve	10.0	—	10.0
Urinal	Public	$^3/_4''$ flushometer valve	5.0	—	5.0
Urinal	Public	Flush tank	3.0	—	3.0
Washing machine (8 lb)	Private	Automatic	1.0	1.0	1.4
Washing machine (8 lb)	Public	Automatic	2.25	2.25	3.0
Washing machine (15 lb)	Public	Automatic	3.0	3.0	4.0
Water closet	Private	Flushometer valve	6.0	—	6.0
Water closet	Private	Flush tank	2.2	—	2.2
Water closet	Public	Flushometer valve	10.0	—	10.0
Water closet	Public	Flush tank	5.0	—	5.0
Water closet	Public or private	Flushometer tank	2.0	—	2.0

For SI: 1 inch = 25.4 mm, 1 pound = 0.454 kg.

a. For fixtures not listed , loads should be assumed by comparing the fixture to one listed using water in similar quantities and at similar rates. The assigned loads for fixtures with both hot and cold water supplies are given for separate hot and cold water loads and for total load. The separate hot and cold water loads being three-fourths of the total load for the fixture in each case.

TABLE E103.3(3)
TABLE FOR ESTIMATING DEMAND

SUPPLY SYSTEMS PREDOMINANTLY FOR FLUSH TANKS			SUPPLY SYSTEMS PREDOMINANTLY FOR FLUSHOMETER VALVES		
Load	Demand		Load	Demand	
(Water supply fixture units)	(Gallons per minute)	(Cubic feet per minute)	(Water supply fixture units)	(Gallons per minute)	(Cubic feet per minute)
1	3.0	0.04104	—	—	—
2	5.0	0.0684	—	—	—
3	6.5	0.86892	—	—	—
4	8.0	1.06944	—	—	—
5	9.4	1.256592	5	15.0	2.0052
6	10.7	1.430376	6	17.4	2.326032
7	11.8	1.577424	7	19.8	2.646364
8	12.8	1.711104	8	22.2	2.967696
9	13.7	1.831416	9	24.6	3.288528
10	14.6	1.951728	10	27.0	3.60936
11	15.4	2.058672	11	27.8	3.716304
12	16.0	2.13888	12	28.6	3.823248
13	16.5	2.20572	13	29.4	3.930192
14	17.0	2.27256	14	30.2	4.037136
15	17.5	2.3394	15	31.0	4.14408
16	18.0	2.90624	16	31.8	4.241024
17	18.4	2.459712	17	32.6	4.357968
18	18.8	2.513184	18	33.4	4.464912
19	19.2	2.566656	19	34.2	4.571856
20	19.6	2.620128	20	35.0	4.6788
25	21.5	2.87412	25	38.0	5.07984
30	23.3	3.114744	30	42.0	5.61356
35	24.9	3.328632	35	44.0	5.88192
40	26.3	3.515784	40	46.0	6.14928
45	27.7	3.702936	45	48.0	6.41664
50	29.1	3.890088	50	50.0	6.684
60	32.0	4.27776	60	54.0	7.21872
70	35.0	4.6788	70	58.0	7.75344
80	38.0	5.07984	80	61.2	8.181216
90	41.0	5.48088	90	64.3	8.595624
100	43.5	5.81508	100	67.5	9.0234
120	48.0	6.41664	120	73.0	9.75864
140	52.5	7.0182	140	77.0	10.29336
160	57.0	7.61976	160	81.0	10.82808
180	61.0	8.15448	180	85.5	11.42964
200	65.0	8.6892	200	90.0	12.0312
225	70.0	9.3576	225	95.5	12.76644
250	75.0	10.026	250	101.0	13.50168

(continued)

TABLE E103.3(3)-continued
TABLE FOR ESTIMATING DEMAND

SUPPLY SYSTEMS PREDOMINANTLY FOR FLUSH TANKS			SUPPLY SYSTEMS PREDOMINANTLY FOR FLUSHOMETER VALVES		
Load	Demand		Load	Demand	
(Water supply fixture units)	(Gallons per minute)	(Cubic feet per minute)	(Water supply fixture units)	(Gallons per minute)	(Cubic feet per minute)
275	80.0	10.6944	275	104.5	13.96956
300	85.0	11.3628	300	108.0	14.43744
400	105.0	14.0364	400	127.0	16.97736
500	124.0	16.57632	500	143.0	19.11624
750	170.0	22.7256	750	177.0	23.66136
1,000	208.0	27.80544	1,000	208.0	27.80544
1,250	239.0	31.94952	1,250	239.0	31.94952
1,500	269.0	35.95992	1,500	269.0	35.95992
1,750	297.0	39.70296	1,750	297.0	39.70296
2,000	325.0	43.446	2,000	325.0	43.446
2,500	380.0	50.7984	2,500	380.0	50.7984
3,000	433.0	57.88344	3,000	433.0	57.88344
4,000	525.0	70.182	4,000	525.0	70.182
5,000	593.0	79.27224	5,000	593.0	79.27224

For SI: 1 inch = 25.4 mm, 1 gallon per minute = 3.785 L/m, 1 cubic foot per minute = 0.28 m³ per minute.

TABLE E103.3(4)
LOSS OF PRESSURE THROUGH TAPS AND TEES IN POUNDS PER SQUARE INCH (psi)

GALLONS PER MINUTE	SIZE OF TAP OR TEE (inches)						
	$5/_8$	$3/_4$	1	$1^1/_4$	$1^1/_2$	2	3
10	1.35	0.64	0.18	0.08	—	—	—
20	5.38	2.54	0.77	0.31	0.14	—	—
30	12.10	5.72	1.62	0.69	0.33	0.10	—
40	—	10.20	3.07	1.23	0.58	0.18	—
50	—	15.90	4.49	1.92	0.91	0.28	—
60	—	—	6.46	2.76	1.31	0.40	—
70	—	—	8.79	3.76	1.78	0.55	0.10
80	—	—	11.50	4.90	2.32	0.72	0.13
90	—	—	14.50	6.21	2.94	0.91	0.16
100	—	—	17.94	7.67	3.63	1.12	0.21
120	—	—	25.80	11.00	5.23	1.61	0.30
140	—	—	35.20	15.00	7.12	2.20	0.41
150	—	—	—	17.20	8.16	2.52	0.47
160	—	—	—	19.60	9.30	2.92	0.54
180	—	—	—	24.80	11.80	3.62	0.68
200	—	—	—	30.70	14.50	4.48	0.84
225	—	—	—	38.80	18.40	5.60	1.06
250	—	—	—	47.90	22.70	7.00	1.31
275	—	—	—	—	27.40	7.70	1.59
300	—	—	—	—	32.60	10.10	1.88

For SI: 1 inch = 25.4 mm, 1 pound per square inch = 6.895 kpa, 1 gallon per minute = 3.785 L/m.

TABLE E103.3(5)
ALLOWANCE IN EQUIVALENT LENGTHS OF PIPE FOR FRICTION LOSS IN VALVES AND THREADED FITTINGS (feet)

FITTING OR VALVE	PIPE SIZE (inches)							
	$^1/_2$	$^3/_4$	1	$1^1/_4$	$1^1/_2$	2	$2^1/_2$	3
45-degree elbow	1.2	1.5	1.8	2.4	3.0	4.0	5.0	6.0
90-degree elbow	2.0	2.5	3.0	4.0	5.0	7.0	8.0	10.0
Tee, run	0.6	0.8	0.9	1.2	1.5	2.0	2.5	3.0
Tee, branch	3.0	4.0	5.0	6.0	7.0	10.0	12.0	15.0
Gate valve	0.4	0.5	0.6	0.8	1.0	1.3	1.6	2.0
Balancing valve	0.8	1.1	1.5	1.9	2.2	3.0	3.7	4.5
Plug-type cock	0.8	1.1	1.5	1.9	2.2	3.0	3.7	4.5
Check valve, swing	5.6	8.4	11.2	14.0	16.8	22.4	28.0	33.6
Globe valve	15.0	20.0	25.0	35.0	45.0	55.0	65.0	80.0
Angle valve	8.0	12.0	15.0	18.0	22.0	28.0	34.0	40.0

For SI: 1 inch = 25.4 mm, 1 foot = 304.8 mm, 1 degree = 0.0175 rad.

TABLE E103.3(6)
PRESSURE LOSS IN FITTINGS AND VALVES EXPRESSED AS EQUIVALENT LENGTH OF TUBE[a] (feet)

NOMINAL OR STANDARD SIZE (inches)	FITTINGS					VALVES			
	Standard Ell		90-Degree Tee						
	90 Degree	45 Degree	Side Branch	Straight Run	Coupling	Ball	Gate	Butterfly	Check
$^3/_8$	0.5	—	1.5	—	—	—	—	—	1.5
$^1/_2$	1	0.5	2	—	—	—	—	—	2
$^5/_8$	1.5	0.5	2	—	—	—	—	—	2.5
$^3/_4$	2	0.5	3	—	—	—	—	—	3
1	2.5	1	4.5	—	—	0.5	—	—	4.5
$1^1/_4$	3	1	5.5	0.5	0.5	0.5	—	—	5.5
$1^1/_2$	4	1.5	7	0.5	0.5	0.5	—	—	6.5
2	5.5	2	9	0.5	0.5	0.5	0.5	7.5	9
$2^1/_2$	7	2.5	12	0.5	0.5	—	1	10	11.5
3	9	3.5	15	1	1	—	1.5	15.5	14.5
$3^1/_2$	9	3.5	14	1	1	—	2	—	12.5
4	12.5	5	21	1	1	—	2	16	18.5
5	16	6	27	1.5	1.5	—	3	11.5	23.5
6	19	7	34	2	2	—	3.5	13.5	26.5
8	29	11	50	3	3	—	5	12.5	39

For SI: 1 inch = 25.4 mm, 1 foot = 304.8 mm, 1 degree = 0.01745 rad.

a. Allowances are for streamlined soldered fittings and recessed threaded fittings. For threaded fittings, double the allowances shown in the table. The equivalent lengths presented above are based on a C factor of 150 in the Hazen-Williams friction loss formula. The lengths shown are rounded to the nearest half-foot.

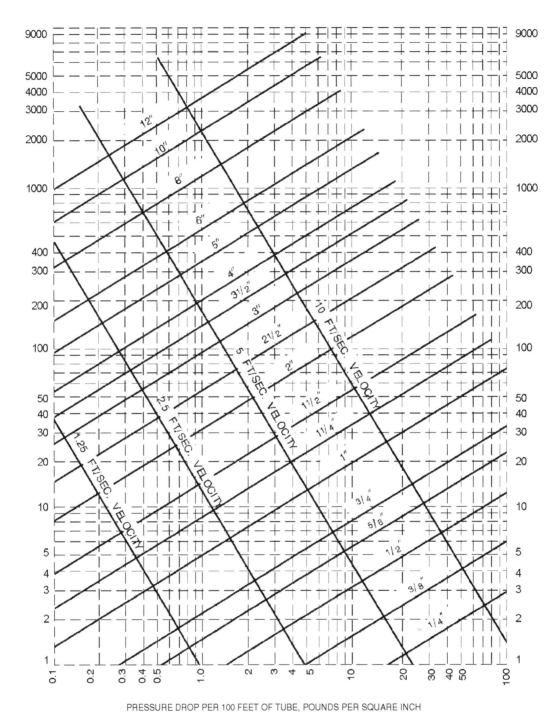

WATER FLOW RATE, GALLONS PER MINUTE

PRESSURE DROP PER 100 FEET OF TUBE, POUNDS PER SQUARE INCH

Note: Fluid velocities in excess of 5 to 8 feet/second are not usually recommended.

For SI: 1 inch = 25.4 mm, 1 foot = 304.8 mm, 1 gpm = 3.785 L/m, 1 psi = 6.895 kPa, 1 foot per second = 0.305 m/s.

a. This chart applies to smooth new copper tubing with recessed (streamline) soldered joints and to the actual sizes of types indicated on the diagram.

FIGURE E103.3(2)
FRICTION LOSS IN SMOOTH PIPE[a] (TYPE K, ASTM B 88 COPPER TUBING)

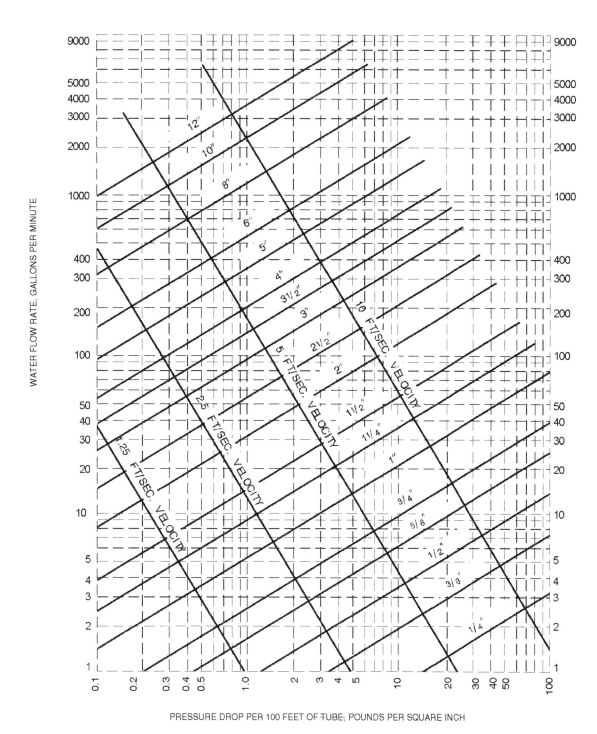

PRESSURE DROP PER 100 FEET OF TUBE, POUNDS PER SQUARE INCH

Note: Fluid velocities in excess of 5 to 8 feet/second are not usually recommended.

For SI: 1 inch = 25.4 mm, 1 foot = 304.8 mm, 1 gpm = 3.785 L/m, 1 psi = 6.895 kPa, 1 foot per second = 0.305 m/s.

a. This chart applies to smooth new copper tubing with recessed (streamline) soldered joints and to the actual sizes of types indicated on the diagram.

FIGURE E103.3(3)
FRICTION LOSS IN SMOOTH PIPE[a] (TYPE L, ASTM B 88 COPPER TUBING)

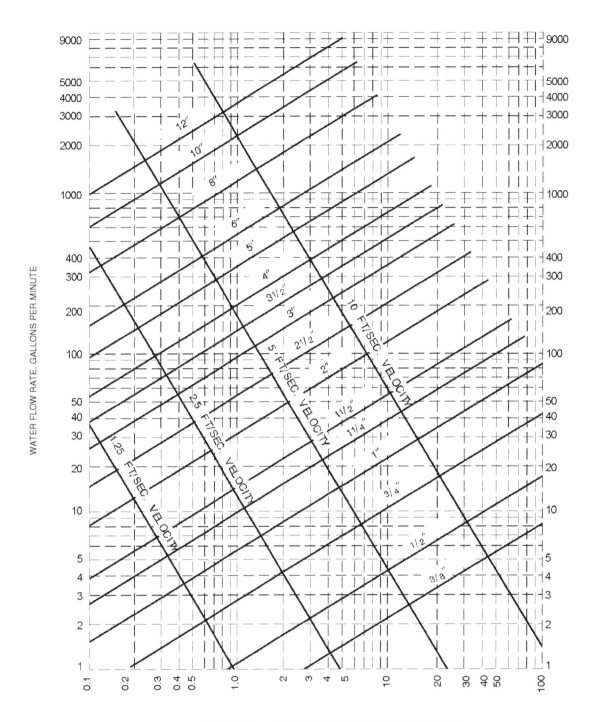

PRESSURE DROP PER 100 FEET OF TUBE, POUNDS PER SQUARE INCH

Note: Fluid velocities in excess of 5 to 8 feet/second are not usually recommended.

For SI: 1 inch = 25.4 mm, 1 foot = 304.8 mm, 1 gpm = 3.785 L/m, 1 psi = 6.895 kPa, 1 foot per second = 0.305 m/s.

a. This chart applies to smooth new copper tubing with recessed (streamline) soldered joints and to the actual sizes of types indicated on the diagram.

FIGURE E103.3(4)
FRICTION LOSS IN SMOOTH PIPE[a] (TYPE M, ASTM B 88 COPPER TUBING)

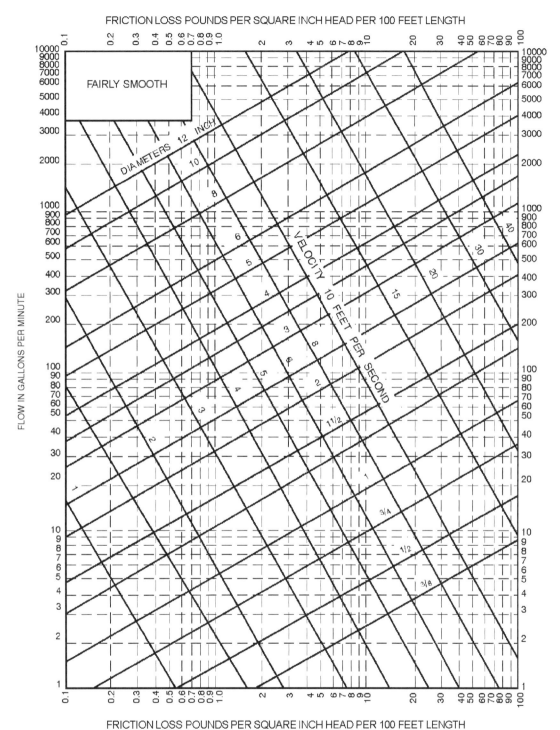

FRICTION LOSS POUNDS PER SQUARE INCH HEAD PER 100 FEET LENGTH

FRICTION LOSS POUNDS PER SQUARE INCH HEAD PER 100 FEET LENGTH

For SI: 1 inch = 25.4 mm, 1 foot = 304.8 mm, 1 gpm = 3.785 L/m, 1 psi = 6.895 kPa, 1 foot per second = 0.305 m/s

a. This chart applies to smooth new steel (fairly smooth) pipe and to actual diameters of standard-weight pipe.

FIGURE E103.3(5)
FRICTION LOSS IN FAIRLY SMOOTH PIPE[a]

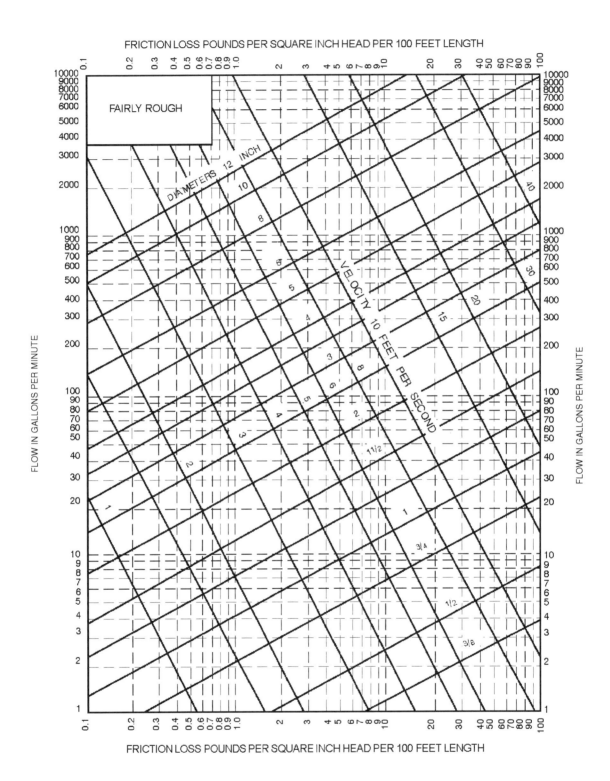

FRICTION LOSS POUNDS PER SQUARE INCH HEAD PER 100 FEET LENGTH

FAIRLY ROUGH

FLOW IN GALLONS PER MINUTE

FRICTION LOSS POUNDS PER SQUARE INCH HEAD PER 100 FEET LENGTH

For SI: 1 inch = 25.4 mm, 1 foot = 304.8 mm, 1 gpm = 3.785 L/m, 1 psi = 6.895 kPa, 1 foot per second = 0.305 m/s.

a. This chart applies to fairly rough pipe and to actual diameters which in general will be less than the actual diameters of the new pipe of the same kind.

FIGURE E103.3(6)
FRICTION LOSS IN FAIRLY ROUGH PIPE[a]

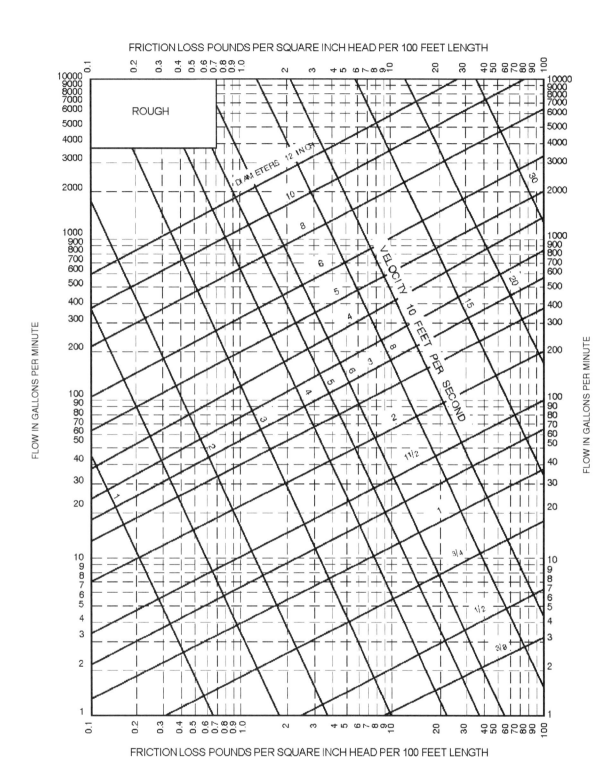

FRICTION LOSS POUNDS PER SQUARE INCH HEAD PER 100 FEET LENGTH

FLOW IN GALLONS PER MINUTE

FRICTION LOSS POUNDS PER SQUARE INCH HEAD PER 100 FEET LENGTH

For SI: 1 inch = 25.4 mm, 1 foot = 304.8 mm, 1 gpm = 3.785 L/m, 1 psi = 6.895 kPa, 1 foot per second = 0.305 m/s.

a. This chart applies to very rough pipe and existing pipe and to their actual diameters.

FIGURE E103.3(7)
FRICTION LOSS IN FAIRLY ROUGH PIPE[a]

SECTION E201
SELECTION OF PIPE SIZE

E201.1 Size of water-service mains, branch mains and risers. The minimum size water service pipe shall be $^3/_4$ inch (19.1 mm). The size of water service mains, *branch* mains and risers shall be determined according to water supply demand [gpm (L/m)], available water pressure [psi (kPa)] and friction loss due to the water meter and *developed length* of pipe [feet (m)], including equivalent length of fittings. The size of each water distribution system shall be determined according to the procedure outlined in this section or by other design methods conforming to acceptable engineering practice and *approved* by the code official:

1. Supply load in the building water-distribution system shall be determined by total load on the pipe being sized, in terms of water-supply fixture units (w.s.f.u.), as shown in Table E103.3(2). For fixtures not listed, choose a w.s.f.u. value of a fixture with similar flow characteristics.

2. Obtain the minimum daily static service pressure [psi (kPa)] available (as determined by the local water authority) at the water meter or other source of supply at the installation location. Adjust this minimum daily static pressure [psi (kPa)] for the following conditions:

    2.1. Determine the difference in elevation between the source of supply and the highest water supply outlet. Where the highest water supply outlet is located above the source of supply, deduct 0.5 psi (3.4 kPa) for each foot (0.3 m) of difference in elevation. Where the highest water supply outlet is located below the source of supply, add 0.5 psi (3.4 kPa) for each foot (0.3 m) of difference in elevation.

    2.2. Where a water pressure reducing valve is installed in the water distribution system, the minimum daily static water pressure available is 80 percent of the minimum daily static water pressure at the source of supply or the set pressure downstream of the pressure reducing valve, whichever is smaller.

    2.3. Deduct all pressure losses due to special equipment such as a backflow preventer, water filter and water softener. Pressure loss data for each piece of equipment shall be obtained through the manufacturer of such devices.

    2.4. Deduct the pressure in excess of 8 psi (55 kPa) due to installation of the special plumbing fixture, such as temperature controlled shower and flushometer tank water closet. Using the resulting minimum available pressure, find the corresponding pressure range in Table E201.1.

3. The maximum *developed length* for water piping is the actual length of pipe between the source of supply and the most remote fixture, including either hot (through the water heater) or cold water branches multiplied by a factor of 1.2 to compensate for pressure loss through fittings. Select the appropriate column in Table E201.1

equal to or greater than the calculated maximum *developed length*.

4. To determine the size of water service pipe, meter and main distribution pipe to the building using the appropriate table, follow down the selected "maximum *developed length*" column to a fixture unit equal to, or greater than the total installation demand calculated by using the "combined" water supply fixture unit column of Table E103.3(2). Read the water service pipe and meter sizes in the first left-hand column and the main distribution pipe to the building in the second left-hand column on the same row.

5. To determine the size of each water distribution pipe, start at the most remote outlet on each *branch* (either hot or cold *branch*) and, working back toward the main distribution pipe to the building, add up the water supply fixture unit demand passing through each segment of the distribution system using the related hot or cold column of Table E103.3(2). Knowing demand, the size of each segment shall be read from the second left-hand column of the same table and maximum *developed length* column selected in Steps 1 and 2, under the same or next smaller size meter row. In no case does the size of any *branch* or main need to be larger that the size of the main distribution pipe to the building established in Step 4.

SECTION E202
DETERMINATION OF PIPE VOLUMES

E202.1 Determining volume of piping systems. Where required for engineering design purposes, Table E202.1 shall be used to determine the approximate internal volume of water distribution piping.

TABLE E201.1
MINIMUM SIZE OF WATER METERS, MAINS AND DISTRIBUTION PIPING
BASED ON WATER SUPPLY FIXTURE UNIT VALUES (w.s.f.u.)

METER AND SERVICE PIPE (inches)	DISTRIBUTION PIPE (inches)	MAXIMUM DEVELOPMENT LENGTH (feet)									
Pressure Range 30 to 39 psi		40	60	80	100	150	200	250	300	400	500
$3/4$	$1/2$ [a]	2.5	2	1.5	1.5	1	1	0.5	0.5	0	0
$3/4$	$3/4$	9.5	7.5	6	5.5	4	3.5	3	2.5	2	1.5
$3/4$	1	32	25	20	16.5	11	9	7.8	6.5	5.5	4.5
1	1	32	32	27	21	13.5	10	8	7	5.5	5
$3/4$	$1^1/_4$	32	32	32	32	30	24	20	17	13	10.5
1	$1^1/_4$	80	80	70	61	45	34	27	22	16	12
$1^1/_2$	$1^1/_4$	80	80	80	75	54	40	31	25	17.5	13
1	$1^1/_2$	87	87	87	87	84	73	64	56	45	36
$1^1/_2$	$1^1/_2$	151	151	151	151	117	92	79	69	54	43
2	$1^1/_2$	151	151	151	151	128	99	83	72	56	45
1	2	87	87	87	87	87	87	87	87	87	86
$1^1/_2$	2	275	275	275	275	258	223	196	174	144	122
2	2	365	365	365	365	318	266	229	201	160	134
2	$2^1/_2$	533	533	533	533	533	495	448	409	353	311

METER AND SERVICE PIPE (inches)	DISTRIBUTION PIPE (inches)	MAXIMUM DEVELOPMENT LENGTH (feet)									
Pressure Range 40 to 49 psi		40	60	80	100	150	200	250	300	400	500
$3/4$	$1/2$ [a]	3	2.5	2	1.5	1.5	1	1	0.5	0.5	0.5
$3/4$	$3/4$	9.5	9.5	8.5	7	5.5	4.5	3.5	3	2.5	2
$3/4$	1	32	32	32	26	18	13.5	10.5	9	7.5	6
1	1	32	32	32	32	21	15	11.5	9.5	7.5	6.5
$3/4$	$1^1/_4$	32	32	32	32	32	32	32	27	21	16.5
1	$1^1/_4$	80	80	80	80	65	52	42	35	26	20
$1^1/_2$	$1^1/_4$	80	80	80	80	75	59	48	39	28	21
1	$1^1/_2$	87	87	87	87	87	87	87	78	65	55
$1^1/_2$	$1^1/_2$	151	151	151	151	151	130	109	93	75	63
2	$1^1/_2$	151	151	151	151	151	139	115	98	77	64
1	2	87	87	87	87	87	87	87	87	87	87
$1^1/_2$	2	275	275	275	275	275	275	264	238	198	169
2	2	365	365	365	365	365	349	304	270	220	185
2	$2^1/_2$	533	533	533	533	533	533	533	528	456	403

(continued)

TABLE E201.1—continued
MINIMUM SIZE OF WATER METERS, MAINS AND DISTRIBUTION PIPING
BASED ON WATER SUPPLY FIXTURE UNIT VALUES (w.s.f.u.)

METER AND SERVICE PIPE (inches)	DISTRIBUTION PIPE (inches)	MAXIMUM DEVELOPMENT LENGTH (feet)									
Pressure Range 50 to 60 psi		40	60	80	100	150	200	250	300	400	500
$3/_4$	$1/_2$ a	3	3	2.5	2	1.5	1	1	1	0.5	0.5
$3/_4$	$3/_4$	9.5	9.5	9.5	8.5	6.5	5	4.5	4	3	2.5
$3/_4$	1	32	32	32	32	25	18.5	14.5	12	9.5	8
1	1	32	32	32	32	30	22	16.5	13	10	8
$3/_4$	$1^1/_4$	32	32	32	32	32	32	32	32	29	24
1	$1^1/_4$	80	80	80	80	80	68	57	48	35	28
$1^1/_2$	$1^1/_4$	80	80	80	80	80	75	63	53	39	29
1	$1^1/_2$	87	87	87	87	87	87	87	87	82	70
$1^1/_2$	$1^1/_2$	151	151	151	151	151	151	139	120	94	79
2	$1^1/_2$	151	151	151	151	151	151	146	126	97	81
1	2	87	87	87	87	87	87	87	87	87	87
$1^1/_2$	2	275	275	275	275	275	275	275	275	247	213
2	2	365	365	365	365	365	365	365	329	272	232
2	$2^1/_2$	533	533	533	533	533	533	533	533	533	486

METER AND SERVICE PIPE (inches)	DISTRIBUTION PIPE (inches)	MAXIMUM DEVELOPMENT LENGTH (feet)									
Pressure Range Over 60		40	60	80	100	150	200	250	300	400	500
$3/_4$	$1/_2$ a	3	3	3	2.5	2	1.5	1.5	1	1	0.5
$3/_4$	$3/_4$	9.5	9.5	9.5	9.5	7.5	6	5	4.5	3.5	3
$3/_4$	1	32	32	32	32	32	24	19.5	15.5	11.5	9.5
1	1	32	32	32	32	32	28	28	17	12	9.5
$3/_4$	$1^1/_4$	32	32	32	32	32	32	32	32	32	30
1	$1^1/_4$	80	80	80	80	80	80	69	60	46	36
$1^1/_2$	$1^1/_4$	80	80	80	80	80	80	76	65	50	38
1	$1^1/_2$	87	87	87	87	87	87	87	87	87	84
$1^1/_2$	$1^1/_2$	151	151	151	151	151	151	151	144	114	94
2	$1^1/_2$	151	151	151	151	151	151	151	151	118	97
1	2	87	87	87	87	87	87	87	87	87	87
$1^1/_2$	2	275	275	275	275	275	275	275	275	275	252
2	2	365	368	368	368	368	368	368	368	318	273
2	$2^1/_2$	533	533	533	533	533	533	533	533	533	533

For SI: 1 inch = 25.4, 1 foot = 304.8 mm.
a. Minimum size for building supply is $3/_4$-inch pipe.

TABLE E202.1
INTERNAL VOLUME OF VARIOUS WATER DISTRIBUTION TUBING

OUNCES OF WATER PER FOOT OF TUBE						
Size Nominal, Inch	Copper Type M	Copper Type L	Copper Type K	CPVC CTS SDR 11	CPVC SCH 40	Composite ASTM F 1281	PEX CTS SDR 9
$^3/_8$	1.06	0.97	0.84	N/A	1.17	0.63	0.64
$^1/_2$	1.69	1.55	1.45	1.25	1.89	1.31	1.18
$^3/_4$	3.43	3.22	2.90	2.67	3.38	3.39	2.35
1	5.81	5.49	5.17	4.43	5.53	5.56	3.91
$1^1/_4$	8.70	8.36	8.09	6.61	9.66	8.49	5.81
$1^1/_2$	12.18	11.83	11.45	9.22	13.20	13.88	8.09
2	21.08	20.58	20.04	15.79	21.88	21.48	13.86

For SI: 1 ounce = 0.030 liter.

APPENDIX F

STRUCTURAL SAFETY

The provisions contained in this appendix are not mandatory unless specifically referenced in the adopting ordinance.

SECTION F101
CUTTING, NOTCHING AND
BORING IN WOOD MEMBERS

[B] F101.1 Joist notching. Notches on the ends of joists shall not exceed one-fourth the joist depth. Holes bored in joists shall not be within 2 inches (51 mm) of the top or bottom of the joist, and the diameter of any such hole shall not exceed one-third the depth of the joist. Notches in the top or bottom of joists shall not exceed one sixth the depth and shall not be located in the middle third of the span.

[B] F101.2 Stud cutting and notching. In exterior walls and bearing partitions, any wood stud is permitted to be cut or notched to a depth not exceeding 25 percent of its width. Cutting or notching of studs to a depth not greater than 40 percent of the width of the stud is permitted in nonbearing partitions supporting no loads other than the weight of the partition.

[B] F101.3 Bored holes. A hole not greater in diameter than 40 percent of the stud width is permitted to be bored in any wood stud. Bored holes not greater than 60 percent of the width of the stud are permitted in nonbearing partitions or in any wall where each bored stud is doubled, provided not more than two such successive doubled studs are so bored. In no case shall the edge of the bored hole be nearer than 0.625 inch (15.9 mm) to the edge of the stud. Bored holes shall not be located at the same section of stud as a cut or notch.

[B] F101.4 Cutting, notching and boring holes in structural steel framing. The cutting, notching and boring of holes in structural steel framing members shall be as prescribed by the registered design professional.

[B] F101.5 Cutting, notching and boring holes in cold-formed steel framing. Flanges and lips of load-bearing cold-formed steel framing members shall not be cut or notched. Holes in webs of load-bearing cold-formed steel framing members shall be permitted along the centerline of the web of the framing member and shall not exceed the dimensional limitations, penetration spacing or minimum hole edge distance as prescribed by the registered design professional. Cutting, notching and boring holes of steel floor/roof decking shall be as prescribed by the registered design professional.

[B] F101.6 Cutting, notching and boring holes in nonstructural cold-formed steel wall framing. Flanges and lips of nonstructural cold-formed steel wall studs shall not be cut or notched. Holes in webs of nonstructural cold-formed steel wall studs shall be permitted along the centerline of the web of the framing member, shall not exceed $1^{1}/_{2}$ inches (38 mm) in width or 4 inches (102 mm) in length, and the holes shall not be spaced less than 24 inches (610 mm) center to center from another hole or less than 10 inches (254 mm) from the bearing end.

INDEX

A

ABS PIPE

H

HANGERS AND SUPPORTS
HEALTH AND SAFETY 101.3
HEALTH CARE PLUMBING 422, 609, 713
HEAT EXCHANGER PROTECTION 608.16.3
HORIZONTAL
HOSPITAL PLUMBING FIXTURES 422
HOT WATER
HOUSE TRAP. 1002.6

I

INDIRECT WASTE 802
INDIVIDUAL VENT 910
INSPECTION 107.1
INSPECTOR (See CODE OFFICIAL)
INSULATION 505, 607.2.1
INTERCEPTORS AND SEPARATORS 1003
IRRIGATION, LAWN. 608.16.5
ISLAND FIXTURE VENT 913

J

JOINTS AND CONNECTIONS. 605, 705

K

KITCHEN 418

L

LABELING 606.7
LAUNDRIES 1003.6
LAUNDRY TRAY. 415

W

2012

INTERNATIONAL
PRIVATE SEWAGE
DISPOSAL
CODE®

A Member of the International Code Family®

2IPSDC®

2012 International Private Sewage Disposal Code®

First Printing: April 2011

ISBN: 978-1-60983-055-7 (soft-cover edition)

COPYRIGHT © 2011
by
INTERNATIONAL CODE COUNCIL, INC.

PRINTED IN THE U.S.A.

PREFACE

Introduction

Internationally, code officials recognize the need for a modern, up-to-date code addressing the safe and sanitary installation of individual sewage disposal systems. The *International Private Sewage Disposal Code*®, in this 2012 edition, is designed to meet these needs through model code regulations that safeguard the public health and safety in all communities, large and small.

This comprehensive sewage disposal code establishes minimum regulations for sewage disposal systems using prescriptive and performance-related provisions. It is founded on broad-based principles that make possible the use of new materials and new sewage disposal designs. This 2012 edition is fully compatible with all of the *International Codes*® (I-Codes®) published by the International Code Council (ICC)®, including the *International Building Code*®, *International Energy Conservation Code*®, *International Existing Building Code*®, *International Fire Code*®, *International Fuel Gas Code*®, *International Green Construction Code*™ (to be available March 2012), *International Mechanical Code*®, ICC *Performance Code*®, *International Plumbing Code*®, *International Property Maintenance Code*®, *International Residential Code*®, *International Swimming Pool and Spa Code*™ (to be available March 2012), *International Wildland-Urban Interface Code*® and *International Zoning Code*®.

The *International Private Sewage Disposal Code* provisions provide many benefits, among which is the model code development process that offers an international forum for plumbing professionals to discuss performance and prescriptive code requirements. This forum provides an excellent arena to debate proposed revisions. This model code also encourages international consistency in the application of provisions.

Development

The first edition of the *International Private Sewage Disposal Code* (1995) was the culmination of an effort initiated in 1994 by a development committee appointed by the ICC and consisting of representatives of the three statutory members of the International Code Council at that time, including: Building Officials and Code Administrators International, Inc. (BOCA), International Conference of Building Officials (ICBO) and Southern Building Code Congress International (SBCCI). The intent was to draft a comprehensive set of regulations for sewage disposal systems consistent with and inclusive of the scope of the existing model codes. Technical content of the latest model codes promulgated by BOCA, ICBO and SBCCI was used as the basis for the development. This 2012 edition presents the code as originally issued, with changes reflected in the 1997 through 2009 editions and further changes approved through the ICC Code Development Process through 2010. A new edition such as this is promulgated every three years.

This code is founded on principles intended to establish provisions consistent with the scope of a sewage disposal code that adequately protects public health, safety and welfare; provisions that do not unnecessarily increase construction costs; provisions that do not restrict the use of new materials, products or methods of construction; and provisions that do not give preferential treatment to particular types or classes of materials, products or methods of construction.

Adoption

The *International Private Sewage Disposal Code* is available for adoption and use by jurisdictions internationally. Its use within a governmental jurisdiction is intended to be accomplished through adoption by reference in accordance with proceedings establishing the jurisdiction's laws. At the time of adoption, jurisdictions should insert the appropriate information in provisions requiring specific local information, such as the name of the adopting jurisdiction. These locations are shown in bracketed words in small capital letters in the code and in the sample ordinance. The sample adoption ordinance on page xi addresses several key elements of a code adoption ordinance, including the information required for insertion into the code text.

Maintenance

The *International Private Sewage Disposal Code* is kept up to date through the review of proposed changes submitted by code enforcing officials, industry representatives, design professionals and other interested parties. Proposed changes are carefully considered through an open code development process in which all interested and affected parties may participate.

The contents of this work are subject to change both through the Code Development Cycles and the governmental body that enacts the code into law. For more information regarding the code development process, contact the Codes and Standards Development Department of the International Code Council.

While the development procedure of the *International Private Sewage Disposal Code* assures the highest degree of care, ICC, its members and those participating in the development of this code do not accept any liability resulting from compliance or noncompliance with the provisions because ICC and its founding members do not have the power or authority to police or enforce compliance with the contents of this code. Only the governmental body that enacts the code into law has such authority.

Code Development Committee Responsibilities
(Letter Designations in Front of Section Numbers)

In each code development cycle, proposed changes to this code are considered at the Code Development Hearing by the International Plumbing Code Development Committee, whose action constitutes a recommendation to the voting membership for final action on the proposed change. Proposed changes to a code section that has a number beginning with a letter in brackets are considered by a different code development committee. For example, proposed changes to code sections that have [B] in front of them (e.g., [B] 309.2) are considered by one of the International Building Code Development Committees (IBC-General) at the Code Development Hearings.

The content of sections in this code that begin with a letter designation is maintained by another code development committee in accordance with the following:

[A] = Administrative Code Development Committee;

[B] = International Building Code Development Committee (IBC-Fire Safety, General, Means of Egress or Structural); and

[P] = International Plumbing Code Development Committee.

Note that, for the development of the 2015 edition of the I-Codes, there will be two groups of code development committees and they will meet in separate years. The groupings are as follows:

Group A Codes (Heard in 2012, Code Change Proposals Deadline: January 3, 2012)	Group B Codes (Heard in 2013, Code Change Proposals Deadline: January 3, 2013)
International Building Code	Administrative Provisions (Chapter 1 all codes except IRC and ICC PC, administrative updates to currently referenced standards, and designated definitions)
International Fuel Gas Code	International Energy Conservation Code
International Mechanical Code	International Existing Building Code
International Plumbing Code	International Fire Code
International Private Sewage Disposal Code	International Green Construction Code
	ICC Performance Code
	International Property Maintenance Code
	International Residential Code
	International Swimming Pool and Spa Code
	International Wildland-Urban Interface Code
	International Zoning Code

Code change proposals submitted for code sections that have a letter designation in front of them will be heard by the respective committee responsible for such code sections. Because different committees will hold its code development hearings in different years, it is possible that some proposals for this code will be heard by a committee in a different year than the year in which the primary committee for this code meets.

For instance, every section of Chapter 1 of this code is designated as the responsibility of the Administrative Code Development Committee, and that committee is part of the Group B portion of the hearings. This committee will hold its code development hearings in 2013 to consider all code change proposals for Chapter 1 of this code and proposals for Chapter 1 of all I-Codes except the *International Residential Code* and ICC *Performance Code*. Therefore, any proposals received for Chapter 1 of this code will be assigned to the Administrative Code Development Committee for consideration in 2013.

It is very important that anyone submitting code change proposals understand which code development committee is responsible for the section of the code that is the subject of the code change proposal. For further information on the code development committee responsibilities, please visit the ICC web site at www.iccsafe.org/scoping.

Marginal Markings

Solid vertical lines in the margins within the body of the code indicate a technical change from the requirements of the 2009 edition. Deletion indicators in the form of an arrow (➡) are provided in the margin where an entire section, paragraph, exception or table has been deleted or an item in a table or list of items has been deleted.

A single asterisk [*] placed in the margin indicates that text or a table has been relocated within the code. A double asterisk [**] placed in the margin indicates that the text or table immediately following it has been relocated there from elsewhere in the code. The following table indicates such relocations in the 2012 *International Private Sewage Disposal Code*.

2012 LOCATION	2009 LOCATION
304.1 through 304.6	105.4 through 105.4.6

Italicized Terms

Selected terms set forth in Chapter 2, Definitions, are italicized where they appear in code text. Such terms are not italicized where the definition set forth in Chapter 2 does not impart the intended meaning in the use of the term. The terms selected have definitions which the user should read carefully to facilitate better understanding of the code.

EFFECTIVE USE OF THE INTERNATIONAL PRIVATE SEWAGE DISPOSAL CODE

The *International Private Sewage Disposal Code* (IPSDC) is a model code that regulates minimum requirements for the installation of new or the alteration of existing private sewage disposal systems. Where a building cannot be served by a public sewer system, the building site must be provided with a system for treating the waste water generated from the use of plumbing fixtures in the building. The IPSDC addresses site evaluations, materials, various soil absorption systems, holding tanks, cesspools and onsite waste water treatment systems. The IPSDC provides a total approach for the onsite, safe disposal of the waste flow discharged to the plumbing fixtures in a building.

The IPSDC is a specification- (prescriptive-) oriented code with very few occurrences of performance-oriented text. The site soil must be evaluated in a prescribed manner to determine its ability to accept the waste flow. The chosen waste treatment method must be designed in a prescribed manner for the soil conditions at the building site, constructed using prescribed materials and installed according to prescribed dimensions. The IPSDC sets forth the minimum acceptable requirements for private sewage disposal systems in order to protect humans and the environment from insanitary conditions that would develop if waste flows were not rendered harmless.

Arrangement and Format of the 2009 IPSDC

The format of the IPSDC allows each chapter to be devoted to a particular subject with the exception of Chapter 3 which contains general subject matters that are not extensive enough to warrant their own independent chapter. The IPSDC is divided into 11 different parts:

Chapters	Subjects
1-2	Administration and Definitions
3	General Regulations
4	Site Evaluation and Requirements
5	Materials
6, 7, 9 &10	Effluent Absorption and Distribution Systems
8	Tanks
11	Waste Water Treatment Systems
12	Inspections
13	Nonliquid Saturated Treatment Systems
14	Referenced Standards
Appendices A & B	Appendices

The following is a chapter-by-chapter synopsis of the scope and intent of the provisions of the *International Private Sewage Disposal Code*:

Chapter 1 Scope and Administration. This chapter contains provisions for the application, enforcement and administration of subsequent requirements of the code. In addition to establishing the scope of the code, Chapter 1 identifies which buildings and structures come under its purview. Chapter 1 is largely concerned with maintaining "due process of law" in enforcing the requirements contained in the body of this code. Only through careful observation of the administrative provisions can the building official reasonably expect to demonstrate that "equal protection under the law" has been provided.

Chapter 2 Definitions. Chapter 2 is the repository of the definitions of terms used in the body of the code. Codes are technical documents and every word, term and punctuation mark can impact the meaning of the code text and the intended results. The code often uses terms that have a unique meaning in the code and the code meaning can differ substantially from the ordinarily understood meaning of the term as used outside of the code.

The terms defined in Chapter 2 are deemed to be of prime importance in establishing the meaning and intent of the code text. The user of the code should be familiar with and consult this chapter because the definitions are essential to the correct interpretation of the code and the user may not be aware that a term is defined.

Where understanding of a term's definition is especially key to or necessary for understanding of a particular code provision, the term is shown in *italics* wherever it appears in the code. This is true only for those terms that have a meaning that is unique to the code. In other words, the generally understood meaning of a term or phrase might not be sufficient or consistent with the meaning prescribed by the code; therefore, it is essential that the code-defined meaning be known.

Guidance regarding tense, gender and plurality of defined terms, as well as guidance regarding terms not defined in this code, is provided.

Chapter 3 General Regulations. The content of Chapter 3 is often referred to as "miscellaneous," rather than general regulations. Chapter 3 received that label because it is the only chapter in the code whose requirements do not interrelate. If a requirement cannot be located in another chapter, it can be found in this chapter. Specific requirements concerning flood hazard areas are in this chapter.

Chapter 4 Site Evaluation and Requirements. A private sewage disposal system has an effluent which cannot be directly discharged into waterways or open ponds. Soil of the right consistency and water content provides a natural filtering and treatment of this discharge. Because soil conditions vary widely, even on the same building site, tests and inspections of the soils must be performed to evaluate the degree to which the soil can accept these liquids. The results of the tests provide necessary information to design an adequate private sewage disposal system. Chapter 4 provides the methods for evaluating the building site.

Chapter 5 Materials. Private sewage disposal systems depend on the strength, quality and chemical resistance of the components that make up the system. To that end, the purpose of Chapter 5 is to specify the minimum material and component standards to ensure that the private sewage disposal system will correctly perform for its intended life.

Chapter 6 Soil Absorption Systems. The design of soil absorption systems depends heavily on the result of the tests and evaluation of the site soil conditions required in Chapter 4. Where soil is less permeable, the area of the soil absorption must be large as compared to that required for soils that are highly permeable. The type of building that is being served by the private sewage disposal system also affects the size of the planned soil absorption area. This chapter provides the methods for computing the required absorption area and details for the proper installation of the soil absorption systems.

Chapter 7 Pressure Distribution Systems. Chapter 6 deals with gravity-type soil absorption systems or systems where the effluent is allowed to drain out of the distribution piping by gravity. This chapter offers an alternative method of discharging the effluent into the ground by pressure means. As such, Chapter 7 provides the necessary details for designing the piping and pumping systems for pressure distribution systems.

Chapter 8 Tanks. Tanks are an integral part of any private sewage disposal system whether they serve as treatment (septic) tanks or merely just holding tanks for leveling the peaks in flow to the system. Where tanks are used for treatment, the dimensions, volume and location of internal features are very important to ensure that the solid wastes are kept within the tank so as to not clog the effluent distribution system. Where tanks are used for holding purposes, they must be sized large enough to accommodate the total of peak flows coming from a building. Chapter 8 provides the necessary requirements for tanks.

Chapter 9 Mound Systems. Mound systems are another method for applying the effluent from a private sewage disposal system to the soil. This type of system may be advantageous in some localities due to the existing soil conditions. Chapter 9 has specific requirements for soil and site evaluations for mound systems.

Chapter 10 Cesspools. Although prohibited from being installed as a permanent private sewage disposal system, cesspools may be necessary where permanent systems are under repair, or are being built. Chapter 10 provides the details for constructing a cesspool.

Chapter 11 Residential Waste Water Systems. Another method of private sewage disposal is a small waste water treatment plant. Where permitted, these systems can discharge effluent directly to streams and rivers. Chapter 11 specifies the standard to which waste water treatment plants must conform.

Chapter 12 Inspections. The best soil and site analysis along with the best design will be rendered useless if the system is not installed according to the plans for the system. Chapter 12 provides requirements for inspection of private sewage disposal systems.

Chapter 13 Nonliquid Saturated Treatment Systems. In some locations, water for the flushing of wastes into and through a sanitary piping system is not available. For example, a toilet facility provided for a remote campground without running water would require such a system. Chapter 13 specifies the standard to which nonliquid saturated treatment systems must conform.

Chapter 14 Referenced Standards. The code contains numerous references to standards that are used to regulate materials and methods of construction. Chapter 14 contains a comprehensive list of all standards that are referenced in the code. The standards are part of the code to the extent of the reference to the standard. Compliance with the referenced standard is necessary for compliance with this code. By providing specifically adopted standards, the construction and installation requirements necessary for compliance with the code can be readily determined. The basis for code compliance is, therefore, established and available on an equal basis to the code official, contractor, designer and owner.

Chapter 14 is organized in a manner that makes it easy to locate specific standards. It lists all of the referenced standards, alphabetically, by acronym of the promulgating agency of the standard. Each agency's standards are then listed in either alphabetical or numeric order based upon the standard identification. The list also contains the title of the standard; the edition (date) of the standard referenced; any addenda included as part of the ICC adoption; and the section or sections of this code that reference the standard.

Appendix A System Layout Illustrations. Because each chapter of this code uses only words to describe requirements, illustrations can offer greater insight as to what the words mean. Appendix A has a number of illustrations referenced to specific sections of the code to help the reader gain a better understanding of the code's requirements.

Appendix B Tables for Pressure Distribution Systems. The design of a pressure distribution system is accomplished by the use of several complex formulas found in Chapter 7. Because a user of the code may not have the necessary experience to manipulate the formulas, a tabular approach for designing pressure distribution systems is provided in Appendix B.

LEGISLATION

The *International Codes* are designed and promulgated to be adopted by reference by legislative action. Jurisdictions wishing to adopt the 2012 *International Private Sewage Disposal Code* as an enforceable regulation governing individual sewage disposal systems should ensure that certain factual information is included in the adopting legislation at the time adoption is being considered by the appropriate governmental body. The following sample adoption legislation addresses several key elements, including the information required for insertion into the code text.

SAMPLE LEGISLATION FOR ADOPTION OF THE *INTERNATIONAL PRIVATE SEWAGE DISPOSAL CODE* ORDINANCE NO._____

A[N] [ORDINANCE/STATUTE/REGULATION] of the [JURISDICTION] adopting the 2012 edition of the *International Private Sewage Disposal Code*, regulating and governing the design, construction, quality of materials, erection, installation, alteration, repair, location, relocation, replacement, addition to, use or maintenance of individual sewage disposal systems in the [JURISDICTION]; providing for the issuance of permits and collection of fees therefor; repealing [ORDINANCE/STATUTE/REGULATION] No. _____ of the [JURISDICTION] and all other ordinances or parts of laws in conflict therewith.

The [GOVERNING BODY] of the [JURISDICTION] does ordain as follows:

Section 1. That a certain document, three (3) copies of which are on file in the office of the [TITLE OF JURISDICTION'S KEEPER OF RECORDS] of [NAME OF JURISDICTION], being marked and designated as the *International Private Sewage Disposal Code*, 2012 edition, including Appendix Chapters [FILL IN THE APPENDIX CHAPTERS BEING ADOPTED] (see *International Private Sewage Disposal Code* Section 101.2.1, 2012 edition), as published by the International Code Council, be and is hereby adopted as the Private Sewage Disposal Code of the [JURISDICTION], in the State of [STATE NAME] regulating and governing the design, construction, quality of materials, erection, installation, alteration, repair, location, relocation, replacement, addition to, use or maintenance of individual sewage disposal systems as herein provided; providing for the issuance of permits and collection of fees therefor; and each and all of the regulations, provisions, penalties, conditions and terms of said Private Sewage Disposal Code on file in the office of the [JURISDICTION] are hereby referred to, adopted, and made a part hereof, as if fully set out in this legislation, with the additions, insertions, deletions and changes, if any, prescribed in Section 2 of this ordinance.

Section 2. The following sections are hereby revised:

Section 101.1. Insert: [NAME OF JURISDICTION]

Section 106.4.2. Insert: [APPROPRIATE SCHEDULE]

Section 106.4.3. Insert: [PERCENTAGES IN TWO LOCATIONS]

Section 108.4. Insert: [OFFENSE, DOLLAR AMOUNT, NUMBER OF DAYS]

Section 108.5. Insert: [DOLLAR AMOUNT IN TWO LOCATIONS]

Section 405.2.5. Insert: [DATE IN THREE LOCATIONS]

Section 405.2.6. Insert: [DATE IN TWO LOCATIONS]

Section 3. That [ORDINANCE/STATUTE/REGULATION] No. _____ of [JURISDICTION] entitled [FILL IN HERE THE COMPLETE TITLE OF THE LEGISLATION OR LAWS IN EFFECT AT THE PRESENT TIME SO THAT THEY WILL BE REPEALED BY DEFINITE MENTION] and all other ordinances or parts of laws in conflict herewith are hereby repealed.

Section 4. That if any section, subsection, sentence, clause or phrase of this legislation is, for any reason, held to be unconstitutional, such decision shall not affect the validity of the remaining portions of this ordinance. The [GOVERNING BODY] hereby declares that it would have passed this law, and each section, subsection, clause or phrase thereof, irrespective of the fact that any one or more sections, subsections, sentences, clauses and phrases be declared unconstitutional.

Section 5. That nothing in this legislation or in the Private Sewage Disposal Code hereby adopted shall be construed to affect any suit or proceeding impending in any court, or any rights acquired, or liability incurred, or any cause or causes of action acquired or existing, under any act or ordinance hereby repealed as cited in Section 3 of this law; nor shall any just or legal right or remedy of any character be lost, impaired or affected by this legislation.

Section 6. That the [JURISDICTION'S KEEPER OF RECORDS] is hereby ordered and directed to cause this legislation to be published. (An additional provision may be required to direct the number of times the legislation is to be published and to specify that it is to be in a newspaper in general circulation. Posting may also be required.)

Section 7. That this law and the rules, regulations, provisions, requirements, orders and matters established and adopted hereby shall take effect and be in full force and effect [TIME PERIOD] from and after the date of its final passage and adoption.

TABLE OF CONTENTS

CHAPTER 1

SCOPE AND ADMINISTRATION

PART 1—SCOPE AND APPLICATION

SECTION 101
GENERAL

[A] 101.1 Title. These regulations shall be known as the Private Sewage Disposal Code of [NAME OF JURISDICTION] hereinafter referred to as "this code."

[A] 101.2 Scope. Septic tank and effluent absorption systems or other treatment tank and effluent disposal systems shall be permitted where a public sewer is not available to the property served. Unless specifically approved, the *private sewage disposal system* of each building shall be entirely separate from and independent of any other building. The use of a common system or a system on a parcel other than the parcel where the structure is located shall be subject to the full requirements of this code as for systems serving public buildings.

> **[A] 101.2.1 Appendices.** Provisions in the appendices shall not apply unless specifically adopted.

[A] 101.3 Public sewer connection. Where public sewers become available to the premises served, the use of the *private sewage disposal system* shall be discontinued within that period of time required by law, but such period shall not exceed 1 year. The building sewer shall be disconnected from the *private sewage disposal system* and connected to the public sewer.

[A] 101.4 Abandoned systems. Abandoned *private sewage disposal systems* shall be plugged or capped in an approved manner. Abandoned treatment tanks and *seepage pits* shall have the contents pumped and discarded in an approved manner. The top or entire tank shall be removed and the remaining portion of the tank or excavation shall be filled immediately.

[A] 101.5 Failing system. When a *private sewage disposal system* fails or malfunctions, the system shall be corrected or use of the system shall be discontinued within that period of time required by the code official, but such period shall not exceed 1 year.

> **[A] 101.5.1 Failure.** A failing *private sewage disposal system* shall be one causing or resulting in any of the following conditions:
>
> 1. The failure to accept sewage discharges and backup of sewage into the structure served by the *private sewage disposal system.*
>
> 2. The discharge of sewage to the surface of the ground or to a drain tile.
>
> 3. The discharge of sewage to any surface or ground waters.

4. The introduction of sewage into saturation zones adversely affecting the operation of a *private sewage disposal system.*

[A] 101.6 Intent. The purpose of this code is to provide minimum standards to safeguard life or limb, health, property and public welfare by regulating and controlling the design, construction, installation, quality of materials, location, operation and maintenance or use of *private sewage disposal systems.*

[A] 101.7 Severability. If any section, subsection, sentence, clause or phrase of this code is for any reason held to be unconstitutional, such decision shall not affect the validity of the remaining portions of this code.

SECTION 102
APPLICABILITY

[A] 102.1 General. Where there is a conflict between a general requirement and a specific requirement, the specific requirement shall govern. Where, in any specific case, different sections of this code specify different materials, methods of construction or other requirements, the most restrictive shall govern.

[A] 102.2 Other laws. The provisions of this code shall not be deemed to nullify any provisions of local, state or federal law.

[A] 102.3 Application of references. Reference to chapter section numbers, or to provisions not specifically identified by number, shall be construed to refer to such chapter, section or provision of this code.

[A] 102.4 Existing installations. *Private sewage disposal systems* lawfully in existence at the time of the adoption of this code shall be permitted to have their use and maintenance continued if the use, maintenance or repair is in accordance with the original design and no hazard to life, health or property is created by the system.

[A] 102.5 Maintenance. *Private sewage disposal systems,* materials and appurtenances, both existing and new, and all parts thereof shall be maintained in proper operating condition in accordance with the original design in a safe and sanitary condition. Devices or safeguards that are required by this code shall be maintained in compliance with the code edition under which they were installed. The owner or the owner's designated agent shall be responsible for maintenance of *private sewage disposal systems.* To determine compliance with this provision, the code official shall have the authority to require reinspection of any *private sewage disposal system.*

[A] 102.6 Additions, alterations or repairs. Additions, alterations, renovations or repairs to any *private sewage disposal system* shall conform to that required for a new system without requiring the existing system to comply with all the requirements of this code. Additions, alterations or repairs shall not cause an existing system to become unsafe, insanitary or overloaded.

Minor additions, alterations, renovations and repairs to existing systems shall meet the provisions for new construction, unless such work is done in the same manner and arrangement as was in the existing system, is not hazardous and is approved.

[A] 102.7 Change in occupancy. It shall be unlawful to make any change in the occupancy of any structure that will subject the structure to any special provision of this code applicable to the new occupancy without approval of the code official. The code official shall certify that such structure meets the intent of the provisions of law governing building construction for the proposed new occupancy and that such change of occupancy does not result in any hazard to the public health, safety or welfare.

[A] 102.8 Historic buildings. The provisions of this code relating to the construction, alteration, repair, enlargement, restoration, relocation or moving of buildings or structures shall not be mandatory for existing buildings or structures identified and classified by the state or local jurisdiction as historic buildings when such buildings or structures are judged by the code official to be safe and in the public interest of health, safety and welfare regarding any proposed construction, alteration, repair, enlargement, restoration, relocation or moving of buildings.

[A] 102.9 Moved buildings. Except as determined by Section 102.4, *private sewage disposal systems* that are a part of buildings or structures moved into or within the jurisdiction shall comply with the provisions of this code for new installations.

[A] 102.10 Referenced codes and standards. The codes and standards referenced in this code shall be those that are listed in Chapter 14 and such codes and standards shall be considered to be part of the requirements of this code to the prescribed extent of each such reference and as further regulated in Sections 102.10.1 and 102.10.2.

> **Exception:** Where enforcement of a code provision would violate the conditions of the listing of the equipment or appliance, the conditions of the listing and the manufacturer's installation instructions shall apply.

> **[A] 102.10.1 Conflicts.** Where conflicts occur between provisions of this code and the referenced standards, the provisions of this code shall apply.

> **[A] 102.10.2 Provisions in referenced codes and standards.** Where the extent of the reference to a referenced code or standard includes subject matter that is within the scope of this code, the provisions of this code, as applicable, shall take precedence over the provisions in the referenced code or standard.

[A] 102.11 Requirements not covered by code. Any requirements necessary for the proper operation of an existing or proposed *private sewage disposal system*, or for the public safety, health and general welfare, not specifically covered by this code, shall be determined by the code official.

PART 2—ADMINISTRATION AND ENFORCEMENT

SECTION 103
DEPARTMENT OF PRIVATE SEWAGE DISPOSAL INSPECTION

[A] 103.1 General. The Department of Private Sewage Disposal Inspection is hereby created and the executive official in charge thereof shall be known as the code official.

[A] 103.2 Appointment. The code official shall be appointed by the chief appointing authority of the jurisdiction.

[A] 103.3 Deputies. In accordance with the prescribed procedures of the jurisdiction and with the concurrence of the appointing authority, the code official shall have the authority to appoint a deputy code official, other related technical officers, inspectors and other employees. Such employees shall have powers as delegated by the code official.

[A] 103.4 Liability. The code official, member of the board of appeals or employee charged with the enforcement of this code, while acting for the jurisdiction in good faith and without malice in the discharge of the duties required by this code or other pertinent law or ordinance, shall not thereby be rendered liable personally, and is hereby relieved from all personal liability for any damage accruing to persons or property as a result of any act or by reason of an act or omission in the discharge of official duties.

Any suit instituted against any officer or employee because of an act performed by that officer or employee in the lawful discharge of duties and under the provisions of this code shall be defended by the legal representative of the jurisdiction until the final termination of the proceedings. The code official or any subordinate shall not be liable for costs in any action, suit or proceeding that is instituted in pursuance of the provisions of this code.

SECTION 104
DUTIES AND POWERS OF THE CODE OFFICIAL

[A] 104.1 General. The code official is hereby authorized and directed to enforce the provisions of this code. The code official shall have the authority to render interpretations of this code and to adopt policies and procedures in order to clarify the application of its provisions. Such interpretations, policies and procedures shall be in compliance with the intent and purpose of this code. Such policies and procedures shall not have the effect of waiving requirements specifically provided for in this code.

[A] 104.2 Applications and permits. The code official shall receive applications, review construction documents and issue permits for the installation and alteration of *private sewage disposal systems*, inspect the premises for which such permits have been issued and enforce compliance with the provisions of this code.

[A] 104.3 Inspections. The code official shall make all of the required inspections, or shall accept reports of inspection by approved agencies or individuals. All reports of such inspections shall be in writing and be certified by a responsible officer of such approved agency or by the responsible individual. The code official is authorized to engage such expert opinion as deemed necessary to report on unusual technical issues that arise, subject to the approval of the appointing authority.

[A] 104.4 Right of entry. Whenever it is necessary to make an inspection to enforce the provisions of this code, or whenever the code official has reasonable cause to believe that there exists in any building or upon any premises any conditions or violations of this code that make the building or premises unsafe, insanitary, dangerous or hazardous, the code official shall have the authority to enter the building or premises at all reasonable times to inspect or to perform the duties imposed on the code official by this code. If such building or premises is occupied, the code official shall present credentials to the occupant and request entry. If such building or premises is unoccupied, the code official shall first make a reasonable effort to locate the owner or other person having charge or control of the building or premises and request entry. If entry is refused, the code official has recourse to every remedy provided by law to secure entry.

When the code official shall have first obtained a proper inspection warrant or other remedy provided by law to secure entry, no owner or occupant or person having charge, care or control of any building or premises shall fail or neglect, after proper request is made as herein provided, to promptly permit entry therein by the code official for the purpose of inspection and examination pursuant to this code.

[A] 104.5 Identification. The code official shall carry proper identification when inspecting structures or premises in the performance of duties under this code.

[A] 104.6 Notices and orders. The code official shall issue all necessary notices or orders to ensure compliance with this code.

[A] 104.7 Department records. The code official shall keep official records of applications received, permits and certificates issued, fees collected, reports of inspections, and notices and orders issued. Such records shall be retained in the official records for the period required for retention of public records.

SECTION 105
APPROVAL

[A] 105.1 Modifications. Whenever there are practical difficulties involved in carrying out the provisions of this code, the code official shall have the authority to grant modifications for individual cases, upon application of the owner or owner's representative provided that the code official shall first find that special individual reason makes the strict letter of this code impractical, the modification is in conformity with the intent and purpose of this code and such modification does not lessen health and fire- and life-safety requirements. The details of action granting modifications shall be recorded and entered in the files of the Private Sewage Disposal Inspection Department.

[A] 105.2 Alternative materials, methods and equipment. The provisions of this code are not intended to prevent the installation of any material or to prohibit any method of construction not specifically prescribed by this code, provided that any such alternative has been approved. An alternative material or method of construction shall be approved where the code official finds that the proposed design is satisfactory and complies with the intent of the provisions of this code, and that the material, method or work offered is, for the purpose intended, at least the equivalent of that prescribed in this code in quality, strength, effectiveness, fire resistance, durability and safety.

[A] 105.2.1 Research reports. Supporting data, where necessary to assist in the approval of materials or assemblies not specifically provided for in this code, shall consist of valid research reports from approved sources.

[A] 105.3 Required testing. Whenever there is insufficient evidence of compliance with the provisions of this code, or evidence that a material or method does not conform to the requirements of this code, or in order to substantiate claims for alternate materials or methods, the code official shall have the authority to require testing as evidence of compliance at no expense to the jurisdiction.

[A] 105.3.1 Test methods. Test methods shall be as specified in this code or by other recognized test standards. In the absence of recognized and accepted test methods, the code official shall approve the testing procedures.

[A] 105.3.2 Testing agency. All tests shall be performed by an approved agency.

[A] 105.3.3 Test reports. Reports of tests shall be retained by the code official for the period required for retention of public records.

[A] 105.4 Used material and equipment. The use of used materials which meet the requirements of this code for new materials is permitted. Materials, equipment and devices shall not be reused unless such elements have been reconditioned, tested and placed in good and proper working condition and approved by the code official.

[A] 105.5 Approved materials and equipment. Materials, equipment and devices approved by the code official shall be constructed and installed in accordance with such approval.

SECTION 106
PERMITS

[A] 106.1 When required. Work on a *private sewage disposal system* shall not commence until a permit for such work has been issued by the code official.

[A] 106.2 Application for permit. Each application for a permit, with the required fee, shall be filed with the code official on a form furnished for that purpose and shall contain a general description of the proposed work and its location. The application shall contain a description of the type of system, the system location, the occupancy of all parts of the structure and all portions of the site or lot not covered by the structure, and such additional information as is required by the code official. The maximum number of bedrooms for residential occupancies shall be indicated.

[A] 106.2.1 Construction documents. An application for a permit shall be accompanied by not less than two copies of construction documents drawn to scale, with sufficient clarity and detail dimensions showing the nature and character of the work to be performed. Specifications shall include pumps and controls, dose volume, elevation differences (vertical lift), pipe friction loss, pump performance curve, pump model and pump manufacturer. The code official is permitted to waive the requirements for filing construction documents where the work involved is of a minor nature. Where the quality of the materials is essential for conformity to this code, specific information shall be given to establish such quality, and this code shall not be cited, or the term "legal" or its equivalent used as a substitute for specific information.

[A] 106.2.2 Preliminary inspection. Before a permit is issued, the code official is authorized to inspect and evaluate the systems, equipment, buildings, devices, premises and spaces or areas to be used.

[A] 106.2.3 Time limitation of application. An application for a permit for any proposed work shall be deemed to have been abandoned 180 days after the date of filing, unless such application has been pursued in good faith or a permit has been issued; except that the code official shall have the authority to grant one or more extensions of time for additional periods not exceeding 180 days each. The extension shall be requested in writing and justifiable cause demonstrated.

[A] 106.2.4 Previous approvals. This code shall not require changes in the construction documents, construction or designated occupancy of a structure for which a lawful permit has been heretofore issued or otherwise lawfully authorized, and the construction of which has been pursued in good faith within 180 days after the effective date of this code and has not been abandoned.

[A] 106.2.5 Soil data. Soil test reports shall be submitted indicating *soil boring* and percolation test data related to the undisturbed and finished grade elevations, vertical elevation reference point and horizontal reference point. Surface elevations shall be given for all *soil borings*. Soil reports shall bear the signature of a soil tester.

[A] 106.2.6 Site plan. A site plan shall be filed showing to scale the location of all septic tanks, holding tanks or other treatment tanks; building sewers; wells; water mains; water service; streams and lakes; *flood hazard areas*; dosing or pumping chambers; distribution boxes; effluent systems; dual disposal systems; replacement system areas; and the location of all buildings or structures. All separating distances and dimensions shall be shown, including any distance to adjoining property. A vertical elevation reference point and a horizontal reference point shall be indicated. For other than single-family dwellings, grade slope with contours shall be shown for the grade elevation of the entire area of the soil absorption system and the area on all sides for a distance of 25 feet (7620 mm).

[A] 106.3 Permit issuance. The application, construction documents and other data filed by an applicant for permit shall be reviewed by the code official. If the code official

finds that the proposed work conforms to the requirements of this code and all laws and ordinances applicable thereto, and that the fees specified in Section 106.4 have been paid, a permit shall be issued to the applicant. A *private sewage disposal system* permit shall not be transferable.

[A] 106.3.1 Approved construction documents. When the code official issues the permit where construction documents are required, the construction documents shall be endorsed in writing and stamped "APPROVED." Such approved construction documents shall not be changed, modified or altered without authorization from the code official. All work shall be done in accordance with the approved construction documents.

The code official shall have the authority to issue a permit for the construction of a part of a *private sewage disposal system* before the construction documents for the whole system have been submitted or approved, provided adequate information and detailed statements have been filed complying with all pertinent requirements of this code. The holder of such permit shall proceed at his or her own risk without assurance that the permit for the entire system will be granted.

[A] 106.3.2 Validity. The issuance of a permit or approval of construction documents shall not be construed to be a permit for, or an approval of, any violation of any of the provisions of this code or of other ordinances of the jurisdiction. No permit presuming to give authority to violate or cancel the provisions of this code shall be valid.

The issuance of a permit based on construction documents and other data shall not prevent the code official from thereafter requiring the correction of errors in said construction documents and other data or from preventing building operations being carried on thereunder when in violation of this code or of other ordinances of the jurisdiction.

[A] 106.3.3 Expiration. Every permit issued by the code official under the provisions of this code shall expire by limitation and become null and void if the work authorized by such permit is not commenced within 180 days from the date of the permit, or if the work authorized by such permit is suspended or abandoned at any time after the work is commenced for a period of 180 days. Before such work can be recommenced, a new permit shall first be obtained and the fee therefor shall be one-half the amount required for a new permit for such work, provided no changes have been or will be made in the original construction documents for such work, and provided further that such suspension or abandonment has not exceeded 1 year.

[A] 106.3.4 Extensions. Any permittee holding an unexpired permit shall have the right to apply for an extension of the time within which the permittee will commence work under that permit when work cannot be commenced within the time required by this section for good and satisfactory reasons. The code official shall extend the time for action by the permittee for a period not exceeding 180 days if there is reasonable cause. No permit shall be extended more than once. The fee for an extension shall be one-half the amount required for a new permit for such work.

[A] 106.3.5 Suspension or revocation of permit. The code official shall have the authority to suspend or revoke a permit issued under the provisions of this code wherever the permit is issued in error or on the basis of incorrect, inaccurate or incomplete information, or in violation of any ordinance or regulation or any of the provisions of this code.

[A] 106.3.6 Retention of construction documents. One set of approved construction documents shall be retained by the code official for a period of not less than 180 days from date of completion of the permitted work, or as required by state or local laws. One set of approved construction documents shall be returned to the applicant, and said set shall be kept on the site of the building or work at all times during which the work authorized thereby is in progress.

[A] 106.3.7 Posting of permit. The permit or a copy shall be kept on the site of the work until the completion of the project.

[A] 106.4 Fees. A permit shall not be issued until the fees prescribed in Section 106.4.2 have been paid, and an amendment to a permit shall not be released until the additional fee, if any, due to an increase of the *private sewage disposal system*, has been paid.

[A] 106.4.1 Work commencing before permit issuance. Any person who commences any work on a *private sewage disposal system* before obtaining the necessary permits shall be subject to 100 percent of the usual permit fee in addition to the required permit fees.

[A] 106.4.2 Fee schedule. The fees for all private sewage disposal work shall be as indicated in the following schedule: [JURISDICTION TO INSERT APPROPRIATE SCHEDULE].

[A] 106.4.3 Fee refunds. The code official shall authorize the refunding of fees as follows:

1. The full amount of any fee paid hereunder that was erroneously paid or collected.

2. Not more than [SPECIFY PERCENTAGE] percent of the permit fee paid when no work has been done under a permit issued in accordance with this code.

3. Not more than [SPECIFY PERCENTAGE] percent of the plan review fee paid when an application for a permit for which a plan review fee has been paid is withdrawn or canceled before any plan review effort has been expended.

The code official shall not authorize the refunding of any fee paid except upon written application filed by the original permittee no later than 180 days after the date of fee payment.

SECTION 107
INSPECTIONS

[A] 107.1 Required inspections. After issuing a permit, the code official shall conduct inspections from time to time during and upon completion of the work for which a permit has been issued. A record of all such examinations and inspections and of all violations of this code shall be maintained by the code official.

[A] 107.1.1 Concealed work. It shall be the duty of the permit applicant to cause the work to remain accessible and exposed for inspection purposes. Neither the code official nor the jurisdiction shall be liable for expense entailed in the removal or replacement of any material required to allow inspection.

[A] 107.1.2 Other inspections. The code official is authorized to make or require other inspections to ascertain compliance with the provisions of this code and other laws that are enforced by the department.

[A] 107.1.3 Approved inspection agencies. The code official shall accept reports of approved inspection agencies provided such agencies satisfy the requirements as to qualifications and reliability.

[A] 107.2 Special inspections. Special inspections of alternative engineered design *private sewage disposal systems* shall be conducted in accordance with Sections 107.2.1 and 107.2.2.

[A] 107.2.1 Periodic inspection. The registered design professional or designated inspector shall periodically inspect and observe the alternative engineered design to determine that the installation is in accordance with the approved plans. All discrepancies shall be brought to the immediate attention of the *private sewage disposal system* contractor for correction. Records shall be kept of all inspections.

[A] 107.2.2 Written report. The registered design professional shall submit a final report in writing to the code official upon completion of the installation, certifying that the alternative engineered design conforms to the approved construction documents. A notice of approval for the *private sewage disposal system* shall not be issued until a written certification has been submitted.

[A] 107.3 Contractor's responsibilities. It shall be the duty of every contractor who enters into contracts for the installation or repair of *private sewage disposal systems* for which a permit is required to comply with adopted state and local rules and regulations concerning licensing.

[A] 107.3.1 Inspection requests. It shall be the duty of the holder of the permit or their duly authorized agent to notify the code official when work is ready for inspection. It shall be the duty of the permit holder to provide access to and means for inspections of such work that are required by this code.

[A] 107.4 Approval required. Work shall not be done beyond the point indicated in each successive inspection

code official, upon notification, shall make the requested inspections and shall either indicate the portion of the construction that is satisfactory as completed, or notify the permit holder or his or her agent wherein the same fails to comply with this code. Any portions that do not comply shall be corrected and such portion shall not be covered or concealed until authorized by the code official.

[A] 107.5 Evaluation and follow-up inspection services. Prior to the approval of a prefabricated construction assembly having concealed work and the issuance of a permit, the code official shall require the submittal of an evaluation report on each prefabricated construction assembly, indicating the complete details of the *private sewage disposal system*, including a description of the system and its components, the basis upon which the system is being evaluated, test results and similar information and other data as necessary for the code official to determine conformance to this code.

[A] 107.5.1 Evaluation service. The code official shall designate the evaluation service of an approved agency as the evaluation agency, and review such agency's evaluation report for adequacy and conformance to this code.

[A] 107.5.2 Follow-up inspection. Except where ready access is provided to *private sewage disposal systems*, service equipment and accessories for complete inspection at the site without disassembly or dismantling, the code official shall conduct the in-plant inspections as frequently as necessary to ensure conformance to the approved evaluation report or shall designate an independent, approved inspection agency to conduct such inspections. The inspection agency shall furnish the code official with the follow-up inspection manual and a report of inspections upon request, and the installation shall have an identifying label permanently affixed to the system indicating that factory inspections have been performed.

[A] 107.5.3 Test and inspection records. Required test and inspection records shall be available to the code official at all times during the fabrication of the installation and the erection of the building; or such records as the code official designates shall be filed.

[A] 107.6 Testing. Installations shall be tested as required in this code and in accordance with Sections 107.6.1 through 107.6.3. Tests shall be made by the permit holder and observed by the code official.

[A] 107.6.1 New, altered, extended or repaired installations. New installations and parts of existing installations, which have been altered, extended, renovated or repaired, shall be tested as prescribed herein to disclose leaks and defects.

[A] 107.6.2 Apparatus, instruments, material and labor for tests. Apparatus, instruments, material and labor required for testing an installation or part thereof shall be furnished by the permit holder.

[A] 107.6.3 Reinspection and testing. Where any work or installation does not pass an initial test or inspection, the necessary corrections shall be made so as to achieve compliance with this code. The work or installation shall then

be resubmitted to the code official for inspection and testing.

[A] 107.7 Approval. After the prescribed inspections indicate that the work complies in all respects with this code, a notice of approval shall be issued by the code official.

[A] 107.7.1 Revocation. The code official is authorized to, in writing, suspend or revoke a notice of approval issued under the provisions of this code wherever the notice is issued in error, on the basis of incorrect information supplied, or where it is determined that the building or structure, premise or portion thereof is in violation of any ordinance or regulation or any of the provisions of this code.

[A] 107.8 Temporary connection. The code official shall have the authority to allow the temporary connection of an installation to the sources of energy for the purpose of testing the installation or for use under a temporary certificate of occupancy.

[A] 107.9 Connection of service utilities. No person shall make connections from a utility, source of energy, fuel or power to any building or system that is regulated by this code for which a permit is required until authorized by the code official.

SECTION 108
VIOLATIONS

[A] 108.1 Unlawful acts. It shall be unlawful for any person, firm or corporation to erect, construct, alter, repair, remove, demolish or use any *private sewage disposal system*, or cause same to be done, in conflict with or in violation of any of the provisions of this code.

[A] 108.2 Notice of violation. The code official shall serve a notice of violation or order to the person responsible for the erection, installation, alteration, extension, repair, removal or demolition of private sewage disposal work in violation of the provisions of this code; in violation of a detailed statement or the approved construction documents thereunder or in violation of a permit or certificate issued under the provisions of this code. Such order shall direct the discontinuance of the illegal action or condition and the abatement of the violation.

[A] 108.3 Prosecution of violation. If the notice of violation is not complied with promptly, the code official shall request the legal counsel of the jurisdiction to institute the appropriate proceeding at law or in equity to restrain, correct or abate such violation, or to require the removal or termination of the unlawful system in violation of the provisions of this code or of the order or direction made pursuant thereto.

[A] 108.4 Violation penalties. Any person who shall violate a provision of this code or fail to comply with any of the requirements thereof or who shall erect, install, alter or repair private sewage disposal work in violation of the approved construction documents or directive of the code official, or of a permit or certificate issued under the provisions of this code, shall be guilty of a [SPECIFY OFFENSE], punishable by a fine of not more than [AMOUNT] dollars or by imprisonment not exceeding [NUMBER OF DAYS], or both such fine and imprisonment. Each day that a violation continues after due notice has been served shall be deemed a separate offense.

[A] 108.5 Stop work orders. Upon notice from the code official, work on any *private sewage disposal system* that is being done contrary to the provisions of this code or in a dangerous or unsafe manner shall immediately cease. Such notice shall be in writing and shall be given to the owner of the property, to the owner's agent or to the person doing the work. The notice shall state the conditions under which work is authorized to resume. Where an emergency exists, the code official shall not be required to give a written notice prior to stopping the work. Any person who shall continue any work on the system after having been served with a stop work order, except such work as that person is directed to perform to remove a violation or unsafe condition, shall be liable to a fine of not less than [AMOUNT] dollars or more than [AMOUNT] dollars.

[A] 108.6 Abatement of violation. The imposition of the penalties herein prescribed shall not preclude the legal officer of the jurisdiction from instituting appropriate action to prevent unlawful construction or to restrain, correct or abate a violation; to prevent illegal occupancy of a building, structure or premises or to stop an illegal act, conduct, business or use of the *private sewage disposal system* on or about any premises.

[A] 108.7 Unsafe systems. Any *private sewage disposal system* regulated by this code that is unsafe or constitutes a health hazard, insanitary condition or is otherwise dangerous to human life is hereby declared unsafe. Any use of *private sewage disposal systems* regulated by this code constituting a hazard to safety, health or public welfare by reason of inadequate maintenance, dilapidation, obsolescence, disaster, damage or abandonment is hereby declared an unsafe use. Any such unsafe equipment is hereby declared to be a public *nuisance* and shall be abated by repair, rehabilitation, demolition or removal.

[A] 108.7.1 Authority to condemn equipment. Whenever the code official determines that any *private sewage disposal system*, or portion thereof, regulated by this code has become hazardous to life, health or property or has become insanitary, the code official shall order in writing that such system be either removed or restored to a safe or sanitary condition. A time limit for compliance with such order shall be specified in the written notice. No person shall use or maintain a defective *private sewage disposal system* after receiving such notice. When such system is to be disconnected, written notice as prescribed in Section 108.2 shall be given. In cases of immediate danger to life or property, such disconnection shall be made immediately without such notice.

[A] 108.7.2 Authority to disconnect service utilities. The code official shall have the authority to authorize disconnection of utility service to the building, structure or system regulated by the technical codes in case of emergency, where necessary, to eliminate an immediate danger to life or property. Where possible, the owner and occupant of the building, structure or service system shall be notified of the decision to disconnect utility service prior to taking such action. If not notified prior to disconnecting, the owner or occupant of the building, structure or service systems shall be notified in writing as soon as is practical thereafter.

SECTION 109
MEANS OF APPEAL

[A] 109.1 Application for appeal. Any person shall have the right to appeal a decision of the code official to the board of appeals. An application for appeal shall be based on a claim that the true intent of this code or the rules legally adopted thereunder has been incorrectly interpreted, the provisions of this code do not fully apply or an equally good or better form of construction is proposed. The application shall be filed on a form obtained from the code official within 20 days after the notice was served.

[A] 109.2 Membership of board. The board of appeals shall consist of five members appointed by the chief appointing authority as follows: one for 5 years, one for 4 years, one for 3 years, one for 2 years and one for 1 year. Thereafter, each new member shall serve for 5 years or until a successor has been appointed.

[A] 109.2.1 Qualifications. The board of appeals shall consist of five individuals, one from each of the following professions or disciplines.

1. Registered design professional that is a registered architect; or a builder or superintendent of building construction with at least 10 years' experience, 5 years of which shall have been in responsible charge of work.

2. Registered design professional with structural engineering or architectural experience.

3. Registered design professional with mechanical and plumbing engineering experience; or a mechanical and plumbing contractor with at least 10 years' experience, 5 years of which shall have been in responsible charge of work.

4. Registered design professional with electrical engineering experience; or an electrical contractor with at least 10 years' experience, 5 years of which shall have been in responsible charge of work.

5. Registered design professional with fire protection engineering experience; or a fire-protection contractor with at least 10 years' experience, 5 years of which shall have been in responsible charge of work.

[A] 109.2.2 Alternate members. The chief appointing authority shall appoint two alternate members who shall be called by the board chairman to hear appeals during the absence or disqualification of a member. Alternate members shall possess the qualifications required for board membership, and shall be appointed for 5 years or until a successor has been appointed.

[A] 109.2.3 Chairman. The board shall annually select one of its members to serve as chairman.

[A] 109.2.4 Disqualification of a member. A member shall not hear an appeal in which that member has any personal, professional or financial interest.

[A] 109.2.5 Secretary. The chief administrative officer shall designate a qualified clerk to serve as secretary to the board. The secretary shall file a detailed record of all proceedings in the office of the chief administrative officer.

[A] 109.2.6 Compensation of members. Compensation of members shall be determined by law.

[A] 109.3 Notice of meeting. The board shall meet upon notice from the chairman, within 10 days of the filing of an appeal or at stated periodic meetings.

[A] 109.4 Open hearing. Hearings before the board shall be open to the public. The appellant, the appellant's representative, the code official and any person whose interests are affected shall be given an opportunity to be heard.

[A] 109.4.1 Procedure. The board shall adopt and make available to the public through the secretary procedures under which a hearing will be conducted. The procedures shall not require compliance with strict rules of evidence, but shall mandate that only relevant information be received.

[A] 109.5 Postponed hearing. When five members are not present to hear an appeal, either the appellant or the appellant's representative shall have the right to request a postponement of the hearing.

[A] 109.6 Board decision. The board shall modify or reverse the decision of the code official by a concurring vote of three members.

[A] 109.6.1 Resolution. The decision of the board shall be by resolution. Certified copies shall be furnished to the appellant and to the code official.

[A] 109.6.2 Administration. The code official shall take immediate action in accordance with the decision of the board.

[A] 109.7 Court review. Any person, whether or not a previous party of the appeal, shall have the right to apply to the appropriate court for a writ of certiorari to correct errors of law. Application for review shall be made in the manner and time required by law following the filing of the decision in the office of the chief administrative officer.

SECTION 110
TEMPORARY EQUIPMENT, SYSTEMS AND USES

[A] 110.1 General. The code official is authorized to issue a permit for temporary equipment, systems and uses. Such permits shall be limited as to time of service, but shall not be permitted for more than 180 days. The code official is authorized to grant extensions for demonstrated cause.

[A] 110.2 Conformance. Temporary equipment, systems and uses shall conform to the structural strength, fire safety, means of egress, accessibility, light, ventilation and sanitary requirements of this code as necessary to ensure the public health, safety and general welfare.

[A] 110.3 Temporary utilities. The code official is authorized to give permission to temporarily supply utilities before an installation has been fully completed and the final certificate of completion has been issued. The part covered by the temporary certificate shall comply with the requirements specified for temporary lighting, heat or power in the code.

[A] 110.4 Termination of approval. The code official is authorized to terminate such permit for a temporary structure or use and to order the temporary structure or use to be discontinued.

CHAPTER 2

DEFINITIONS

SECTION 201
GENERAL

201.1 Scope. Unless otherwise expressly stated, the following words and terms shall, for the purposes of this code, have the meanings indicated in this chapter.

201.2 Interchangeability. Words used in the present tense include the future; words in the masculine gender include the feminine and neuter; the singular number includes the plural and the plural, the singular.

201.3 Terms defined in other codes. Where terms are not defined in this code and are defined in the *International Building Code* or the *International Plumbing Code*, such terms shall have meanings ascribed to them as in those codes.

201.4 Terms not defined. Where terms are not defined through the methods authorized by this section, such terms shall have ordinarily accepted meanings such as the context implies.

SECTION 202
GENERAL DEFINITIONS

AGGREGATE. Graded hard rock that has been washed with water under pressure over a screen during or after grading to remove fine material and with a hardness value of 3 or greater on Mohs' Scale of Hardness. Aggregate that will scratch a copper penny without leaving any residual rock material on the coin has a hardness value of 3 or greater on Mohs' Scale of Hardness.

[P] AIR BREAK (Drainage System). A piping arrangement in which a drain from a fixture, appliance or device discharges indirectly into another fixture, receptacle or interceptor at a point below the flood level rim and above the trap seal.

ALLUVIUM. Soil deposited by floodwaters.

BEDROCK. The rock that underlies soil material or is located at the earth's surface. Bedrock is encountered when the weathered in-place consolidated material, larger than 0.08 inch (2 mm) in size, is more than 50 percent by volume.

CESSPOOL. A covered excavation in the ground receiving sewage or other organic wastes from a drainage system that is designed to retain the organic matter and solids, permitting the liquids to seep into the soil cavities.

CLEAR-WATER WASTES. Cooling water and condensate drainage from refrigeration compressors and air-conditioning equipment, water used for equipment chilling purposes, liquid having no impurities or where impurities have been reduced below a minimum concentration considered harmful, and cooled condensate from steam-heating systems or other equipment.

[A] CODE OFFICIAL. The officer or other designated authority charged with administration and enforcement of this code or a duly authorized representative.

COLLUVIUM. Soil transported under the influence of gravity.

COLOR. The moist color of the soil based on Munsell soil color charts.

[A] CONSTRUCTION DOCUMENTS. All the written, graphic and pictorial documents prepared or assembled for describing the design, location and physical characteristics of the elements of the project necessary for obtaining a building permit. The construction drawings shall be drawn to an appropriate scale.

CONVENTIONAL SOIL ABSORPTION SYSTEM. A system employing gravity flow from the septic or other treatment tank and applying effluent to the soil through the use of a *seepage trench*, bed or pit.

[B] DESIGN FLOOD ELEVATION. The elevation of the "design flood," including wave height, relative to the datum specified on the community's legally designated flood hazard map.

DETAILED SOIL MAP. A map prepared by or for a state or federal agency participating in the National Cooperative Soil Survey showing soil series, type and phases at a scale of not more than 2,000 feet to the inch (24 m/mm) and which includes related explanatory information.

DOSING SOIL ABSORPTION SYSTEM. A system employing a pump or automatic siphon to elevate or distribute effluent to the soil through the use of a *seepage trench* or bed.

EFFLUENT. Liquid discharged from a septic or other treatment tank.

[B] FLOOD HAZARD AREA. The greater of the following two areas:

1. The area within a flood plain subject to a 1-percent or greater chance of flooding in any given year.

2. The area designated as a flood hazard area on a community's flood hazard map or as otherwise legally designated.

HIGH GROUND WATER. Soil saturation zones, including perched water tables, shallow regional ground water tables or aquifers, or zones seasonally, periodically or permanently saturated.

HOLDING TANK. An approved water tight receptacle for collecting and holding sewage.

HORIZONTAL REFERENCE POINT. A stationary, easily identifiable point to which horizontal dimensions are related.

LEGAL DESCRIPTION. An accurate metes and bounds description, a lot and block number in a recorded subdivision, a recorded assessor's plat or a public land survey description to the nearest 40 acres (16 ha).

MANHOLE. An opening of sufficient size to permit a person to gain access to a sewer or any portion of a private sewage disposal system.

MOBILE UNIT. A structure of vehicular, portable design, built on a chassis and designed to be moved from one site to another and to be used with or without a permanent foundation.

MOBILE UNIT PARK. Any plot or plots of ground owned by a person, state or local government upon which two or more units, occupied for dwelling or sleeping purposes regardless of mobile unit ownership, are located and whether or not a charge is made for such accommodation.

[P] NUISANCE. Public *nuisance* as known in common law or equity jurisprudence; whatever is dangerous to human life or detrimental to health; whatever building, structure or premises is not sufficiently ventilated, sewered, drained, cleaned or lighted, in reference to its intended use; and whatever renders the air, human food, drink or water supply unwholesome.

PAN. A soil horizon cemented with any one of a number of cementing agents such as iron, organic matter, silica, calcium, carbonate, gypsum or a combination of chemicals. Pans will resist penetration from a knife blade and are slowly permeable horizons or are impermeable.

PERCOLATION TEST. The method of testing absorption qualities of the soil (see Section 404).

PERMEABILITY. The ease with which liquids move through the soil. One of the soil qualities listed in soil survey reports.

PRESSURE DISTRIBUTION SYSTEM. A soil absorption system using a pump or automatic siphon and smaller diameter distribution piping with small-diameter perforations to introduce effluent into the soil.

PRIVATE SEWAGE DISPOSAL SYSTEM. A sewage treatment and disposal system serving a single structure with a septic tank and soil absorption field located on the same parcel as the structure. This term also means an alternative sewage disposal system, including a substitute for the septic tank or soil absorption field, a holding tank, a system serving more than one structure or a system located on a different parcel than the structure. A private sewage disposal system is permitted to be owned by the property owner or a special-purpose district.

PRIVY. A structure not connected to a plumbing system and which is used by persons for the deposition of human body waste.

[A] REGISTERED DESIGN PROFESSIONAL. An individual who is registered or licensed to practice their respective design profession, as defined by the statutory requirements of the professional registration laws of the state or jurisdiction in which the project is to be constructed.

SEEPAGE BED. An excavated area more than 5 feet (1524 mm) wide that contains a bedding of aggregate and has more than one distribution line.

SEEPAGE PIT. An underground receptacle constructed to permit disposal of effluent or clear wastes by soil absorption through its floor and walls.

SEEPAGE TRENCH. An area excavated 1 foot to 5 feet (305 mm to 1524 mm) wide containing a bedding of aggregate and a single distribution line.

SEPTAGE. All sludge, scum, liquid and any other material removed from a private sewage treatment and disposal system.

SEPTIC TANK. A tank that receives and partially treats sewage through processes of sedimentation, flotation and bacterial action to separate solids from the liquid in the sewage, and which discharges the liquid to a soil absorption system.

SOIL. The unconsolidated material over bedrock, 0.08 inch (2 mm) and smaller.

SOIL BORING. An observation pit dug by hand or backhoe, a hole dug by augering or a soil core taken intact and undisturbed with a probe.

SOIL MOTTLES. Spots, streaks or contrasting soil colors usually caused by soil saturation for one period of a normal year, with a color value of 4 or more and a chroma of 2 or less. Gray-colored mottles are called low chroma; reddish-brown, red- and yellow-colored mottles are called high chroma.

SOIL SATURATION. The state in which all pores in a soil are filled with water. Water will flow from saturated soil into a bore hole.

VENT CAP. An approved appurtenance used for covering the vent terminal of an effluent disposal system to avoid closure by mischief or debris and still permit circulation of air within the system.

VERTICAL ELEVATION REFERENCE POINT. An easily identifiable stationary point or object of constant elevation for establishing the relative elevation of percolation tests, soil borings and other locations.

WATERCOURSE. A stream usually flowing in a particular direction, though it need not flow continually and is sometimes dry. A *watercourse* flows in a definite channel, with a bed, sides or banks, and usually discharges itself into some other stream or body of water. It must be something more than mere surface drainage over the entire face of a tract of land, occasioned by unusual freshets or other extraordinary cause. It does not include the water flowing in the hollows or ravines in land, which is the mere surface water from rains or melting snows, and is discharged through them from a higher to a lower level, but which at other times are destitute of water. Such hollows or ravines are not, in legal contemplation, *watercourses*.

WORKMANSHIP. Work of such character that will fully secure the results sought in all the sections of this code as intended for the health, safety and welfare protection of all individuals.

CHAPTER 3

GENERAL REGULATIONS

SECTION 301
GENERAL

301.1 Scope. The provisions of this chapter shall govern the general regulations of *private sewage disposal systems*, including specific limitations and *flood hazard areas*.

SECTION 302
SPECIFIC LIMITATIONS

302.1 Domestic waste. All wastes and sewage derived from ordinary living uses shall enter the septic or treatment tank unless otherwise specifically exempted by the code official or this code.

302.2 Cesspools and privies. Privies shall be prohibited. Cesspools shall be prohibited, except where approved by the code official. Where approved, cesspools shall be designed and installed in accordance with Chapter 10.

302.3 Industrial wastes. The code official shall approve the method of treatment and disposal of all waste products from manufacturing or industrial operations, including combined industrial and domestic waste.

302.4 Detrimental or dangerous waste. Material such as ashes, cinders or rags; flammable, poisonous or explosive liquids or gases; oil, grease or other insoluble material that is capable of obstructing, damaging or overloading the private sewage disposal system, or is capable of interfering with the normal operation of the private sewage disposal system, shall not be deposited, by any means, into such systems. The code official shall approve the method of treatment and disposal.

302.5 Clear water. The discharge of surface, rain or other clear water into a private sewage disposal system shall be prohibited.

302.6 Water softener and iron filter backwash. Water softener or iron filter discharge shall be indirectly connected by means of an air gap to the private sewage disposal system or discharge onto the ground surface, provided that a nuisance is not created.

302.7 Food waste disposals. Where a food waste disposal connects to a private sewage disposal system, the system shall be designed to accommodate the solids loading from the disposal unit.

SECTION 303
FLOOD HAZARD AREAS

[B] 303.1 General. Soil absorption systems shall be located outside of flood hazard areas.

> **Exception:** Where suitable soil absorption sites outside of the flood hazard area are not available, the soil absorption site is permitted to be located within the flood hazard area.

The soil absorption site shall be located to minimize the effects of inundation under conditions of the design flood.

[B] 303.2 Tanks. In flood hazard areas, tanks shall be anchored to counter buoyant forces during condition of the design flood. The vent termination and service manhole of the tank shall be not less than 2 feet (610 mm) above the design flood elevation or fitted with covers designed to prevent the inflow of floodwater or outflow of the contents of the tanks during conditions of the design flood.

[B] 303.3 Mound systems. Mound systems shall be prohibited in flood hazard areas.

SECTION 304
ALTERNATIVE ENGINEERED DESIGN

304.1 Alternative engineered design. The design, documentation, inspection, testing and approval of an alternative engineered design *private sewage disposal system* shall comply with Sections 304.1.1 through 304.6.

> **304.1.1 Design criteria.** An alternative engineered design shall conform to the intent of the provisions of this code and shall provide an equivalent level of quality, strength, effectiveness, fire resistance, durability and safety. Material, equipment or components shall be designed and installed in accordance with the manufacturer's instructions.

304.2 Submittal. The registered design professional shall indicate on the permit application that the *private sewage disposal system* is an alternative engineered design. The permit and permanent permit records shall indicate that an alternative engineered design was part of the approved installation.

304.3 Technical data. The registered design professional shall submit sufficient technical data to substantiate the proposed alternative engineered design and to prove that the performance meets the intent of this code.

304.4 Construction documents. The registered design professional shall submit to the code official two complete sets of signed and sealed construction documents for the alternative engineered design.

304.5 Design approval. Where the code official determines that the alternative engineered design conforms to the intent of this code, the *private sewage disposal system* shall be approved. If the alternative engineered design is not approved, the code official shall notify the registered design professional in writing, stating the reasons therefor.

304.6 Inspection and test. The alternative engineered design shall be inspected in accordance with the requirements of Section 107.

CHAPTER 4

SITE EVALUATION AND REQUIREMENTS

SECTION 401
GENERAL

401.1 Scope. The provisions of this chapter shall govern the evaluation of and requirements for private sewage disposal system sites.

401.2 Site evaluation. Site evaluation shall include soil conditions, properties and permeability, depth to zones of soil saturation, depth to bedrock, slope, landscape position, all setback requirements and the presence of flood hazard areas. Soil test data shall relate to the undisturbed elevations, and a vertical elevation reference point or benchmark shall be established. Evaluation data shall be reported on approved forms. Reports shall be filed within 30 days of the completion of testing for all sites investigated.

401.3 Replacement system area. On each parcel of land being initially developed, sufficient area of suitable soils based on the soil tests and system location and site requirements of this code for one replacement system shall be established. Where bore hole test data in the replacement system area are equivalent to data in the proposed system area, the percolation test is not required.

401.3.1 Nonconforming site conditions. Where site conditions do not permit replacement systems in accordance with this code and an alternative system is used, the alternative system shall be approved in accordance with Section 105.

401.3.2 Undisturbed site. The replacement system shall not be disturbed to the extent that the site area is no longer suitable. The replacement system area shall not be used for construction of buildings, parking lots or parking areas, below-ground swimming pools or any other use that will adversely affect the replacement area.

SECTION 402
SLOPE

402.1 General. A *conventional soil absorption system* shall not be located on land with a slope greater than 20 percent. A *conventional soil absorption system* shall be located a minimum of 20 feet (6096 mm) from the crown of land with a slope greater than 20 percent, except where the top of the aggregate of a system is at or below the bottom of an adjacent roadside ditch. Where a more restrictive land slope is to be observed for a soil absorption system, other than a *conventional soil absorption system*, the more restrictive land slope specified in the design sections of this code shall apply.

SECTION 403
SOIL BORINGS AND EVALUATION

403.1 Soil borings and profile descriptions. *Soil borings* shall be conducted on all sites, regardless of the type of pri-

vate sewage system planned to serve the parcel. Borings shall extend at least 3 feet (914 mm) below the bottom of the proposed system. Borings shall be of sufficient size and extent to determine the soil characteristics important to an on-site liquid waste disposal system. Borehole data shall be used to determine the suitability of soils at the site with respect to zones of seasonal or permanent soil saturation and the depth to bedrock. Borings shall be conducted prior to percolation tests to determine whether the soils are suitable to warrant such tests and, if suitable, at what depth percolation tests shall be conducted. The use of power augers for *soil borings* is prohibited. *Soil borings* shall be conducted and reported in accordance with Sections 403.1.1 through 403.1.5. Where it is not practical to have borings made with a backhoe, such borings shall be augered or dug by hand.

403.1.1 Number. There shall be a minimum of three borings per soil absorption site. Where necessary, more *soil borings* shall be made for an accurate evaluation of a site. Borings shall be constructed to a depth of at least 3 feet (914 mm) below the proposed depth of the system.

Exception: On new parcels, the requirement of six borings (three for initial area and three for replacement area) shall be reduced to five where the initial and replacement system areas are contiguous and one boring is made on each outer corner of the contiguous area and the fifth boring is made between the system areas (see Appendix A, Figure A-1).

403.1.2 Location. Each borehole shall be accurately located and referenced to the vertical elevation and horizontal reference points. Reports of boring location shall either be drawn to scale or have the horizontal dimensions clearly indicated between the borings and the horizontal reference point.

403.1.3 Soil description. Soil profile descriptions shall be written for all borings. The thickness in inches (mm) of the different soil horizons observed shall be indicated. Horizons shall be differentiated on the basis of color, texture, soil mottles or bedrock. Depths shall be measured from the ground surface.

403.1.4 Soil mottles. Seasonal or periodic soil saturation zones shall be estimated at the highest level of soil mottles. The code official shall require, where deemed necessary, a detailed description of the soil mottling on a marginal site. The abundance, size, contrast and color of the soil mottles shall be described in the following manner:

Abundance shall be described as "few" if the mottled color occupies less than 2 percent of the exposed surface; "common" if the mottled color occupies from 2 to 20 percent of the exposed surface; or "many" if the mottled color occupies more than 20 percent of the exposed surface. Size refers to length of the mottle measured along the longest dimension and shall be described as "fine" if the mottle is less than 0.196 inch (5 mm); medium if the mottle is from 0.196 inch to 1.590 inches (5 mm to 40 mm); or coarse if the mottle is larger than 1.590 inches (40 mm). Contrast refers to the difference in color between the soil mottle and the background color of the soil and is described as "faint" if the mottle is evident but recognizable with close examination; "distinct" if the mottle is readily seen but not striking; or "prominent" if the mottle is obvious and one of the outstanding features of the horizon. The color(s) of the mottle(s) shall be indicated.

403.1.5 Observed ground water. The depth to ground water, if present, shall be reported. Observed ground water shall be reported at the level that ground water reaches in the soil borehole or the highest level of sidewall seepage into the boring. Measurements shall be made from ground level. Soil located above the water level in the boring shall be checked for the presence of soil mottles.

403.2 Color patterns not indicative of soil saturation. The following soil conditions shall be reported, but shall not be interpreted as color patterns caused by wetness or saturation. Soil profiles with an abrupt textural change with finer-textured soils overlying more than 4 feet (1219 mm) of unmottled, loamy sand or coarser soils can have a mottled zone for the finer textured material. Where the mottled zone is less than 12 inches (305 mm) thick and located immediately above the textural change, a soil absorption system shall be permitted in the loamy sand or coarser material below the mottled layer. The site shall be considered unsuitable where any soil mottles occur within the sandy material. The code official shall consider certain coarse sandy loam soils to be included as a coarse material.

403.2.1 Other soil color patterns. Soil mottles occur that are not caused by seasonal or periodic soil saturation zones. Examples of such soil conditions not limited by enumeration are soil mottles formed from residual sandstone deposits; soil mottles formed from uneven weathering of glacially deposited material or glacially deposited material that is naturally gray in color, including any concretionary material in various stages of decomposition; deposits of lime in a profile derived from highly calcareous parent material; light-colored silt coats deposited on soil bed faces; and soil mottles usually vertically oriented along old or decayed root channels with a dark organic stain usually present in the center of the mottled area.

403.2.2 Reporting exceptions. The site evaluator shall report any mottled soil condition. The observation of soil mottles not caused by soil saturation shall be reported. Upon request, the code official shall make a determination of the acceptability of the site.

403.3 Bedrock. The depth of the bedrock, except sandstone, shall be established at the depth in a *soil boring* where more than 50 percent of the weathered-in-place material is consolidated. Sandstone bedrock shall be established at the depth where an increase in resistance to penetration of a knife blade occurs.

403.4 Alluvial and colluvial deposits. Subsurface soil absorption systems shall not be placed in alluvial and collu-

vial deposits with shallow depths, extended periods of saturation or possible flooding.

SECTION 404
PERCOLATION OR PERMEABILITY EVALUATION

404.1 General. The permeability of the soil in the proposed absorption system shall be determined by percolation tests or permeability evaluation.

404.2 Percolation tests and procedures. At least three percolation tests in each system area shall be conducted. The holes shall be spaced uniformly in relation to the bottom depth of the proposed absorption system. More percolation tests shall be made where necessary, depending on system design.

404.2.1 Percolation test hole. The test hole shall be dug or bored. The test hole shall have vertical sides and a horizontal dimension of 4 inches to 8 inches (102 mm to 203 mm). The bottom and sides of the hole shall be scratched with a sharp-pointed instrument to expose the natural soil. All loose material shall be removed from the hole, and the bottom shall be covered with 2 inches (51 mm) of gravel or coarse sand.

404.2.2 Test procedure, sandy soils. The hole shall be filled with clear water to a minimum of 12 inches (305 mm) above the bottom of the hole for tests in sandy soils. The time for this amount of water to seep away shall be determined and this procedure shall be repeated if the water from the second filling of the hole seeps away in 10 minutes or less. The test shall proceed as follows: Water shall be added to a point not more than 6 inches (152 mm) above the gravel or coarse sand. Thereupon, from a fixed reference point, water levels shall be measured at 10-minute intervals for a period of 1 hour. Where 6 inches (152 mm) of water seeps away in less than 10 minutes, a shorter interval between measurements shall be used, but in no case shall the water depth exceed 6 inches (152 mm). Where 6 inches (152 mm) of water seeps away in less than 2 minutes, the test shall be stopped and a rate of less than 3 minutes per inch (7.2 s/mm) shall be reported. The final water level drop shall be used to calculate the percolation rate. Soils not meeting the above requirements shall be tested in accordance with Section 404.2.3.

404.2.3 Test procedure, other soils. The hole shall be filled with clear water, and a minimum water depth of 12 inches (305 mm) shall be maintained above the bottom of the hole for a 4-hour period by refilling whenever necessary or by use of an automatic siphon. Water remaining in the hole after 4 hours shall not be removed. Thereafter, the soil shall be allowed to swell not less than 16 hours or more than 30 hours. Immediately after the soil swelling period, the measurements for determining the percolation rate shall be made as follows: Any soil sloughed into the hole shall be removed, and the water level shall be adjusted to 6 inches (152 mm) above the gravel or coarse sand. Thereupon, from a fixed reference point, the water level shall be measured at 30-minute intervals for a period of 4 hours, unless two successive water level drops do not vary by more than $^1/_{16}$-inch (1.59 mm). At least three water level drops shall be observed and recorded. The hole shall be filled with clear water to a point not more than 6 inches (152 mm) above the gravel or coarse sand whenever it becomes nearly empty. The water level shall not be adjusted during the three measurement periods except to the limits of the last measured water level drop. When the first 6 inches (152 mm) of water seeps away in less than 30 minutes, the time interval between measurements shall be 10 minutes and the test run for 1 hour. The water depth shall not exceed 5 inches (127 mm) at any time during the measurement period. The drop that occurs during the final measurement period shall be used in calculating the percolation rate.

404.2.4 Mechanical test equipment. Mechanical percolation test equipment shall be of an approved type.

404.3 Permeability evaluation. Soil shall be evaluated for estimated percolation based on structure and texture in accordance with accepted soil evaluation practices. Borings shall be made in accordance with Section 404.2 for evaluating the soil.

SECTION 405
SOIL VERIFICATION

405.1 Verification. Where required by the code official, depth to soil mottles, depth to high ground water, soil textures, depth to bedrock and land slope shall be verified by the code official. The code official shall require, where necessary, backhoe pits to be provided for verification of *soil boring* data. Where required by the code official, the results of percolation tests or permeability evaluation shall be subject to verification. The code official shall require, where necessary, that percolation tests be conducted under supervision. Where the natural soil condition has been altered by filling or other methods used to improve wet areas, the code official shall require, where necessary, observation of high ground water levels under saturated soil conditions. Detailed soil maps, or other adequate information, shall be used for determining estimated percolation rates and other soil characteristics.

405.2 Monitoring ground water levels. A property owner or developer shall have the option to provide documentation that soil mottling or other color patterns at a particular site are not an indication of seasonally saturated soil conditions of high ground water levels. Direct observation shall be used to document ground water levels. Monitoring shall be in accordance with the procedures cited in Sections 405.2.1 through 405.2.6.

405.2.1 Precipitation. Monitoring shall be performed at a time of the year when maximum ground water elevation occurs. In determining whether a near-normal season has occurred where sites are subject to broad regional water tables, such as large areas of sandy soils, the fluctuation over the several-year cycle shall be considered. In such cases, data obtained from the United States Geological Survey (USGS) shall be used to determine if a regional water table was at or near its normal level.

405.2.2 Artificial drainage. Areas to be monitored shall be checked for drainage tile and open ditches that alter nat-

ural high ground water levels. Where such factors are involved, information on the location, design, ownership and maintenance responsibilities for such drainage shall be provided. Documentation shall be provided to show that the drainage network has an adequate outlet and will be maintained. Sites affected by agricultural drain tile shall not be acceptable for system installation.

405.2.3 Procedures. The owner or the owner's agent shall notify the code official in writing of the intent to monitor. Where necessary, the code official shall field check the monitoring at least once during the time of expected saturated soil conditions.

At least three wells shall be monitored at a site for a proposed system and replacement. Where necessary, the code official shall require more than three monitoring sites, and the site evaluator shall be so advised in writing.

405.2.4 Monitoring well design. At least two wells shall extend to a depth of at least 6 feet (1829 mm) below the ground surface and shall be a minimum of 3 feet (914 mm) below the designed system depth. However, with layered mottled soil over permeable unmottled soil, at least one well shall terminate within the mottled layer. Monitoring at greater depths shall be required, where necessary, due to site conditions. The site evaluator shall determine the depth of the monitoring wells for each specific site. All depths shall be approved. The monitoring well shall be a solid pipe installed in a bore hole. The pipe size shall be a minimum of 1 inch (25 mm) and a maximum of 4 inches (102 mm). The bore hole shall be a minimum of 4 inches (102 mm) and a maximum of 8 inches (203 mm) larger than the pipe (see Appendix A, Figure A-2).

405.2.5 Observations. The first observation shall be made on or before [DATE]. Observations shall be made thereafter every 7 days or less until [DATE] or until the site is determined to be unacceptable, whichever occurs first. Where water is observed above the critical depth at any time, an observation shall be made 1 week later. Where water is present above the critical depth at both observations, monitoring shall cease and the site shall be considered unacceptable. Where water is not present above the critical depth at the second observation, monitoring shall continue until [DATE]. Where any two observations 7 days apart show the presence of water above the critical depth, the site shall be considered unacceptable and the code official shall be notified in writing. When rainfall of 0.5 inch (12.7 mm) or more occurs in a 24-hour period during monitoring, observations shall be made at more frequent intervals, where necessary.

405.2.6 Reporting data. Where monitoring shows saturated conditions, the following data shall be submitted in writing: test locations; ground elevations at the wells; soil profile descriptions; soil series, if available from soil maps; dates observed; depths to observed water; and local precipitation data-monthly from [DATE] and daily during monitoring.

Where monitoring discloses that the site is acceptable, the following data shall be submitted in writing: location and depth of test holes, ground elevations at the wells and soil profile descriptions; soil series, if available from soil maps; dates observed; results of observations; information on artificial drainage; and local precipitation data-monthly from [DATE] and daily during monitoring. A request to install a soil absorption system shall be made in accordance with Section 106.

SECTION 406
SITE REQUIREMENTS

406.1 Soil absorption site location. The surface grade of all soil absorption systems shall be located at a point lower than the surface grade of any nearby water well or reservoir on the same or adjoining property. Where this is not possible, the site shall be located so surface water drainage from the site is not directed toward a well or reservoir. The soil absorption system shall be located with a minimum horizontal distance between various elements as indicated in Table 406.1. Private sewage disposal systems in compacted areas, such as parking lots and driveways, are prohibited. Surface water shall be diverted away from any soil absorption site on the same or neighboring lots.

TABLE 406.1
MINIMUM HORIZONTAL SEPARATION DISTANCES
FOR SOIL ABSORPTION SYSTEMS

ELEMENT	DISTANCE (feet)
Cistern	50
Habitable building, below-grade foundation	25
Habitable building, slab-on-grade	15
Lake, high-water mark	50
Lot line	5
Reservoir	50
Roadway ditches	10
Spring	100
Streams or watercourse	50
Swimming pool	15
Uninhabited building	10
Water main	50
Water service	10
Water well	50

For SI: 1 foot = 304.8 mm.

406.1.1 Flood hazard areas. The site shall be located outside of flood hazard areas.

Exception: Where suitable sites outside of the *flood hazard area* are not available, it is permitted for the site to be located within the *flood hazard area*. The site shall be located to minimize the effects of inundation under conditions of the design flood.

406.2 Ground water, bedrock or slowly permeable soils. There shall be a minimum of 3 feet (914 mm) of soil between the bottom of the soil absorption system and high ground water or bedrock. Soil with a percolation rate of 60 minutes per 1 inch (25 mm) or faster shall exist for the depth of the proposed soil absorption system and at least 3 feet (914 mm)

below the proposed bottom of the soil absorption system. There shall be 56 inches (1422 mm) of suitable soil from original grade for a *conventional soil absorption system.*

406.3 Percolation rate, trench or bed. A subsurface soil absorption system of the trench or bed type shall not be installed where the percolation rate for any one of the three tests is slower than 60 minutes for water to fall 1 inch (25 mm). The slowest percolation rate shall be used to determine the absorption area.

406.4 Percolation rate, seepage pit. Percolation tests shall be made in each horizon penetrated below the inlet pipe for a *seepage pit.* Soil strata in which the percolation rates are slower than 30 minutes per 1 inch (25 mm) shall not be included in computing the absorption area. The slowest percolation rate shall be used to determine the absorption area.

406.5 Soil maps. When a parcel of land consists entirely of soils with very severe or severe limitations for on-site liquid-waste disposal as determined by use of a detailed soil map and supporting data, that map and supporting data shall be permitted to be used as a basis for denial for an on-site waste disposal system. However, the property owner shall be permitted to present evidence that a suitable site for an on-site liquid-waste disposal system does exist.

406.6 Filled area. A soil absorption system shall not be installed in a filled area unless written approval is received.

406.6.1 Placement of fill. The approval of a *conventional soil absorption system* shall be based on evidence indicating its conformance to code requirements for area, percolation and elevation.

406.6.2 Bedrock. Sites with less than 56 inches (1422 mm) but at least 30 inches (762 mm) of soil over bedrock, where the original soil texture is sand or loamy sand, are permitted to be filled with the same soil texture as the natural soil or coarser material up to and including medium sand to overcome the site limitations. The fill material shall not be of a finer texture than the natural soil.

406.6.3 High ground water. Sites with less than 56 inches (1422 mm) of soil over high ground water or estimated high ground water, where the original soil texture is sand or loamy sand, are permitted to be filled in accordance with Section 406.6.1 or 406.6.2.

406.6.4 Natural soil. Sites with soils finer than sand or loamy sand shall not be approved for systems in fill.

406.6.5 Monitoring. Sites that will have 36 inches (762 mm) or less of soil above high ground water after the topsoil is removed shall be monitored for high ground water levels in the filled area in accordance with Section 405.2.

406.6.6 Inspection of fill. Placement of the fill material shall be inspected by the code official.

406.6.7 Design requirements. Filled areas shall be large enough to accommodate a shallow trench system and a replacement system. The site of the area to be filled shall be determined by the percolation rate of the natural soil and use of the building. Where any portion of the trench system or its replacement is in the fill, the fill shall extend 20 feet (6096 mm) beyond all sides of both systems before the slope begins. *Soil borings* and percolation tests shall be conducted before filling to determine soil textures and depth to high ground water or bedrock. Vegetation and topsoil shall be removed prior to filling. Slopes at the edge of the filled areas shall have a maximum ratio of one unit vertical to three units horizontal (33-percent slope), provided the 20-foot (6096 mm) separating distance is maintained (see Appendix A, Figure A-3).

406.7 Altering slopes. Areas with slopes exceeding those specified in Section 402.1 shall not be used unless graded and reshaped in accordance with Sections 406.7.1 through 406.7.3.

406.7.1 Site investigation. Soil test data shall show that a sufficient depth of suitable soil material is present to provide the required amount of soil over bedrock and ground water after alteration. A complete site evaluation as specified in this section shall be performed after alteration of the site.

406.7.2 System location. A soil absorption system shall be installed in the cut area of an altered site. A soil absorption system shall not be installed in the fill area of an altered site. The area of fill on an altered site is permitted to be used as a portion of the required 20-foot (6096 mm) separating distance from the crown of a critical slope. There shall be a minimum of 6 feet (1829 mm) of natural soil between the edge of a system area and the downslope side of the altered area.

406.7.3 Site protection. Altered slope areas shall be positioned so that surface water drainage will be diverted away from the system areas. Disturbed areas shall be seeded or sodded with grass, and appropriate steps shall be taken to control erosion (see Figure 406.7.3).

A. EXCAVATION OF COMPLETE HILLTOP

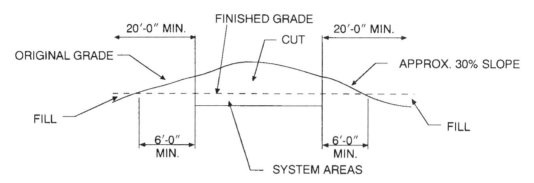

B. EXCAVATION INTO HILLSIDE

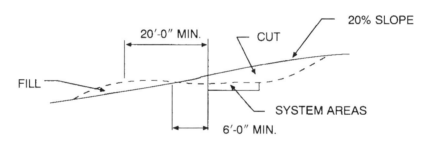

A SURFACE WATER DIVERSION
MAY BE NEEDED AT ONE OF
THESE POINTS IF LONG
SLOPES ARE PRESENT.

C. REGRADE OF HILLSIDE

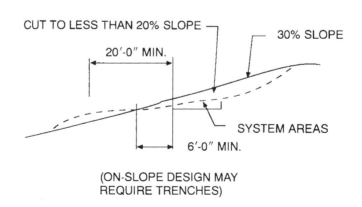

(ON-SLOPE DESIGN MAY
REQUIRE TRENCHES)

For SI: 1 inch = 25.4 mm, 1 foot = 304.8 mm.

**FIGURE 406.7.3
CONCEPTUAL DESIGN SKETCH FOR ALTERING SLOPES**

CHAPTER 5

MATERIALS

SECTION 501
GENERAL

501.1 Scope. The provisions of this chapter shall govern the requirements for materials for *private sewage disposal systems*.

501.2 Minimum standards. Materials shall conform to the standards referenced in this code for the construction, installation, alteration or repair of *private sewage disposal systems* or parts thereof.

> **Exception:** The extension, addition to or relocation of existing pipes with materials of like grade or quality in accordance with Sections 102.6 and 105.

SECTION 502
IDENTIFICATION

502.1 General. The manufacturer's mark or name and the quality of the product or identification shall be cast, embossed, stamped or indelibly marked on each length of pipe and each pipe fitting, fixture, tank, material and device used in a *private sewage disposal system* in accordance with the approved standard. Tanks shall indicate their capacity.

SECTION 503
PERFORMANCE REQUIREMENTS

503.1 Approved materials required. All materials, fixtures or equipment used in the installation, repair or alteration of any *private sewage disposal system* shall conform to the standards referenced in this code, except as otherwise approved in accordance with Section 105.

503.2 Care in installation. All materials installed in *private sewage disposal systems* shall be handled and installed so as to avoid damage. The quality of the material shall not be impaired.

503.3 Defective materials prohibited. Defective or damaged materials, equipment or apparatus shall not be installed or maintained.

SECTION 504
TANKS

504.1 Approval. All tanks shall be of an approved type. The design of tanks shall conform to the requirements of Chapter 8. All tanks shall be designed to withstand the pressures to which they are subjected.

504.2 Precast concrete and site-constructed tanks. Precast concrete tanks shall conform to ASTM C 913. The floor and sidewalls of a site-constructed concrete tank shall be monolithic, except a construction joint is permitted in the lower 12 inches (305 mm) of the sidewalls of the tank. The construction joint shall have a keyway in the lower section of the joint. The width of the keyway shall be approximately 30 percent of the thickness of the sidewall with a depth equal to the width. A continuous water stop or baffle at least 56 inches (1422 mm) wide shall be set vertically in the joint, embedded one-half its width in the concrete below the joint with the remaining width in the concrete above the joint. The water stop or baffle shall be copper, neoprene, rubber or polyvinyl chloride designed for this specific purpose. Joints between the concrete septic tank and the tank cover and between the septic tank cover and manhole riser shall be tongue and groove or shiplap-type and sealed water tight using cement, mortar or bituminous compound.

504.3 Steel tanks. Steel tanks shall conform to UL 70. Any damage to the bituminous coating shall be repaired by recoating. The gage of the steel shall be in accordance with Table 504.3.

TABLE 504.3
TANK CAPACITY

TANK DESIGN AND CAPACITY		MINIMUM GAGE THICKNESS	MINIMUM DIAMETER
Vertical cylindrical			
500 to 1,000 gallons	Bottom and sidewalls	12 gage	None
	Cover	12 gage	
	Baffles	12 gage	
1,001 to 1,250 gallons	Complete tank	10 gage	None
1,251 to 1,500 gallons	Complete tank	7 gage	None
Horizontal cylindrical			
500 to 1,000 gallons	Complete tank	12 gage	54-inch diameter
1,001 to 1,500 gallons	Complete tank	12 gage	64-inch diameter
1,501 to 2,500 gallons	Complete tank	10 gage	76-inch diameter
2,501 to 9,000 gallons	Complete tank	7 gage	76-inch diameter
9,001 to 12,000 gallons	Complete tank	$^1/_4$-inch plate	None
Over 12,000 gallons	Complete tank	$^5/_{16}$ inch	None

For SI: 1 inch = 25.4 mm, 1 gallon = 3.785 L.

504.4 Fiberglass tanks. Fiberglass tanks shall conform to ASTM D 4021.

504.5 Manholes. Manhole collars and extensions shall be of the same material as the tank. Manhole covers shall be of concrete, steel, cast iron or other approved material.

SECTION 505
PIPE, JOINTS AND CONNECTIONS

505.1 Pipe. Pipe for *private sewage disposal systems* shall have a smooth wall and conform to one of the standards listed in Table 505.1.

505.1.1 Distribution pipe. Perforated pipe for distribution systems shall conform to one of the standards listed in Table 505.1 or Table 505.1.1.

505.2 Joints and connection approval. All joints and connections shall be of an approved type.

505.3 ABS plastic pipe. Joints between acrylonitrile butadiene styrene (ABS) plastic pipe or fittings shall be in accordance with Sections 505.3.1 and 505.3.2.

505.3.1 Mechanical joints. Mechanical joints on drainage pipes shall be made with an elastomeric seal conforming to ASTM C 1173, ASTM D 3212 or CSA B 602. Mechanical joints shall be installed only in underground systems, except as otherwise approved. Joints shall be installed in accordance with the manufacturer's installation instructions.

505.3.2 Solvent cementing. Joint surfaces shall be clean and free from moisture. Solvent cement conforming to ASTM D 2235 or CSA B181.1 shall be applied to all joint surfaces. The joint shall be made while the cement is wet. Joints shall be made in accordance with ASTM D 2235, ASTM D 2661, ASTM F 628 or CSA B181.1. Solvent cement joints shall be permitted above or below ground.

505.4 Asbestos-cement pipe. Joints between asbestos-cement pipe or fittings shall be made with a sleeve coupling of the same composition as the pipe and sealed with an elastomeric ring conforming to ASTM D 1869.

505.5 Coextruded composite ABS pipe and joints. Joints between coextruded composite pipe with an ABS outer layer or ABS fittings shall comply with Sections 505.5.1 and 505.5.2.

505.5.1 Mechanical joints. Mechanical joints on drainage pipe shall be made with an elastomeric seal conforming to ASTM C 1173, ASTM D 3212, or CSA B 602. Mechanical joints shall not be installed in above-ground systems, except as otherwise approved. Joints shall be installed in accordance with the manufacturer's installation instructions.

505.5.2 Solvent cementing. Joint surfaces shall be clean and free from moisture. Solvent cement conforming to ASTM D 2235 or CSA B 181.1 shall be applied to all joint surfaces. The joint shall be made while the cement is wet. Joints shall be made in accordance with ASTM D 2235, ASTM D 2661, ASTM F 628 or CSA B181.1. Solvent cement joints shall be permitted above or below ground.

505.6 Cast-iron pipe. Joints between cast-iron pipe or fittings shall be in accordance with Sections 505.6.1 through 505.6.3.

505.6.1 Caulked joints. Joints for hub and spigot pipe shall be firmly packed with oakum or hemp. Molten lead shall be poured in one operation to a depth of not less than 1 inch (25 mm). The lead shall not recede more than 0.125

TABLE 505.1
PRIVATE SEWAGE DISPOSAL SYSTEM PIPE

MATERIAL	STANDARD
Acrylonitrile butadiene styrene (ABS) plastic pipe	ASTM D 2661; ASTM D 2751; ASTM F 628
Asbestos-cement pipe	ASTM C 428
Cast-iron pipe	ASTM A 74; ASTM A 888; CISPI 301
Coextruded composite ABS DWV Schedule 40 IPS pipe (solid)	ASTM F 1488; ASTM F 1499
Coextruded composite ABS DWV Schedule 40 IPS pipe (cellular core)	ASTM F 1488; ASTM F 1499
Coextruded composite ABS sewer and drain DR-PS in PS35, PS50, PS100, PS140 and PS200	ASTM F 1488; ASTM F 1499
Coextruded composite PVC DWV Schedule 40 IPS pipe (solid)	ASTM F 1488
Coextruded composite PVC DWV Schedule 40 IPS pipe (cellular core)	ASTM F 1488
Coextruded composite PVC-IPS-DR of PS140, PS200, DWV	ASTM F 1488
Coextruded composite PVC 3.25 OD DWV pipe	ASTM F 1488
Coextruded composite PVC sewer and drain DR-PS in PS35, PS50, PS100, PS140 and PS200	ASTM F 1488
Concrete pipe	ASTM C 14; ASTM C 76; CSA A257.1M; CSA A257.2M
Copper or copper-alloy tubing (Type K or L)	ASTM B 75; ASTM B 88; ASTM B 251
Polyvinyl chloride (PVC) plastic pipe (Type DWV, SDR26, SDR35, SDR41, PS50 or PS100)	ASTM D 2665; ASTM D 2949; ASTM D 3034; ASTM F 891; CSA B182.2; CSA B182.4
Vitrified clay pipe	ASTM C 4; ASTM C 700

TABLE 505.1.1
DISTRIBUTION PIPE

MATERIAL	STANDARD
Polyethylene (PE) plastic pipe	ASTM F 405
Polyvinyl chloride (PVC) plastic pipe	ASTM D 2729
Polyvinyl chloride (PVC) plastic pipe with pipe stiffness of PS35 and PS50	ASTM F 1488

inch (3.2 mm) below the rim of the hub, and shall be caulked tight. Paint, varnish or other coatings shall not be applied to the joining material until after the joint has been tested and approved. Lead shall be run in one pouring and shall be caulked tight. Acid-resistant rope and acidproof cement shall be permitted.

505.6.2 Mechanical compression joints. Compression gaskets for hub and spigot pipe and fittings shall conform to ASTM C 564. Gaskets shall be compressed when the pipe is fully inserted.

505.6.3 Mechanical joint coupling. Mechanical joint couplings for hubless pipe and fittings shall comply with CISPI 310 or ASTM C 1277. The elastomeric sealing sleeve shall conform to ASTM C 564 or CSA B 602 and shall be provided with a center stop. Mechanical joint couplings shall be installed in accordance with the manufacturer's installation instructions.

505.7 Concrete pipe. Joints between concrete pipe or fittings shall be made by the use of an elastomeric seal conforming to ASTM C 443, ASTM C 1173, CSA A 257.3M or CSA B 602.

505.8 Copper or copper-alloy tubing or pipe. Joints between copper or copper-alloy tubing, pipe or fittings shall be in accordance with Sections 505.8.1 and 505.8.2.

505.8.1 Mechanical joints. Mechanical joints shall be installed in accordance with the manufacturer's installation instructions.

505.8.2 Soldered joints. Solder joints shall be made in accordance with the methods of ASTM B 828. All cut ends shall be reamed to the full inside diameter of the tube end. All joint surfaces shall be cleaned. A flux conforming to ASTM B 813 shall be applied. The joint shall be soldered with a solder conforming to ASTM B 32.

505.9 Polyethylene plastic pipe and tubing. Joints between polyethylene plastic pipe and tubing or fittings shall be in accordance with Sections 505.9.1 and 505.9.2.

505.9.1 Heat-fusion joints. Joint surfaces shall be clean and free from moisture. All joint surfaces shall be heated to melting temperature and joined. The joint shall be undisturbed until cool. Joints shall be made in accordance with ASTM D 2657.

505.9.2 Mechanical joints. Mechanical joints shall be installed in accordance with the manufacturer's instructions.

505.10 PVC plastic pipe. Joints between polyvinyl chloride (PVC) plastic pipe and fittings shall be in accordance with Sections 505.10.1 and 505.10.2.

505.10.1 Mechanical joints. Mechanical joints shall be made with an elastomeric seal conforming to ASTM C 1173, ASTM D 3212 or CSA B602. Mechanical joints shall not be installed in above-ground systems, except as otherwise approved. Joints shall be installed in accordance with the manufacturer's installation instructions.

505.10.2 Solvent cementing. Joint surfaces shall be clean and free from moisture. A purple primer that conforms to ASTM F 656 shall be applied. Solvent cement not purple in color and conforming to ASTM D 2564, CSA B137.3,

CSA B181.2 or CSA B182.1 shall be applied to all joint surfaces. The joint shall be made while the cement is wet, and shall be in accordance with ASTM D 2855. Solvent cement joints shall be permitted above or below ground.

505.11 Coextruded composite PVC pipe. Joints between coextruded composite pipe with a PVC outer layer or PVC fittings shall comply with Sections 505.11.1 and 505.11.2.

505.11.1 Mechanical joints. Mechanical joints on drainage pipe shall be made with an elastomeric seal conforming to ASTM D 3212. Mechanical joints shall not be installed in above-ground systems, except as otherwise approved. Joints shall be installed in accordance with the manufacturer's installation instructions.

505.11.2 Solvent cementing. Joint surfaces shall be clean and free from moisture. A purple primer that conforms to ASTM F 656 shall be applied. Solvent cement not purple in color and conforming to ASTM D 2564, CSA B137.3, CSA B181.2 or CSA B 182.1 shall be applied to all joint surfaces. The joint shall be made while the cement is wet, and shall be in accordance with ASTM D 2855. Solvent cement joints shall be permitted above or below ground.

505.12 Vitrified clay pipe. Joints between vitrified clay pipe or fittings shall be made by the use of an elastomeric seal conforming to ASTM C 425, ASTM C 1173 or CSA B602.

505.13 Different piping materials. Joints between different piping materials shall be made with a mechanical joint of the compression or mechanical-sealing type conforming to ASTM C 1173, ASTM C 1460 or ASTM C 1461. Connectors or adapters shall be approved for the application and such joints shall have an elastomeric seal conforming to ASTM C 425, ASTM C 443, ASTM C 564, ASTM C 1440, ASTM D 1869, ASTM F 477, CSA A257.3M or CSA B602 or as required in Sections 505.13.1 and 505.13.2. Joints shall be installed in accordance with the manufacturer's instructions.

505.13.1 Copper or copper-alloy tubing to cast-iron hub pipe. Joints between copper or copper-alloy tubing and cast-iron hub pipe shall be made with a brass ferrule or compression joint. The copper or copper-alloy tubing shall be soldered to the ferrule in an approved manner, and the ferrule shall be joined to the cast-iron hub by a caulked joint or a mechanical compression joint.

505.13.2 Plastic pipe or tubing to other piping material. Joints between different grades of plastic pipe or between plastic pipe and other piping material shall be made with an approved adapter fitting. Joints between plastic pipe and cast-iron hub pipe shall be made by a caulked joint or a mechanical compression joint.

505.14 Pipe installation. Pipe shall be installed in accordance with the *International Plumbing Code*.

SECTION 506
PROHIBITED JOINTS AND CONNECTIONS

506.1 General. The following types of joints and connections shall be prohibited:

1. Cement or concrete joints.

2. Mastic or hot-pour bituminous joints.

3. Joints made with fittings not approved for the specific installation.

4. Joints between different diameter pipes made with elastomeric rolling O-rings.

5. Solvent-cement joints between different types of plastic pipe.

CHAPTER 6

SOIL ABSORPTION SYSTEMS

SECTION 601
GENERAL

601.1 Scope. The provisions of this chapter shall govern the sizing and installation of soil absorption systems.

SECTION 602
SIZING SOIL ABSORPTION SYSTEMS

602.1 General. Effluent from septic tanks and other approved treatment tanks shall be disposed of by soil absorption or an approved manner. Sizing shall be in accordance with this chapter for systems with a daily effluent application of 5,000 gallons (18 925 L) or less. Two systems of equal size shall be required for systems receiving effluents exceeding 5,000 gallons (18 925 L) per day. Each system shall have a minimum capacity of 75 percent of the area required for a single system. An approved means of alternating waste application shall be provided. A dual system shall be considered as one system.

602.2 Pressure system. A *pressure distribution system* shall be permitted in place of a conventional or dosing *conventional soil absorption system* where a site is suitable for a conventional *private sewage disposal system*. A *pressure distribution system* shall be approved as an alternative *private sewage disposal system* where the site is unsuitable for conventional treatment (for sizing and design criteria, see Chapter 7).

602.3 Method of discharge. Flow from the septic or treatment tank to the soil absorption system shall be by gravity or dosing for facilities with a daily effluent application of 1,500 gallons (5678 L) or less. The tank effluent shall be discharged by pumping or an automatic siphon for systems over 1,500 gallons (5678 L).

SECTION 603
RESIDENTIAL SIZING

603.1 General. The bottom area for *seepage trenches* or beds or the sidewall area for seepage pits required for a soil absorption system serving residential property shall be determined from Table 603.1 using soil percolation test data and type of construction.

TABLE 603.1
MINIMUM ABSORPTION AREA FOR
ONE- AND TWO-FAMILY DWELLINGS

PERCOLATION CLASS	PERCOLATION RATE (minutes required for water to fall 1 inch)	SEEPAGE TRENCHES OR PITS (square feet per bedroom)	SEEPAGE BEDS (square feet per bedroom)
1	0 to less than 10	165	205
2	10 to less than 30	250	315
3	30 to less than 45	300	375
4	45 to 60	330	415

For SI: 1 minute per inch = 2.4 s/mm, 1 square foot = 0.0929 m².

SECTION 604
OTHER BUILDING SIZING

604.1 General. The minimum required soil absorption system area for all occupancies, except one- and two-family dwellings, shall be based on building usage, the percolation rate and system design in accordance with Tables 604.1(1) and 604.1(2). The minimum soil absorption area shall be calculated by the following equation:

$$A = U \times CF \times AA \qquad \text{(Equation 6-1)}$$

where:

A = Minimum system absorption area.

AA = Absorption area from Table 604.1(1).

CF = Conversion factor from Table 604.1(2).

U = Number of units.

TABLE 604.1(1)
MINIMUM ABSORPTION AREA FOR OTHER THAN
ONE- AND TWO-FAMILY DWELLINGS

PERCOLATION CLASS	PERCOLATION RATE (minutes required for water to fall 1 inch)	SEEPAGE TRENCHES OR PITS (square feet per unit)	SEEPAGE BEDS (square feet per unit)
1	0 to less than 10	110	140
2	10 to less than 30	165	205
3	30 to less than 45	220	250
4	45 to 60	220	280

For SI: 1 minute per inch = 2.4 s/mm, 1 square foot = 0.0929 m².

SECTION 605
INSTALLATION OF CONVENTIONAL SOIL
ABSORPTION SYSTEMS

605.1 Seepage trench excavations. *Seepage trench* excavations shall be 1 foot to 5 feet (305 mm to 1524 mm) wide. Trench excavations shall be spaced a minimum of 6 feet (1829 mm) apart. The absorption area of a seepage trench shall be computed by using only the bottom of the trench area. The bottom excavation area of the distribution header shall not be computed as absorption area. Individual *seepage trenches* shall be a maximum of 100 feet (30 480 mm) long, except as otherwise approved.

605.2 Seepage bed excavations. *Seepage bed* excavations shall be a minimum of 5 feet (1524 mm) wide and have more than one distribution pipe. The absorption area of a *seepage bed* shall be computed by using the bottom of the trench area. Distribution piping in a *seepage bed* shall be uniformly spaced a maximum of 5 feet (1524 mm) and a minimum of 3 feet (914 mm) apart, and a maximum of 3 feet (914 mm) and a minimum of 1 foot (305 mm) from the sidewall or headwall.

TABLE 604.1(2)
CONVERSION FACTOR

BUILDING CLASSIFICATION	UNITS	FACTOR
Apartment building	1 per bedroom	1.5
Assembly hall—no kitchen	1 per person	0.02
Auto washer (service buildings, etc.)	1 per machine	6.0
Bar and cocktail lounge	1 per patron space	0.2
Beauty salon	1 per station	2.4
Bowling center	1 per bowling lane	2.5
Bowling center with bar	1 per bowling lane	4.5
Camp, day and night	1 per person	0.45
Camp, day use only	1 per person	0.2
Campground and camping resort	1 per camping space	0.9
Campground and sanitary dump station	1 per camping space	0.085
Car wash	1 per car	1.0
Catch basin—garages, motor fuel-dispensing facility, etc.	1 per basin	2.0
Catch basin—truck wash	1 per truck	5.0
Church—no kitchen	1 per person	0.04
Church—with kitchen	1 per person	0.09
Condominium	1 per bedroom	1.5
Dance hall	1 per person	0.06
Dining hall—kitchen and toilet	1 per meal served	0.2
Dining hall—kitchen and toilet waste with dishwasher or food waste grinder or both	1 per meal served	0.25
Dining hall—kitchen only	1 per meal served	0.06
Drive-in restaurant, inside seating	1 per seat	0.3
Drive-in restaurant, without inside seating	1 per car space	0.3
Drive-in theater	1 per car space	0.1
Employees—in all buildings	1 per person	0.4
Floor drain	1 per drain	1.0
Hospital	1 per bed space	2.0
Hotel or motel and tourist rooming house	1 per room	0.9
Labor camp—central bathhouse	1 per employee	0.25
Medical office buildings, clinics and dental offices 　　Doctors, nurses and medical staff 　　Office personnel 　　Patients	 1 per person 1 per person 1 per person	 0.8 0.25 0.15
Mobile home park	1 per mobile home site	3.0
Motor-fuel-dispensing facility	1 per car served	0.15
Nursing or group homes	1 per bed space	1.0
Outdoor sports facility—toilet waste only	1 per person	0.35
Park—showers and toilets	1 per acre	8.0
Park—toilet waste only	1 per acre	4.0
Restaurant—dishwasher or food waste grinder or both	1 per seating space	0.15
Restaurant—kitchen and toilet	1 per seating space	0.6

(continued)

TABLE 604.1(2)—continued
CONVERSION FACTOR

BUILDING CLASSIFICATION	UNITS	FACTOR
Restaurant—kitchen waste only	1 per seating space	0.18
Restaurant—toilet waste only	1 per seating space	0.42
Restaurant—(24-hour) kitchen and toilet	1 per seating space	1.2
Restaurant—(24-hour) with dishwasher or food waste grinder or both	1 per seating space	1.5
Retail store	1 per customer	0.03
School—meals and showers	1 per classroom	8.0
School—meals served or showers	1 per classroom	6.7
School—no meals, no showers	1 per classroom	5.0
Self-service laundry—toilet wastes only	1 per machine	1.0
Showers—public	1 per shower	0.3
Swimming pool bathhouse	1 per person	0.2

605.3 Seepage pits. A seepage pit shall have a minimum inside diameter of 5 feet (1524 mm) and shall consist of a chamber walled-up with material, such as perforated precast concrete ring, concrete block, brick or other approved material allowing effluent to percolate into the surrounding soil. The pit bottom shall be left open to the soil. Aggregate of $^1/_2$ inch to $2^1/_2$ inches (12.7 mm to 64 mm) in size shall be placed into a 6-inch minimum (152 mm) annular space separating the outside wall of the chamber and sidewall excavation. The depth of the annular space shall be measured from the inlet pipe to the bottom of the chamber. Each seepage pit shall be provided with a 24-inch (610 mm) manhole extending to within 56 inches (1422 mm) of the ground surface and a 4-inch-diameter (102 mm) fresh air inlet. Seepage pits shall be located a minimum of 5 feet (1524 mm) apart. Excavation and scarifying shall be in accordance with Section 605.4. The effective area of a seepage pit shall be the vertical wall area of the walled-up chamber for the depth below the inlet for all strata in which the percolation rates are less than 30 minutes per inch (70 s/mm). The 6-inch (152 mm) annular opening outside the vertical wall area is permitted to be included for determining the effective area. Table 605.3, or an approved method, shall be used for determining the effective sidewall area of circular seepage pits.

TABLE 605.3
EFFECTIVE SQUARE-FOOT ABSORPTION AREA
FOR SEEPAGE PITS

INSIDE DIAMETER OF CHAMBER IN FEET PLUS 1 FOOT FOR WALL THICKNESS PLUS 1 FOOT FOR ANNULAR SPACE	DEPTH IN FEET OF PERMEABLE STRATA BELOW INLET					
	3	4	5	6	7	8
7	47	88	110	132	154	176
8	75	101	126	151	176	201
9	85	113	142	170	198	226
10	94	126	157	188	220	251
11	104	138	173	208	242	277
13	123	163	204	245	286	327

For SI: 1 foot = 304.8 mm.

605.4 Excavation and construction. The bottom of a trench or bed excavation shall be level. *Seepage trenches* or beds shall not be excavated where the soil is so wet that such material rolled between the hands forms a soil wire. All smeared or compacted soil surfaces in the sidewalls or bottom of *seepage trench* or bed excavations shall be scarified to the depth of smearing or compaction and the loose material removed. Where rain falls on an open excavation, the soil shall be left until sufficiently dry so a soil wire will not form when soil from the excavation bottom is rolled between the hands. The bottom area shall then be scarified and loose material removed.

605.5 Aggregate and backfill. A minimum of 6 inches (152 mm) of aggregate ranging in size from $^1/_2$ inch to $2^1/_2$ inches (12.7 mm to 64 mm) shall be laid into the trench or bed below the distribution pipe elevation. The aggregate shall be evenly distributed a minimum of 2 inches (51 mm) over the top of the distribution pipe. The aggregate shall be covered with approved synthetic materials or 9 inches (229 mm) of uncompacted marsh hay or straw. Building paper shall not be used to cover the aggregate. A minimum of 18 inches (457 mm) of soil backfill shall be provided above the covering.

605.6 Distribution piping. Distribution piping for gravity systems shall be not less than 4 inches (102 mm) in diameter. The distribution header (PVC) shall be solid-wall pipe. The top of the distribution pipe shall be not less than 8 inches (203 mm) below the original surface in continuous straight or curved lines. The slope of the distribution pipes shall be 2 inches to 4 inches (51 mm to 102 mm) per 100 feet (30 480 mm). Effluent shall be distributed to all distribution pipes. Distribution of effluent to *seepage trenches* on sloping sites shall be accomplished by using a drop box design or other approved methods. Where dosing is required, the siphon or pump shall discharge a dose of minimum capacity equal to 75 percent of the combined volume of the distribution piping in the absorption system.

605.7 Observation pipes. Observation pipes shall be provided. Such pipes shall be not less than 4 inches (102 mm) in diameter, not less than 12 inches (305 mm) above final grade and shall terminate with an approved vent cap.

The bottom 12 inches (305 mm) of the observation pipe shall be perforated and extend to the bottom of the aggregate. Observation pipes shall be located at least 25 feet (7620 mm) from any window, door or air intake of any building used for human occupancy. A maximum of four distribution pipelines shall be served by one common 4-inch (102 mm) observation pipe when interconnected by a common header pipe (see Appendix-A, Figure A-4).

> **Exception:** Where approved and where the location of the observation pipe is permanently recorded, the observation pipe shall be not more than 2 inches (51 mm) below the finished grade.

605.8 Winter installation. Soil absorption systems shall not be installed during periods of adverse weather conditions unless the installation is approved. A soil absorption system shall not be installed where the soil at the system elevation is frozen. Snow cover shall be removed from the soil absorption area before excavation begins. Snow shall not be placed in a manner that will cause water to pond on the soil absorption system area during snow melt. Excavated soil to be used as backfill shall be protected from freezing. Excavated soil that freezes solid shall not be used as backfill. The first 12 inches (305 mm) of backfill shall be loose, unfrozen soil. Inspection of systems installed during winter conditions shall include inspection of the trench or bed excavation prior to the placement of gravel and inspection of backfill material at the time of placement.

605.9 Evaporation. Soil absorption systems shall not be covered or paved over by material that inhibits the evaporation of the effluent.

CHAPTER 7
PRESSURE DISTRIBUTION SYSTEMS

SECTION 701
GENERAL

701.1 Scope. The provisions of this chapter shall govern the design and installation of *pressure distribution systems*.

SECTION 702
DESIGN LOADING RATE

702.1 General. A *pressure distribution system* shall be permitted for use on any site meeting the conventional *private sewage disposal system* criteria. There shall be not less than 6 inches (152 mm) to the top of the distribution piping from original grade for any *pressure distribution system*. The minimum required suitable soil depths from original grade for *pressure distribution systems* shall be in accordance with Table 702.1.

TABLE 702.1
SOIL REQUIRED

DISTRIBUTION PIPE (inches)	SUITABLE SOIL (inches)
1	49
2	50
3	52
4	53

For SI: 1 inch = 25.4 mm.

702.2 Absorption area. The total absorption area required shall be computed from the estimated daily wastewater flow and the design loading rate based on the percolation rate for the site. The required absorption area equals wastewater flow divided by the design loading rate from Table 702.2. Two systems of equal size shall be required for systems receiving effluents exceeding 5,000 gallons (18 925 L). Each system shall have a minimum capacity of 75 percent of the area required for a single system and shall be provided with a suitable means of alternating waste applications. A dual system shall be considered as one system.

TABLE 702.2
DESIGN LOADING RATE

PERCOLATION RATE (minutes per inch)	DESIGN LOADING FACTOR (gallons per square foot per day)
0 to less than 10	1.2
10 to less than 30	0.8
30 to less than 45	0.72
45 to 60	0.4

For SI: 1 minute per inch = 2.4 s/mm, 1 gallon per square foot = 0.025 L/m².

702.3 Estimated waste-water flow. The estimated waste-water flow from a residence shall be 150 gallons (568 L) per bedroom per day. Waste-water flow rates for other occupancies in a 24-hour period shall be based on the values in Table 802.7.2.

SECTION 703
SYSTEM DESIGN

703.1 General. *Pressure distribution systems* shall discharge effluent into trenches or beds. Each pipe connected to an outlet of a manifold shall be counted as a separate distribution pipe. The horizontal spacing of distribution pipes shall be 30 inches to 72 inches (762 mm to 1829 mm). The system shall be sized in accordance with the formulas listed in this section. Systems using Schedule 40 plastic pipe shall be sized in accordance with the formulas listed in this section or in accordance with the tables listed in Appendix B. Distribution piping shall be installed at the same elevation, unless an approved system provides for a design ensuring equal flow through each of the perforations and the effluent is uniformly applied to the soil infiltrative surface (see Appendix A, Figure A-5).

703.2 Symbols. The following symbols and notations shall apply to the provisions of this chapter:

C_h = Hazen-Williams friction factor.

D = Distribution pipe diameter, inches (mm).

d = Perforation diameter, inches (mm).

D_d = Delivery pipe diameter, inches (mm).

D_m = Manifold pipe diameter, inches (mm).

f = Fraction of total head loss in the manifold segment.

F_D = Friction loss in the delivery pipe, feet of head (mm of head).

F_i = Friction factor for i^{th} manifold segment.

F_N = Friction loss in the network pipe, feet of head (mm of head).

h = Pressure in distribution pipe, feet of head (mm of head).

h_d = In-line pressure at distal end of lateral, feet of head (mm of head).

L_D = Length of delivery pipe, feet (mm).

L_i = Length of ith manifold segment, feet (mm).

N = Number of perforations in the lateral.

q = Perforation discharge rate, gpm (L/min).

Q_i = Flow rate i^{th} manifold segment, gpm (L/min).

Q_m = Flow rate at manifold inlet, gpm (L/min).

703.3 Distribution pipe. Distribution pipe size, hole diameter and hole spacing shall be selected. The hole diameter and spacing shall be equal for each manifold segment. Distribution pipe size shall not be required to be the same for each segment. Changes in pressure in the distribution pipe shall be less than or equal to 10 percent by conforming to the following formula:

$$\sum \Delta h \, 0.2 h_d$$ **(Formula 7-1)**

For SI: 1 foot = 304.8 mm.

where:

$$\Delta h = 4.71 L \left(\frac{q}{C_h D^{2.65}} \right)^{1.85}$$

$$q = 11.79 d^2 \sqrt{h_d}$$

The Hazen-Williams friction factor, C_h, for each pipe material shall be determined in accordance with Table 703.3.

TABLE 703.3
HAZEN-WILLIAMS FRICTION FACTOR

MATERIAL	FRICTION FACTOR, C_h
ABS plastic pipe	150
Asbestos-cement pipe	140
Bituminized fiber pipe	120
Cast-iron pipe	100
Concrete pipe	110
Copper or copper-alloy tubing	150
PVC plastic pipe	150
Vitrified clay pipe	100

703.4 Manifolds. The diameter of the manifold pipe shall be determined by the following equation:

$$D_m = \left(\frac{\sum L_i F_i}{f h_d} \right)^{0.21}$$ **(Equation 7-1)**

For SI: 1 inch = 25.4 mm.

where:

$$F_i = 9.8 \times 10^{-4} Q_i$$

$$q = 11.79 d^2 \sqrt{h_d}$$

$$Q_i = Nq$$

The fraction of the total head loss at the manifold segment, f, shall be less than or equal to 0.1. The in-line pressure at the distal end of the lateral, h_d, shall be a minimum of 2.5 feet (762 mm) of head. Distribution pipes shall be connected to the manifold with tees or 90-degree (1.57 rad) ells. Distribution pipes shall have the ends capped.

703.5 Friction loss. The delivery pipe shall include all pipe between the pump and the supply end of the distribution pipe. The friction loss in the delivery pipe, F_D, shall be determined by the following equation:

$$F_D = L_D \left(\frac{3.55 Q_m}{C_h D_d^{2.63}} \right)^{1.85}$$ **(Equation 7-2)**

For SI: 1 inch of head = 25.4 mm of head.

The Hazen-Williams friction factor, C_h, for each pipe material shall be determined in accordance with Table 703.3.

The friction loss in the network pipe shall be determined by the following equation:

$$F_N = 1.31 \, h_d$$ **(Equation 7-3)**

For SI: 1 inch of head = 25.4 mm of head.

Pipe in the system shall be increased in size if the friction loss is excessive.

703.6 Force main. Size of the force main between the pump and manifold shall be based on the friction loss and velocity of effluent through the pipe. The velocity of effluent in a force main shall be not more than 5 feet per second (1524 mm/sec).

SECTION 704
BED AND TRENCH CONSTRUCTION

704.1 General. The excavation and construction for *pressure distribution system* trenches and beds shall be in accordance with Chapter 6. Aggregate shall be not less than 6 inches (152 mm) beneath the distribution pipe with 2 inches (51 mm) spread evenly above the pipe. The aggregate shall be clean, nondeteriorating 0.5-inch to 2.5-inch (12.7 mm to 64 mm) stone.

SECTION 705
PUMPS

705.1 General. Pump selection shall be based on the discharge rate and total dynamic head of the pump performance curve. The total dynamic head shall be equal to the difference in feet of elevation between the pump and distribution pipe invert plus the friction loss and a minimum of 2.5 feet (762 mm) where using low pressure distribution in the delivery pipe and network pipe.

705.2 Pump and alarm controls. The control system for the pumping chamber shall consist of a control for operating the pump and an alarm system to detect a pump. Pump start and stop depth controls shall be adjustable. Pump and alarm controls shall be of an approved type. Switches shall be resistant to sewage corrosion.

705.3 Alarm system. Alarm systems shall consist of a bell or light, mounted in the structure, and shall be located to be easily seen or heard. The high-water sensing device shall be installed approximately 2 inches (51 mm) above the depth set for the "on" pump control but below the bottom of the inlet to the pumping chamber. Alarm systems shall be installed on a separate circuit from the electrical service.

705.4 Electrical connections. Electrical connections shall be located outside the pumping chamber.

SECTION 706
DOSING

706.1 General. The dosing frequency shall be a maximum of four times daily. A volume per dose shall be established by dividing the daily wastewater flow by the dosing frequency. The dosing volume shall be a minimum of 10 times the capacity of the distribution pipe volume. Table 706.1 provides the estimated volume for various pipe diameters.

TABLE 706.1
ESTIMATED VOLUME FOR VARIOUS DIAMETER PIPES

DIAMETER (inches)	VOLUME (gallons per foot length)
1	0.041
1¼	0.064
1½	0.092
2	0.164
3	0.368
4	0.655
5	1.47

For SI: 1 inch = 25.4 mm, 1 gallon per foot = 0.012 L/mm.

CHAPTER 8
TANKS

SECTION 801
GENERAL

801.1 Scope. The provisions of this chapter shall govern the design, installation, repair and maintenance of septic tanks, treatment tanks and holding tanks.

SECTION 802
SEPTIC TANKS AND OTHER TREATMENT TANKS

802.1 General. Septic tanks shall be fabricated or constructed of welded steel, monolithic concrete, fiberglass or an approved material. Tanks shall be water tight and fabricated to constitute an individual structure, and shall be designed and constructed to withstand anticipated loads. The design of prefabricated septic tanks shall be approved. Plans for site-constructed concrete tanks shall be approved prior to construction.

802.2 Design of septic tanks. Septic tanks shall have not less than two compartments. The inlet compartment shall be not less than two-thirds of the total capacity of the tank, not less than a 500-gallon (1893 L) liquid capacity and not less than 3 feet (914 mm) wide and 5 feet (1524 mm) long. The secondary compartment of a septic tank shall have not less than a capacity of 250 gallons (946 L) and not more than one-third of the total capacity. The secondary compartment of septic tanks having a capacity more than 1,500 gallons (5678 L) shall be not less than 5 feet (1524 mm) long.

The liquid depth shall be not less than 30 inches (762 mm) and a maximum average of 6 feet (1829 mm). The total depth shall be not less than 8 inches (203 mm) greater than the liquid depth.

Rectangular tanks shall be constructed with the longest dimensions parallel to the direction of the flow.

Cylindrical tanks shall be not less than 48 inches (1219 mm) in diameter.

802.3 Inlets and outlets. The inlet and outlet on all tanks or tank compartments shall be provided with open-end coated sanitary tees or baffles made of approved materials constructed to distribute flow and retain scum in the tank or compartments. The inlet and outlet openings on all tanks shall contain a stop or other provision that will prevent the insertion of the sewer piping beyond the inside wall of the tank. The tees or baffles shall extend a minimum of 6 inches (152 mm) above and 9 inches (229 mm) below the liquid level, but shall not exceed one-third the liquid depth. A minimum of 2 inches (51 mm) of clear space shall be provided over the top of the baffles or tees. The bottom of the outlet opening shall be a minimum of 2 inches (51 mm) lower than the bottom of the inlet.

802.4 Manholes. Each compartment of a tank shall be provided with at least one manhole opening located over the inlet or outlet opening, and such opening shall be not less than 24 inches (610 mm) square or 24 inches (610 mm) in diameter. Where the inlet compartment of a septic tank exceeds 12 feet (3658 mm) in length, an additional manhole shall be provided over the baffle wall. Manholes shall terminate a maximum of 6 inches (152 mm) below the ground surface and be of the same material as the tank. Steel tanks shall have not less than a 2-inch (51 mm) collar for the manhole extensions permanently welded to the tank. The manhole extension on fiberglass tanks shall be of the same material as the tank and an integral part of the tank. The collar shall be not less than 2 inches (51 mm) high.

802.5 Manhole covers. Manhole risers shall be provided with a fitted, water-tight cover of concrete, steel, cast iron or other approved material capable of withstanding all anticipated loads. Manhole covers terminating above grade shall have an approved locking device.

802.6 Inspection opening. An inspection opening shall be provided over either the inlet or outlet baffle of every treatment tank. The opening shall be not less than 4 inches (102 mm) in diameter with a tight-fitting cover. Inspection pipes terminating above ground shall be not less than 6 inches (152 mm) above finished grade. Inspection pipes approved for terminating below grade shall be not more than 2 inches (51 mm) below finished grade, and the location shall be permanently recorded.

802.7 Capacity and sizing. The capacity of a septic tank or other treatment tank shall be based on the number of persons using the building to be served or on the volume and type of waste, whichever is greater. The minimum liquid capacity shall be 750 gallons (2839 L). Where the required capacity is to be provided by more than one tank, the minimum capacity of any tank shall be 750 gallons (2839 L). The installation of more than four tanks in series is prohibited.

802.7.1 Sizing of tank. The minimum liquid capacity for one- and two-family dwellings shall be in accordance with Table 802.7.1.

TABLE 802.7.1
SEPTIC TANK CAPACITY
FOR ONE- AND TWO-FAMILY DWELLINGS

NUMBER OF BEDROOMS	SEPTIC TANK (gallons)
1	750
2	750
3	1,000
4	1,200
5	1,425
6	1,650
7	1,875
8	2,100

For SI: 1 gallon = 3.785 L.

802.7.2 Other buildings. For buildings, the liquid capacity shall be increased above the 750-gallon (2839 L) minimum as established in Table 802.7.1. In buildings with kitchen or laundry waste, the tank capacity shall be increased to receive the anticipated volume for a 24-hour period from the kitchen or laundry or both. The liquid capacities established in Table 802.7.2 do not include employees.

Exception: One- or two-family dwellings.

802.8 Installation. Septic and other treatment tanks shall be located with a horizontal distance not less than specified in Table 802.8 between various elements. Tanks installed in ground water shall be securely anchored. A 3-inch-thick (76 mm) compacted bedding shall be provided for all septic and other treatment tank installations. The bedding material shall be sand, gravel, granite, limerock or other noncorrosive materials of such size that the material passes through a 0.5-inch (12.7 mm) screen.

TABLE 802.8
MINIMUM HORIZONTAL SEPARATION DISTANCES FOR
TREATMENT TANKS

ELEMENT	DISTANCE (feet)
Building	5
Cistern	25
Foundation wall	5
Lake, high water mark	25
Lot line	2
Pond	25
Reservoir	25
Spring	50
Stream or watercourse	25
Swimming pool	15
Water service	5
Well	25

For SI: 1 foot = 304.8 mm.

802.9 Backfill. The backfill material for steel and fiberglass tanks shall be specified for bedding and shall be tamped into place without causing damage to the coating. The backfill for concrete tanks shall be soil material, which shall pass a 4-inch (102 mm) screen and be tamped into place.

802.10 Manhole riser joints. Joints on concrete risers and manhole covers shall be tongue-and-groove or shiplap type and sealed water tight using neat cement, mortar or bituminous compound. Joints on steel risers shall be welded or flanged and bolted and water tight. Steel manhole extensions shall be bituminous coated both inside and outside. Methods of attaching fiberglass risers shall be water tight and approved.

802.11 Dosing or pumping chambers. Dosing or pumping chambers shall be fabricated or constructed of welded steel, monolithic concrete, glass-fiber-reinforced polyester or other approved materials. Manholes for dosing or pumping chambers shall terminate not less than 4 inches (102 mm) above the ground surface. Dosing or pumping chambers shall be water tight, and materials and construction specifications shall meet the same criteria specified for septic tanks in this chapter.

802.11.1 Capacity sizing. The working capacity of the dosing or pumping chamber shall be sized to permit automatic discharge of the total daily sewage flow with discharge occurring not more than four times per 24 hours. Minimum capacity of a dosing chamber shall be 500 gallons (1893 L) and a space shall be provided between the bottom of the pump and floor of the dosing or pumping chamber. A dosing chamber shall have a 1-day holding capacity located above the high-water alarm for one- and two-family dwellings based on 100 gallons (379 L) per day per bedroom, or in the case of other buildings, in accordance with Section 802.7. Minimum pump chamber sizes are indicated for one- and two-family dwellings in Table 802.11.1. Where the total developed length of distribution piping exceeds 1,000 feet (305 m), the dosing or pumping chamber shall have two siphons or pumps dosing alternately and serving one-half of the soil absorption system.

TABLE 802.11.1
PUMP CHAMBER SIZES

NUMBER OF BEDROOMS	MINIMUM PUMPING CHAMBER SIZE (gallons)
1	500
2	500
3	750
4	750
5	1,000

For SI: 1 gallon = 3.785 L.

802.12 Design of other treatment tanks. The design of other treatment tanks shall be approved on an individual basis. The capacity, sizing and installation of the tank shall be in accordance with this section except as otherwise approved. Where a treatment tank is preceded by a conventional septic tank, credit shall be given for the capacity of the septic tank.

SECTION 803
MAINTENANCE AND SLUDGE DISPOSAL

803.1 Maintenance. Septic tanks and other treatment tanks shall be cleaned whenever the sludge and scum occupy one-third of the tank's liquid capacity.

803.2 Septage. All septage shall be disposed of at an approved location.

SECTION 804
CHEMICAL RESTORATION

804.1 General. Products for chemical restoration or chemical restoration procedures for *private sewage disposal systems* shall not be used unless approved.

**TABLE 802.7.2
ADDITIONAL CAPACITY FOR OTHER BUILDINGS**

BUILDING CLASSIFICATION	CAPACITY (gallons)
Apartment buildings (per bedroom—includes automatic clothes washer)	150
Assembly halls (per person—no kitchen)	2
Bars and cocktail lounges (per patron space)	9
Beauty salons (per station—includes customers)	140
Bowling centers (per lane)	125
Bowling centers with bar (per lane)	225
Camp, day use only—no meals served (per person)	15
Campgrounds and camping resorts (per camp space)	100
Campground sanitary dump stations (per camp space) (omit camp spaces with sewer connection)	5
Camps, day and night (per person)	40
Car washes (per car handwash)	50
Catch basins—garages, motor-fuel-dispensing facilities, etc. (per basin)	100
Catch basins—truck washing (per truck)	100
Places of religious worship—no kitchen (per person)	3
Places of religious worship—with kitchen (per person)	7.5
Condominiums (per bedroom—includes automatic clothes washer)	150
Dance halls (per person)	3
Dining halls—kitchen and toilet waste—with dishwasher, food waste grinder or both (per meal served)	11
Dining halls—kitchen waste only (per meal served)	3
Drive-in restaurants—all paper service (per car space)	15
Drive-in restaurants—all paper service, inside seating (per seat)	15
Drive-in theaters (per car space)	5
Employees—in all buildings, per employee—total all shifts	20
Floor drains (per drain)	50
Hospitals (per bed space)	200
Hotels or motels and tourist rooming houses	100
Labor camps, central bathhouses (per employee)	30
Medical office buildings, clinics and dental offices Doctors, nurses, medical staff (per person) Office personnel (per person) Patients (per person)	75 20 10
Mobile home parks, homes with bathroom groups (per site)	300
Motor-fuel-dispensing facilities	10
Nursing and rest homes—without laundry (per bed space)	100
Outdoor sports facilities (toilet waste only—per person)	5
Parks, toilet wastes (per person—75 persons per acre)	5
Parks, with showers and toilet wastes (per person—75 persons per acre)	10
Restaurants—dishwasher or food waste grinder or both (per seat)	3
Restaurants—kitchen and toilet wastes (per seating space)	30
Restaurants—kitchen waste only—without dishwasher and food waste grinder (per seat)	9
Restaurants—toilet waste only (per seat)	21
Restaurants (24-hour)—dishwasher or food waste grinder (per seat)	6

(continued)

TABLE 802.7.2—continued
ADDITIONAL CAPACITY FOR OTHER BUILDINGS

BUILDING CLASSIFICATION	CAPACITY (gallons)
Restaurants (24 hour)—kitchen and toilet wastes (per seating space)	60
Retail stores—customers	1.5
Schools (per classroom—25 pupils per classroom)	450
Schools with meals served (per classroom—25 pupils per classroom)	600
Schools with meals served and showers provided (per classroom)	750
Self-service laundries (toilet waste only, per machine) Automatic clothes washers (apartments, service buildings, etc.—per machine)	50 300
Showers—public (per shower taken)	15
Swimming pool bathhouses (per person)	10

For SI: 1 gallon = 3.785 L.

SECTION 805
HOLDING TANKS

805.1 Approval. The installation of a holding tank shall not be approved where the site can accommodate the installation of any other *private sewage disposal system* specified in this code. A pumping and maintenance schedule for each holding tank installation shall be submitted to the code official.

805.2 Sizing. The minimum liquid capacity of a holding tank for one- and two-family dwellings shall be in accordance with Table 805.2. Other buildings shall have a minimum 5-day holding capacity, but not less than 2,000 gallons (7570 L). Sizing shall be in accordance with Table 802.7.2. Not more than four holding tanks shall be installed in series.

TABLE 805.2
MINIMUM LIQUID CAPACITY OF HOLDING TANKS

NUMBER OF BEDROOMS	TANK CAPACITY (gallons)
1	2,000
2	2,000
3	2,000
4	2,500
5	3,000
6	3,500
7	4,000
8	4,500

For SI: 1 gallon = 3.785 L.

805.3 Construction. Holding tanks shall be constructed of welded steel, monolithic concrete, glass-fiber reinforced polyester or other approved materials.

805.4 Installation. Tanks shall be located in accordance with Section 802.8, except the tanks shall be not less than 20 feet (6096 mm) from any part of a building. Holding tanks shall be located so the servicing manhole is located not less than 10 feet (3048 mm) from an all-weather access road or drive.

805.5 Warning device. A high-water warning device shall be installed to activate 1 foot (305 mm) below the inlet pipe. This device shall be either an audible or an approved illuminated alarm. The electrical junction box, including warning equipment junctions, shall be located outside the holding tank or housed in waterproof, explosion proof enclosures. Electrical relays or controls shall be located outside the holding tank.

805.6 Manholes. Each tank shall be provided with either a manhole not less than 24 inches (610 mm) square or a manhole with a 24-inch (610 mm) inside diameter extending not less than 4 inches (102 mm) above ground. Finish grade shall be sloped away from the manhole to divert surface water from the manhole. Each manhole cover shall have an effective locking device. Service ports in manhole covers shall be not less than 8 inches (203 mm) in diameter and shall be 4 inches (102 mm) above finished grade level. The service port shall have an effective locking cover or a brass cleanout plug.

805.7 Septic tank. The outlet shall be sealed where an approved septic tank is installed to serve as a holding tank. Removal of the inlet and outlet baffle shall not be prohibited.

805.8 Vent. Each tank shall be provided with a vent not less than 2 inches (51 mm) in diameter and shall extend not less than 12 inches (305 mm) above finished grade, terminating with a return bend fitting or approved vent cap.

CHAPTER 9

MOUND SYSTEMS

SECTION 901
GENERAL

901.1 Scope. The provisions of this chapter shall govern the design and installation of mound systems.

SECTION 902
SOIL AND SITE REQUIREMENTS

902.1 Soil borings. A minimum of three *soil borings* per site shall be conducted in accordance with Chapter 4 to determine the depth to seasonal or permanent soil saturation or bedrock. Identification of a replacement system area is not required.

902.2 Prohibited locations. A mound system shall be prohibited on sites not having the minimum depths of soil specified in Table 902.2. The installation of a mound in a filled area shall be prohibited. A mound shall not be installed in a compacted area or over a failing conventional system.

TABLE 902.2
MINIMUM SOIL DEPTHS FOR MOUND SYSTEM INSTALLATION

RESTRICTING FACTOR	MINIMUM SOIL DEPTH TO RESTRICTION (inches)
High ground water	24
Impermeable rock strata	60
Pervious rock	24
Rock fragments (50-percent volume)	24

For SI: 1 inch = 25.4 mm.

902.3 Slowly permeable soils with or without high ground water. Percolation tests shall be conducted at a depth of 20 inches to 24 inches (508 mm to 610 mm) from existing grade. Where a more slowly permeable horizon exists at less than 20 inches to 24 inches (508 mm to 610 mm), percolation tests shall be conducted within that horizon. A mound system shall be suitable for such site condition where the percolation rate is greater than 60 minutes per inch and less than or equal to 120 minutes per inch (2.4 min/mm to 4.7 min/mm).

902.4 Shallow permeable soils over creviced bedrock. Percolation tests shall be conducted at a depth of 12 inches to 18 inches (305 mm to 457 mm) from existing grade. Where a more slowly permeable horizon exists within 12 inches to 18 inches (305 mm to 457 mm), percolation tests shall be conducted within that horizon. A mound system shall be suitable for such site condition where the percolation rate is between 3 minutes per inch and 60 minutes per inch (0.12 min/mm and 2.4 min/mm).

902.5 Permeable soils with high ground water. Percolation tests shall be conducted at a depth of 20 inches to 24 inches (508 mm to 610 mm) from existing grade. Where a more slowly permeable horizon exists at less than 20 inches to 24 inches (508 mm to 610 mm), percolation tests shall be conducted within that horizon. A mound system shall be suitable for such site condition where the percolation rate is between 0 minutes per inch and 60 minutes per inch (0 min/mm and 2.4 min/mm).

902.6 Depth to pervious rock. A minimum of 24 inches (610 mm) of unsaturated natural soil shall be over creviced or porous bedrock.

902.7 Depth to high ground water. A minimum of 24 inches (610 mm) of unsaturated natural soil shall be present over high ground water as indicated by soil mottling or direct observation of water in accordance with Chapter 4.

902.8 Slopes. A mound shall not be installed on a slope greater than 6 percent where the percolation rate is between 30 and 120 minutes per inch (1.2 and 4.7 min/mm). The maximum allowable slope shall be 12 percent where there is a complex slope (two directions).

902.9 Location of mound on sloping sites. The mound shall be located so the longest dimension of the mound and the distribution lines are perpendicular to the slope. The mound shall be placed upslope and not at the base of a slope. The mound shall be situated so the effluent is not concentrated in one direction where there is a complex slope (two directions). Surface water runoff shall be diverted around the mound.

902.10 Depth to rock strata or 50 percent by volume rock fragments. A minimum of 60 inches (1524 mm) of soil shall be present over uncreviced, impermeable bedrock. Where the soil contains 50 percent coarse fragments by volume in the upper 24 inches (610 mm), a mound shall not be installed unless there is at least 24 inches (610 mm) of permeable, unsaturated soil with less than 50-percent coarse fragments located beneath this layer.

SECTION 903
SYSTEM DESIGN

903.1 Mound dimensions and design. For one- and two-family dwellings and other buildings with estimated waste water flows less than 600 gallons (2271 L) per day, the mound dimensions shall be determined in accordance with this section or Tables 903.1(1) through 903.1(12). Dimensions and corresponding letter designations listed in the tables and referenced in this section are shown in Appendix A, Figures A-6 through A-10. For buildings with estimated waste water flows exceeding 600 gallons (2271 L) per day, the mound shall be designed in accordance with this section. Daily waste water flow shall be estimated as 150 gallons (568 L) per day per bedroom for one- and two-family dwellings. For other buildings the total daily waste water flow shall be determined in accordance with Table 802.7.2.

903.1.1 Symbols. The following symbols and notations shall apply to the provisions of this section.

A = Bed or trench width, feet (mm).

A_A = Required absorption area, square feet (m^2).

B = Bed or trench length, feet (mm).

B_A = Basal area, square feet (m²).

C = Trench spacing, feet (mm).

C_I = Infiltration capacity of natural soil, gallons per foot per day (L/mm/day).

D = Fill depth, feet (mm).

E = Downslope fill depth, feet (mm).

F = Bed or trench depth, feet (mm).

G = Minimum cap and topsoil depth, feet (mm).

H = Cap and topsoil depth at center of mound, foot (mm).

I = Downslope width, feet (mm).

J = Upslope width, feet (mm).

K = End slope length, feet (mm).

L = Total mound length, feet (mm).

N = Number of trenches.

P = Distribution pipe length, feet (mm).

R = Manifold length, feet (mm).

S = Distribution pipe spacing, feet (mm).

S_D = Downslope correction factor.

S_U = Upslope correction factor.

T_W = Total daily waste-water flow, gallons per day (L/day).

W = Total mound width, feet (mm).

X = Slope, percent.

903.2 Size of absorption area. The absorption area shall be sized based on the daily waste-water flow and the infiltrative capacity of the medium sand texture fill material, equaling 1.2 gallons per square foot (0.03 L/m²) per day. The required absorption area shall be determined by the following equation:

$$A_A = \frac{T_W}{1.2 \text{ gal./ft}^2/\text{day}}$$ **(Equation 9-1)**

For SI: 1 square foot = 0.0929 m², 1 gallon = 3.785 L.

903.3 Trenches. Effluent shall be distributed in the mound through a trench system for slowly permeable soils with or without high ground water. Trench length shall be selected by determining the longest dimension perpendicular to any slope on the site. Trench width and spacing is dependent on specific site conditions. Trenches shall be 2 feet to 4 feet (610 mm to 1219 mm) wide. Trench length (B) shall be not more than 100 feet (2540 mm). Trenches shall be of equal length where more than one trench is required. A mound shall not have more than three trenches. Trench spacing (C) shall be determined by the following equation:

$$C = \frac{T_W}{N \times 0.24 \text{ gal./ft}^2/\text{day} \times B}$$ **(Equation 9-2)**

For SI: 1 gallon = 3.785 L, 1 square foot = 0.0929 m².

The calculated trench spacing (C) shall be measured from center to center of the trenches. Facilities with more than 1,500 gallons (56 775 L) per day shall be specifically engineered and approved for use with a trench system.

903.4 Beds. A long, narrow bed design shall be used for permeable soils with high water tables. The bed shall be square or rectangular for shallow permeable soils over bedrock. The bed length (B) shall be set after determining the longest dimension available and perpendicular to any slope on the site.

903.5 Mound dimensions. The mound height consists of the fill depth, bed or trench depth, the cap and topsoil depth.

903.5.1 Fill depth. The fill depth (D) shall be not less than 1 foot (305 mm) for slowly permeable soils and permeable soils with high water tables and not less than 2 feet (610 mm) of fill shall be required for shallow permeable soils over bedrock. Additional fill shall be placed at the downslope end of the bed or trench where the site is not level so the bottom of the bed or trenches is level. The downslope fill depth for bed systems shall be determined by the following equation:

$$E = D + XA$$ **(Equation 9-3)**

For SI: 1 foot = 304.8 mm.

The downslope fill depth for trench systems shall be determined by the following equation:

$$E = D + X(C + A)$$ **(Equation 9-4)**

For SI: 1 foot = 304.8 mm.

903.5.2 Bed or trench depth. The bed or trench depth (F) shall be not less than 9 inches (229 mm) and not less than 6 inches (152 mm) of aggregate shall be placed under the distribution pipes and not less than 2 inches (51 mm) of aggregate shall be placed over the top of the distribution pipes.

903.5.3 Cap and topsoil depth. The cap and topsoil depth (H) at the center of the mound shall be not less than 18 inches (457 mm), which includes 1 foot (305 mm) of subsoil and 6 inches (152 mm) of topsoil. Outer edges of the mound, G (the minimum cap and topsoil depth), shall be not less than 1 foot (305 mm), which includes 6 inches (152 mm) of subsoil and 6 inches (152 mm) of topsoil. The soil used for the cap shall be topsoil or finer textured subsoil.

903.5.4 Mound lengths. The total mound length (L) shall be determined by the following equation:

$$L = B + 2K$$ **(Equation 9-5)**

For SI: 1 foot = 304.8 mm.

where:

$$K = 3\left[\frac{(D+E)}{2} + F + H\right]$$

TABLE 903.1(1)
DESIGN CRITERIA FOR A MOUND FOR A ONE-BEDROOM HOME ON A 0- TO 6-PERCENT SLOPE
WITH LOADING RATES OF 150 GALLONS PER DAY FOR SLOWLY PERMEABLE SOIL

	DESIGN PARAMETER	SLOPE (percent)			
		0	2	4	6
A	Trench width, feet	3	3	3	3
B	Trench length, feet	42	42	42	42
	Number of trenches	1	1	1	1
D	Mound height, inches	12	12	12	12
F	Mound height, inches	9	9	9	9
G	Mound height, inches	12	12	12	12
H	Mound height, inches	18	18	18	18
I	Mound width, feet[a]	15	15	15	15
J	Mound width, feet[a]	11	8	8	8
K	Mound length, feet	10	10	10	10
L	Mound length, feet	62	62	62	62
P	Distribution pipe length, feet	20	20	20	20
	Distribution pipe diameter, inches	1	1	1	1
	Number of holes per distribution pipe[b]	9	9	9	9
	Hole spacing, inches[b]	30	30	30	30
	Hole diameter, inches[b]	0.25	0.25	0.25	0.25
W	Mound width, feet	25	26	26	26

For SI: 1 inch = 25.4 mm, 1 foot = 304.8 mm, 1 gallon = 3.785 L.

a. Additional width to obtain required basal area.

b. Last hole is located at the end of the distribution pipe, which is 15 inches from the other hole.

TABLE 903.1 (2)
DESIGN CRITERIA FOR A TWO-BEDROOM HOME FOR A MOUND ON A 0- TO 6-PERCENT SLOPE
WITH LOADING RATES OF 300 GALLONS PER DAY FOR SLOWLY PERMEABLE SOIL

	DESIGN PARAMETER	SLOPE (percent)			
		0	2	4	6
A	Trench width, feet	3	3	3	3
B	Trench length, feet	42	42	42	42
	Number of trenches	2	2	2	2
C	Trench spacing, feet	15	15	15	15
D	Mound height, inches	12	12	12	12
E	Mound height, inches	12	17	25	25
F	Mound height, inches	9	9	9	9
G	Mound height, inches	12	12	12	12
H	Mound height, inches	18	18	18	18
I	Mound width, feet[a]	12	20	20	20
J	Mound width, feet	12	8	8	8
K	Mound length, feet	10	10	10	10
L	Mound length, feet	62	62	62	62
P	Distribution pipe length, feet	20	20	20	20
	Distribution pipe diameter, inches	1	1	1	1
	Number of holes per distribution pipes[b]	9	9	9	9
	Hole spacing, inches[b]	30	30	30	30
	Hole diameter, inches	0.25	0.25	0.25	0.25
R	Manifold length, feet	15	15	15	15
	Manifold diameter, inches[c]	2	2	2	2
W	Mound width, feet	42	46	46	46

For SI: 1 inch = 25.4 mm, 1 foot = 304.8 mm, 1 gallon = 3.785 L.

a. Additional width to obtain required basal area.

b. Last hole is located at the end of the distribution pipe, which is 15 inches from the other hole.

c. Diameter dependent on the size of pipe from pump and inlet position.

TABLE 903.1(3)
DESIGN CRITERIA FOR A THREE-BEDROOM HOME FOR A MOUND ON A 0- TO 6-PERCENT SLOPE
WITH LOADING RATES OF 450 GALLONS PER DAY FOR SLOWLY PERMEABLE SOIL

	DESIGN PARAMETER	SLOPE (percent)			
		0	2	4	6
A	Trench width, feet	3	3	3	3
B	Trench length, feet	63	63	63	63
	Number of trenches	2	2	2	2
C	Trench spacing, feet	15	15	15	15
D	Mound height, inches	12	12	12	12
E	Mound height, inches	12	17	20	25
F	Mound height, inches	9	9	9	9
G	Mound height, inches	12	12	12	12
H	Mound height, inches	18	18	18	18
I	Mound width, feet[a]	12	20	20	20
J	Mound width, feet[a]	12	8	8	8
K	Mound length, feet	10	10	10	10
L	Mound length, feet	62	62	62	62
P	Distribution pipe length, feet	31	31	31	31
	Distribution pipe diameter, inches	$1^1/_4$	$1^1/_4$	$1^1/_4$	$1^1/_4$
	Number of holes per distribution pipe[b]	13	13	13	13
	Hole spacing[b], inches	30	30	30	30
	Hole diameter, inches	0.25	0.25	0.25	0.25
R	Manifold length, feet	15	15	15	15
	Manifold diameter, inches[c]	2	2	2	2
W	Mound width, feet	42	46	46	46

For SI: 1 inch = 25.4 mm, 1 foot = 304.8 mm, 1 gallon = 3.785 L.

a. Additional width to obtain required basal area.

b. First hole is located 12 inches from the manifold.

c. Diameter dependent on the size of pipe from pump and inlet position.

TABLE 903.1(4)
DESIGN CRITERIA FOR A FOUR-BEDROOM HOME FOR A MOUND ON A 0- TO 6-PERCENT SLOPE WITH
LOADING RATES OF 600 GALLONS PER DAY FOR SLOWLY PERMEABLE SOIL

	DESIGN PARAMETER	SLOPE (percent)			
		0	2	4	6
A	Trench width, feet	3	3	3	3
B	Trench length, feet	56	56	56	56
	Number of trenches	3	3	3	3
C	Trench spacing, feet	15	15	15	15
D	Mound height, inches	12	12	12	12
E	Mound height, inches	12	20	28	36
F	Mound height, inches	9	9	9	9
G	Mound height, inches	12	12	12	12
H	Mound height, inches	24	24	24	24
I	Mound width, feet[a]	12	20	20	20
J	Mound width, feet[a]	12	8	8	8
K	Mound length, feet	12	12	12	14
L	Mound length, feet	80	80	80	84
P	Distribution pipe length, feet	27.5	27.5	27.5	27.5
	Distribution pipe diameter, inches	$1^1/_4$	$1^1/_4$	$1^1/_4$	$1^1/_4$
	Number of holes per distribution pipe[b]	12	12	12	12
	Hole spacing, inches[b]	30	30	30	30
	Hole diameter, inches	0.25	0.25	0.25	0.25
R	Manifold length, feet	30	30	30	30
	Manifold diameter, inches[c]	2	2	2	2
W	Mound width, feet	57	61	61	61

For SI: 1 inch = 25.4 mm, 1 foot = 304.8 mm, 1 gallon = 3.785 L.

a. Additional width to obtain required basal area.

b. Last hole is located at the end of the distribution pipe, which is 15 inches from the previous hole.

c. Diameter dependent on the size of pipe from pump and inlet position.

TABLE 903.1(5)
DESIGN CRITERIA FOR A ONE-BEDROOM HOME FOR A MOUND ON A 0- TO 12-PERCENT SLOPE WITH
LOADING RATES OF 150 GALLONS PER DAY FOR SHALLOW PERMEABLE SOIL OVER CREVICED BEDROCK

	DESIGN PARAMETER	PERCOLATION RATE (minutes per inch) SLOPE (percent)						
		3 to 60				3 to less than 30		
		0	2	4	6	8	10[a]	12[a]
A	Bed width, feet[b]	10	10	10	10	10	10	10
B	Bed length, feet	13	13	13	13	13	13	13
D	Mound height, inches	24	24	24	24	24	24	24
E	Mound height, inches	24	26	29	31	34	36	38
F	Mound height, inches	9	9	9	9	9	9	9
G	Mound height, inches	12	12	12	12	12	12	12
H	Mound height, inches	18	18	18	18	18	18	18
I	Mound width, feet	12	13	14	17	18	21	26
J	Mound width, feet	12	11	10	10	9	9	9
K	Mound length, feet	12	12	12	13	13	13	15
L	Mound length, feet	37	37	37	39	39	39	43

(continued)

TABLE 903.1(5)—continued
DESIGN CRITERIA FOR A ONE-BEDROOM HOME FOR A MOUND ON A 0- TO 12-PERCENT SLOPE WITH
LOADING RATES OF 150 GALLONS PER DAY FOR SHALLOW PERMEABLE SOIL OVER CREVICED BEDROCK

	DESIGN PARAMETER	PERCOLATION RATE (minutes per inch) SLOPE (percent)						
		3 to 60				3 to less than 30		
		0	2	4	6	8	10[a]	12[a]
P	Distribution pipe length, feet[c]	12.5	12.5	12.5	12.5	12.5	12.5	12.5
	Distribution pipe diameter, inches	1	1	1	1	1	1	1
	Number of distribution pipes	6	6	6	6	6	6	6
R	Manifold length, feet	6	6	6	6	6	6	6
	Manifold diameter, inches[c]	2	2	2	2	2	2	2
S	Distribution pipe spacing, feet	3	3	3	3	3	3	3
	Number of holes per distribution pipe[d]	6	6	6	6	6	6	6
	Hole spacing, inches[d]	30	30	30	30	30	30	30
	Hole diameter, inches	0.25	0.25	0.25	0.25	0.25	0.25	0.25
W	Mound width, feet	34	34	34	37	37	41	45

For SI: 1 inch = 25.4 mm, 1 foot = 304.8 mm, 1 gallon = 3.785 L, 1 minute per inch = 2.4 s/mm.

a. On sites with a 10- to 12-percent slope, the fill depth (D) shall be reduced to a minimum of 1.5 feet or the bed width shall be reduced to decrease E [downslope fill depth, feet (mm)].

b. Bed widths shall not be limited.

c. Use a manifold with distribution pipes on only one side.

d. Last hole is located at the end of the distribution pipe, which is 15 inches from the previous hole.

TABLE 903.1(6)
DESIGN CRITERIA FOR A TWO-BEDROOM HOME FOR A MOUND ON A 0- TO 12-PERCENT SLOPE WITH
LOADING RATES OF 300 GALLONS PER DAY FOR SHALLOW PERMEABLE SOIL OVER CREVICED BEDROCK

	DESIGN PARAMETER	PERCOLATION RATE (minutes per inch) SLOPE (percent)						
		3 to 60				3 to less than 30		
		0	2	4	6	8	10[a]	12[a]
A	Bed width, feet[b]	10	10	10	10	10	10	10
B	Bed length, feet	25	25	25	25	25	25	25
D	Mound height, inches	24	24	24	24	24	24	24
E	Mound height, inches	24	26	29	31	34	36	38
F	Mound height, inches	9	9	9	9	9	9	9
G	Mound height, inches	12	12	12	12	12	12	12
H	Mound height, inches	18	18	18	18	18	18	18
I	Mound width, feet	12	13	14	17	18	21	26
J	Mound width, feet	12	11	10	10	9	9	9
K	Mound length, feet	12	12	12	13	13	13	15
L	Mound length, feet	49	49	49	51	51	51	55
P	Distribution pipe length, feet[c]	12	12	12	12	12	12	12
	Distribution pipe diameter, inches	1	1	1	1	1	1	1
	Number of distribution pipes	6	6	6	6	6	6	6
R	Manifold length, feet	6	6	6	6	6	6	6
	Manifold diameter, inches	2	2	2	2	2	2	2
S	Distribution pipe spacing, feet	3	3	3	3	3	3	3
	Number of holes per distribution pipe[d]	5	5	5	5	5	5	5
	Hole spacing, inches[d]	30	30	30	30	30	30	30
	Hole diameter, inches	0.25	0.25	0.25	0.25	0.25	0.25	0.25
W	Mound width, feet	34	34	34	37	37	41	45

For SI: 1 inch = 25.4 mm, 1 foot = 304.8 mm, 1 gallon = 3.785 L, 1 minute per inch = 2.4 s/mm.

(continued)

40   **2012 INTERNATIONAL PRIVATE SEWAGE DISPOSAL CODE®**

TABLE 903.1(6)—continued
DESIGN CRITERIA FOR A TWO-BEDROOM HOME FOR A MOUND ON A 0- TO 12-PERCENT SLOPE WITH LOADING RATES OF 300 GALLONS PER DAY FOR SHALLOW PERMEABLE SOIL OVER CREVICED BEDROCK

a. On sites with a 10- to 12-percent slope, the fill depth (*D*) shall be reduced to a minimum of 1.5 feet or the bed width shall be reduced to decrease *E* [downslope fill depth, feet (mm)].

b. Bed widths shall not be limited.

c. This design is based on a manifold with distribution pipes on both sides. An alternative design basis is 24-foot distribution pipes, with manifold at the end.

d. Last hole is located 9 inches from the end of the distribution pipe.

TABLE 903.1(7)
DESIGN CRITERIA FOR A THREE-BEDROOM HOME FOR A MOUND ON A 0- TO 12-PERCENT SLOPE WITH LOADING RATES OF 450 GALLONS PER DAY FOR SHALLOW PERMEABLE SOIL OVER CREVICED BEDROCK

	DESIGN PARAMETER	PERCOLATION RATE (minutes per inch) SLOPE (percent)						
		3 to 60		3 to less than 30				
		0	2	4	6	8	10[a]	12[a]
A	Bed width, feet[b]	10	10	10	10	10	10	10
B	Bed length, feet	38	38	38	38	38	38	38
D	Mound height, inches	24	24	24	24	24	24	24
E	Mound height, inches	24	26	29	31	34	36	38
F	Mound height, inches	9	9	9	9	9	9	9
G	Mound height, inches	12	12	12	12	12	12	12
H	Mound height, inches	18	18	18	18	18	18	18
I	Mound width, feet	12	13	14	17	18	21	26
J	Mound width, feet	12	11	10	10	9	9	9
K	Mound length, feet	12	12	12	13	13	13	15
L	Mound length, feet	62	62	62	64	64	64	68
P	Distribution pipe length, feet[c]	18.5	18.5	18.5	18.5	18.5	18.5	18.5
	Distribution pipe diameter, inches	1	1	1	1	1	1	1
	Number of distribution pipes	6	6	6	6	6	6	6
R	Manifold length, feet	6	6	6	6	6	6	6
	Manifold diameter, inches	2	2	2	2	2	2	2
S	Distribution pipe spacing, feet	3	3	3	3	3	3	3
	Number of holes per distribution pipe[d]	8	8	8	8	8	8	8
	Hole spacing, inches[d]	30	30	30	30	30	30	30
	Hole diameter, inches	0.25	0.25	0.25	0.25	0.25	0.25	0.25
W	Mound width, feet	34	34	34	37	37	41	45

For SI: 1 inch = 25.4 mm, 1 foot = 304.8 mm, 1 gallon = 3.785 L, 1 minute per inch = 2.4 s/mm.

a. On sites with a 10- to 12-percent slope, the fill depth (*D*) shall be reduced to a minimum of 1.5 feet or the bed width shall be reduced to decrease *E* [downslope fill depth, feet (mm)].

b. Bed widths shall not be limited.

c. Use a manifold with distribution pipes on only one side.

d. Last hole is located at the end of the distribution pipe, which is 27 inches from the previous hole.

TABLE 903.1(8)
DESIGN CRITERIA FOR A FOUR-BEDROOM HOME FOR A MOUND ON A 0- TO 12-PERCENT SLOPE WITH
LOADING RATES OF 600 GALLONS PER DAY FOR SHALLOW PERMEABLE SOIL OVER CREVICED BEDROCK

	DESIGN PARAMETER	PERCOLATION RATE (minutes per inch) SLOPE (percent)						
		3 to 60				3 to less than 30		
		0	2	4	6	8	10[a]	12[a]
A	Bed width, feet[b]	10	10	10	10	10	10	10
B	Bed length, feet	50	50	50	50	50	50	50
D	Mound height, inches	24	24	24	24	24	24	24
E	Mound height, inches	24	26	29	31	34	36	38
F	Mound height, inches	9	9	9	9	9	9	9
G	Mound height, inches	12	12	12	12	12	12	12
H	Mound height, inches	18	18	18	18	18	18	18
I	Mound width, feet	12	13	14	17	18	21	26
J	Mound width, feet	12	11	10	10	9	9	9
K	Mound length, feet	12	12	12	13	13	13	15
L	Mound length, feet	74	74	74	76	76	76	78
P	Distribution pipe length, feet[c] Distribution pipe diameter, inches Number of distribution pipes	24.5 1 6	24.5 1 6	24.5 1 6	24.5 1 6	24.5 1 6	24.5 1 6	24.5 1 6
R	Manifold length, feet Manifold diameter, inches	6 2	6 2	6 2	6 2	6 2	6 2	6 2
S	Distribution pipe spacing, feet Number of holes per distribution pipe[d] Hole spacing, inches[d] Hole diameter, inches	3 10 30 0.25	3 10 30 0.25	3 10 30 0.25	3 10 30 0.25	3 10 30 0.25	3 10 30 0.25	3 10 30 0.25
W	Mound width, feet	34	34	34	37	37	41	45

For SI: 1 inch 25.4 mm, 1 foot = 304.8 mm, 1 gallon = 3.785 L, 1 minute per inch = 2.4 s/mm.

a. On sites with a 10- to 12-percent slope, the fill depth (*D*) shall be reduced to a minimum of 1.5 feet or the bed width shall be reduced to decrease *E* [downslope fill depth, feet (mm)].

b. Bed widths shall not be limited.

c. Use a manifold with distribution pipes on only one side.

d. Last hole is located 9 inches from the end of the distribution pipe.

TABLE 903.1(9)
DESIGN CRITERIA FOR A ONE-BEDROOM HOME FOR A MOUND ON A 0- TO 12-PERCENT SLOPE WITH LOADING RATES OF 150
GALLONS PER DAY FOR PERMEABLE SOIL WITH A HIGH WATER TABLE

	DESIGN PARAMETER	PERCOLATION RATE (minutes per inch) SLOPE (percent)						
		0 to 60				0 to less than 30		
		0	2	4	6	8	10	12
A	Bed width, feet	4	4	4	4	4	4	4
B	Bed length, feet	32	32	32	32	32	32	32
D	Mound height, inches	12	12	12	12	12	12	12
E	Mound height, inches	12	13	14	14	16	17	18
F	Mound height, inches	9	9	9	9	9	9	9
G	Mound height, inches	12	12	12	12	12	12	12
H	Mound height, inches	18	18	18	18	18	18	18
I	Mound width, feet	9	10	11	12	13	14	15
J	Mound width, feet	9	9	8	8	7	7	6

(continued)

TABLE 903.1(9)—continued
DESIGN CRITERIA FOR A ONE-BEDROOM HOME FOR A MOUND ON A 0- TO 12-PERCENT SLOPE WITH
LOADING RATES OF 150 GALLONS PER DAY FOR PERMEABLE SOIL WITH A HIGH WATER TABLE

	DESIGN PARAMETER	PERCOLATION RATE (minutes per inch) SLOPE (percent)						
		0 to 60				0 to less than 30		
		0	2	4	6	8	10	12
K	Mound length, feet	10	10	10	10	10	11	11
L	Mound length, feet	52	52	52	52	52	53	53
P	Distribution pipe length	15.5	15.5	15.5	15.5	15.5	15.5	15.5
	Distribution pipe diameter, inches	1	1	1	1	1	1	1
	Number of distribution pipes	2	2	2	2	2	2	2
	Number of holes per distribution pipe[a]	7	7	7	7	7	7	7
	Hole spacing, inches[a]	30	30	30	30	30	30	30
	Hole diameter, inches	0.25	0.25	0.25	0.25	0.25	0.25	0.25
W	Mound width, feet	22	23	23	24	24	25	25

For SI: 1 inch = 25.4 mm, 1 foot = 304.8 mm, 1 gallon = 3.785 L, 1 minute per inch = 2.4 s/mm.

a. Last hole is located at the end of the distribution pipe, which is 21 inches from the previous hole.

TABLE 903.1(10)
DESIGN CRITERIA FOR A TWO-BEDROOM HOME FOR A MOUND ON A 0- TO 12-PERCENT SLOPE WITH
LOADING RATES OF 300 GALLONS PER DAY FOR PERMEABLE SOIL WITH A HIGH WATER TABLE

	DESIGN PARAMETER	PERCOLATION RATE (minutes per inch) SLOPE (percent)						
		0 to 60				0 to less than 30		
		0	2	4	6	8	10	12
A	Bed width, feet	6	6	6	6	6	6	6
B	Bed length, feet	42	42	42	42	42	42	42
D	Mound height, inches	12	12	12	12	12	12	12
E	Mound height, inches	12	13	14	17	18	19	22
F	Mound height, inches	9	9	9	9	9	9	9
G	Mound height, inches	12	12	12	12	12	12	12
H	Mound height, inches	18	18	18	18	18	18	18
I	Mound width, feet	9	10	11	12	13	15	16
J	Mound width, feet	9	9	8	8	7	7	6
K	Mound length, feet	10	10	10	10	10	11	11
L	Mound length, feet	62	62	62	62	62	64	64
P	Distribution pipe length, feet[a]	20	20	20	20	20	20	20
	Distribution pipe diameter, inches	1	1	1	1	1	1	1
	Number of distribution pipes	4	4	4	4	4	4	4
R	Manifold length, feet	3	3	3	3	3	3	3
	Manifold diameter, inches	2	2	2	2	2	2	2
S	Distribution pipe spacing, feet	3	3	3	3	3	3	3
	Number of holes per distribution pipe[b]	9	9	9	9	9	9	9
	Hole spacing, inches[b]	30	30	30	30	30	30	30
	Hole diameter, inches	0.25	0.25	0.25	0.25	0.25	0.25	0.25
W	Mound width, feet	24	25	25	26	26	28	29

For SI: 1 inch = 25.4 mm, 1 foot = 304.8 mm, 1 gallon = 3.785 L, 1 minute per inch = 2.4 s/mm.

a. Use a manifold with distribution pipes only on one side.

b. Last hole is located at the end of the distribution pipe, which is 15 inches from the previous hole.

TABLE 903.1(11)
**DESIGN CRITERIA FOR A THREE-BEDROOM HOME FOR A MOUND ON A 0- TO 12-PERCENT SLOPE WITH
LOADING RATES OF 450 GALLONS PER DAY FOR PERMEABLE SOIL WITH A HIGH WATER TABLE**

	DESIGN PARAMETER	PERCOLATION RATE (minutes per inch) SLOPE (percent)						
		0 to 60				0 to less than 30		
		0	2	4	6	8	10	12
A	Bed width, feet	8	8	8	8	8	8	8
B	Bed length, feet	47	47	47	47	47	47	47
D	Mound height, inches	12	12	12	12	12	12	12
E	Mound height, inches	12	12	16	18	19	22	24
F	Mound height, inches	9	9	9	9	9	9	9
G	Mound height, inches	12	12	12	12	12	12	12
H	Mound height, inches	18	18	18	18	18	18	18
I	Mound width, feet	9	11	12	13	15	17	18
J	Mound width, feet	9	9	8	8	7	7	6
K	Mound length, feet	10	10	10	10	10	11	12
L	Mound length, feet	67	67	67	67	69	69	71
P	Distribution pipe length, feet Distribution pipe diameter, inches Number of distribution pipes	23 1 6	23 1 6	23 1 6	23 1 6	23 1 6	23 1 6	23 1 6
R	Manifold length, feet Manifold diameter, inches	64 2	64 2	64 2	64 2	64 2	64 2	64 2
S	Distribution pipe spacing, feet Number of holes per distribution pipe[a] Hole spacing, inches[a] Hole diameter, inches	32 10 30 0.25	32 10 30 0.25	32 10 30 0.25	32 10 30 0.25	32 10 30 0.25	32 10 30 0.25	32 10 30 0.25
W	Mound width, feet	26	28	28	29	30	32	32

For SI: 1 inch = 25.4 mm, 1 foot = 304.8 mm, 1 gallon = 3.785 L, 1 minute per inch = 2.4 s/mm.

a. Last hole is located at the end of the distribution pipe, which is 21 inches from the previous hole.

TABLE 903.1(12)
**DESIGN CRITERIA FOR A FOUR-BEDROOM HOME FOR A MOUND ON A 0- TO 12-PERCENT SLOPE WITH
LOADING RATES OF 600 GALLONS PER DAY FOR PERMEABLE SOIL WITH A HIGH WATER TABLE**

	DESIGN PARAMETER	PERCOLATION RATE (minutes per inch) SLOPE (percent)						
		0 to 60				0 to less than 30		
		0	2	4	6	8	10	12
A	Bed width, feet	10	10	10	10	10	10	10
B	Bed length, feet	50	50	50	50	50	50	50
D	Mound height, inches	12	12	12	12	12	12	12
E	Mound height, inches	12	14	17	19	22	24	26
F	Mound height, inches	9	9	9	9	9	9	9
G	Mound height, inches	12	12	12	12	12	12	12
H	Mound height, inches	18	18	18	18	18	18	18
I	Mound width, feet	9	11	13	14	17	18	19
J	Mound width, feet	9	9	8	8	7	7	6
K	Mound length, feet	10	10	10	10	11	11	12
L	Mound length, feet	70	70	70	70	72	72	74

(continued)

TABLE 903.1(12)—continued
TABLE 903.1(12)—continued
DESIGN CRITERIA FOR A FOUR-BEDROOM HOME FOR A MOUND ON A 0- TO 12-PERCENT SLOPE WITH
LOADING RATES OF 600 GALLONS PER DAY FOR PERMEABLE SOIL WITH A HIGH WATER TABLE

DESIGN PARAMETER		PERCOLATION RATE (minutes per inch) SLOPE (percent)						
		0 to 60				0 to less than 30		
		0	2	4	6	8	10	12
P	Distribution pipe length, feet	24.5	24.5	24.5	24.5	24.5	24.5	24.5
	Distribution pipe diameter, inches	1	1	1	1	1	1	1
	Number of distribution pipes	6	6	6	6	6	6	6
R	Manifold length, feet	6	6	6	6	6	6	6
	Manifold diameter, inches	2	2	2	2	2	2	2
S	Distribution pipe spacing, feet	3	3	3	3	3	3	3
	Number of holes per distribution pipe[a]	10	10	10	10	10	10	10
	Hole spacing, inches[a]	30	30	30	30	30	30	30
	Hole diameter, inches	0.25	0.25	0.25	0.25	0.25	0.25	0.25
W	Mound width, feet	28	29	31	32	34	35	36

For SI: 1 inch = 25.4 mm, 1 foot = 304.8 mm, 1 gallon = 3.785 L, 1 minute per inch = 2.4 s/mm.

a. Last hole is 9 inches from the end of the distribution pipe.

903.5.5 Mound widths. The mound width for a bed system shall be determined by the following equation:

$$W = J + A + I \qquad \textbf{(Equation 9-6)}$$

For SI: 1 foot = 304.8 mm.

The mound width for a trench system shall be determined by the following equation:

$$W = J + \frac{A}{2} + C(N-1) - \frac{A}{2} + I \qquad \textbf{(Equation 9-7)}$$

For SI: 1 foot = 304.8 mm.

where:

$$J = 3(D + F + G)S_U$$

$$I = 3(E + F + G)S_D$$

The upslope correction factor (S_U) and the downslope correction factor (S_D) shall be determined based on the slope in accordance with Table 903.5.5.

TABLE 903.5.5
DOWNSLOPE AND UPSLOPE WIDTH CORRECTIONS
FOR MOUNDS ON SLOPING SITES

SLOPE (percent)	DOWNSLOPE CORRECTION FACTOR (S_D)	UPSLOPE CORRECTION FACTOR (S_U)
0	1	1
1	1.03	0.97
2	1.06	0.94
3	1.10	0.915
4	1.14	0.89
5	1.18	0.875
6	1.22	0.86
7	1.27	0.83
8	1.32	0.80
9	1.38	0.785
10	1.44	0.77
11	1.51	0.75
12	1.57	0.73

903.6 Basal area. The minimum basal area required shall be determined by the following equation:

$$B_A = \frac{T_W}{C_I} \qquad \textbf{(Equation 9-8)}$$

For SI: 1 square foot = 0.0929 m².

The infiltrative capacity of natural soil shall be determined on the percolation rate in accordance with Table 903.6.

TABLE 903.6
INFILTRATIVE CAPACITY OF NATURAL SOIL

PERCOLATION RATE (minutes per inch)	INFILTRATIVE CAPACITY (gallons per foot per day)
Less than 30	1.2
30 to 60	0.74
More than 60 to 120	0.24

For SI: 1 gallon per foot per day = 0.012 L/mm/day,
1 minute per inch = 2.4 s/mm.

903.6.1 Basal area available in bed system. The available basal area for a bed system shall be determined by one of the following equations:

$$B_A = B(A + I) \text{ for sloping sites} \qquad \textbf{(Equation 9-9)}$$

$$B_A = BW \text{ for level sites} \qquad \textbf{(Equation 9-10)}$$

For SI: 1 square foot = 0.0929 m².

903.6.2 Basal area available in trench system. The available basal area for a trench system shall be determined by one of the following equations:

$$B_A = B\left(W + J + \frac{A}{2}\right) \text{ for sloping sites} \qquad \textbf{(Equation 9-11)}$$

$$B_A = BW \qquad \text{for level sites} \qquad \textbf{(Equation 9-12)}$$

For SI: 1 square foot = 0.0929 m².

903.6.3 Adequacy of basal area. The downslope width (*I*) on a sloping site shall be increased or the upslope width (*J*) and downslope (*I*) widths on a level site shall be increased until sufficient area is available if the basal area available is not equal to or greater than the basal area required.

903.7 Dose volume and pump. The dose volume and pump shall conform to the requirements of Chapters 7 and 8.

SECTION 904
CONSTRUCTION TECHNIQUES

904.1 General. Construction shall not commence where the soil is so wet a soil wire forms when the soil is rolled between the hands. Installation of mound systems where the soil on the site is frozen shall be prohibited for new construction.

904.2 Site preparation. Excess vegetation shall be cut and removed from the mound area. Small trees shall be cut to grade surface, leaving the stumps in place.

904.3 Force main. The force main from the pumping chamber shall be installed before the mound site is plowed. The force main shall be sloped uniformly toward the pumping chamber so the force main drains after each dose.

904.4 Plowing. The site shall be plowed with a moldboard plow or chisel plow. The site shall be plowed to a depth of 7 inches to 8 inches (178 mm to 203 mm) with the plowing perpendicular to the slope. Rototillers shall not be used. The sand fill shall be placed immediately after plowing. All foot and vehicular traffic shall be kept off the plowed area.

904.5 Sand fill material. The fill material shall be medium sand texture defined as 25 percent or more very coarse, coarse and medium sand and a maximum of 50 percent fine sand, very fine sand, silt and clay. The percentage of silt plus one and one-half times the percentage of clay shall not exceed 15 percent. Fill materials with higher content of silt and clay shall not be used.

    904.5.1 Placement of sand fill. The medium sand fill shall be moved into place from the upslope and side edges of the plowed area. Vehicular traffic shall be prohibited in the area extending to 25 feet (7620 mm) beyond the downslope edge of the mound. The sand fill shall be moved into place with a track-type tractor and not less than 6 inches (152 mm) of sand shall be kept beneath the tracks at all times.

904.6 Installation of the absorption area. The bed or trenches shall be formed within the sand fill. The bottom of the trenches or bed shall be level. The elevation of the bottom of the trenches or bed shall be checked at the upslope and downslope edges to ensure that the fill has been placed to the proper depth.

904.7 Placement of the aggregate. A minimum of 6 inches (152 mm) of coarse aggregate ranging in size from $^1/_2$ inch to $2^1/_2$ inches (12.7 mm to 64 mm) shall be placed in the bed or trench excavation. The top of the aggregate shall be level.

904.8 Distribution system. Distribution systems shall be placed on the aggregate, with the holes located on the bottom of the distribution pipe. The ends of all distribution pipes shall be marked at the surface, and an observation pipe shall be placed to the bottom of the bed or each trench.

904.9 Cover. The top of the bed or trenches shall be covered with not less than 2 inches (51 mm) of aggregate ranging in size from $^1/_2$ inch to $2^1/_2$ inches (12.7 mm to 64 mm) and not less than 4 inches to 5 inches (102 mm to 127 mm) of uncompacted straw or marsh hay or approved synthetic fabric shall be placed over the aggregate. Cap and topsoil covers shall be in place and the mound shall be seeded immediately and protected from erosion.

904.10 Maintenance. When the septic tank is pumped, the pump chamber shall be inspected and pumped to remove any solids present. Excess traffic in the mound area shall be avoided.

CHAPTER 10

CESSPOOLS

SECTION 1001
GENERAL

1001.1 Scope. The provisions of this chapter shall govern the design and installation of *cesspools*.

1001.2 Application. *Cesspools* shall not be installed, except where approved by the code official. A *cesspool* shall be considered as only a temporary expedient pending the construction of a public sewer; as an overflow facility where installed in conjunction with an existing *cesspool*; or as a means of sewage disposal for limited, minor or temporary applications.

1001.3 Construction. *Cesspools* shall conform to the construction requirements of Section 605.3 for *seepage pits*. The *seepage pit* shall have a minimum sidewall of 20 feet (6096 mm) below the inlet opening. Where a stratum of gravel or equally pervious material of 4 feet (1219 mm) or more in thickness is found, the sidewall need not be more than 10 feet (3048 mm) below the inlet.

CHAPTER 11

RESIDENTIAL WASTE WATER SYSTEMS

SECTION 1101
GENERAL

1101.1 Scope. The provisions of this chapter shall govern residential waste water systems.

1101.2 Residential waste water treatment systems. The regulations for materials, design, construction and performance shall comply with NSF 40.

CHAPTER 12

INSPECTIONS

SECTION 1201
GENERAL

1201.1 Scope. The provisions of this chapter shall govern the inspection of *private sewage disposal systems*.

SECTION 1202
INSPECTIONS

1202.1 Initial inspection procedures. All *private sewage disposal systems* shall be inspected after construction, but before backfilling. The code official shall be notified when the *private sewage disposal system* is ready for inspection.

1202.2 Preparation for inspection. The installer shall make such arrangements as will enable the code official to inspect all parts of the system when a *private sewage disposal system* is ready. The installer shall provide the proper apparatus and equipment for conducting the inspection and furnish such assistance as is necessary to conduct the inspection.

1202.3 Covering of work. A *private sewage disposal system* or part thereof shall not be backfilled until such system has been inspected and approved. Any system that has been covered before being inspected and approved shall be uncovered as required by the code official.

1202.4 Other inspections. In addition to the required inspection prior to backfilling, the code official shall conduct any other inspections deemed necessary to determine compliance with this code.

1202.5 Inspections for additions, alterations or modifications. Additions, alterations or modifications to *private sewage disposal systems* shall be inspected.

1202.6 Defects in materials and workmanship. Where inspection discloses defective material, design or siting or unworkmanlike construction not conforming to the requirements of this code, the nonconforming parts shall be removed, replaced and reinspected.

CHAPTER 13

NONLIQUID SATURATED TREATMENT SYSTEMS

SECTION 1301
GENERAL

1301.1 Scope. The provisions of this chapter shall govern nonliquid saturated treatment systems.

1301.2 Nonliquid saturated treatment systems. The regulations for materials, design, construction and performance shall comply with NSF Standard 41.

CHAPTER 14
REFERENCED STANDARDS

ASTM

ASTM International
100 Barr Harbor Drive
West Conshohocken, PA 19428-2959

Standard reference number	Title	Referenced in code section number
A 74—06	Specification for Cast Iron Soil Pipe and Fittings	Table 505.1
A 888—09	Specification for Hubless Cast Iron Soil Pipe and Fittings for Sanitary and Storm Drain, Waste, and Vent Piping Application	Table 505.1
B 32—08	Specification for Solder Metal	505.8.2
B 75—02	Specification for Seamless Copper Tube	Table 505.1
B 88—03	Specification for Seamless Copper Water Tube	Table 505.1
B 251—02e01	Specification for General Requirements for Wrought Seamless Copper and Copper-alloy Tube	Table 505.1
B 813—00(2009)	Specification for Liquid and Paste Fluxes for Soldering of Copper and Copper-alloy Tube	505.8.2
B 828—02	Practice for Making Capillary Joints by Soldering of Copper and Copper-alloy Tube and Fittings	505.8.2
C 4—04e01	Specification for Clay Drain Tile and Perforated Clay Drain Tile	Table 505.1
C 14—07	Specification for Nonreinforced Concrete Sewer, Storm Drain and Culvert Pipe	Table 505.1
C 76—08a	Specification for Reinforced Concrete Culvert, Storm Drain and Sewer Pipe	Table 505.1
C 425—04	Specification for Compression Joints for Vitrified Clay Pipe and Fittings	505.12, 505.13
C 428—05(2006)	Specification for Asbestos-cement Nonpressure Sewer Pipe	Table 505.1
C 443—05a	Specification for Joints for Concrete Pipe and Manholes, Using Rubber Gaskets	505.7, 505.13
C 564—07e1	Specification for Rubber Gaskets for Cast Iron Soil Pipe and Fittings	505.6.2, 505.6.3, 505.13
C 700—07a	Specification for Vitrified Clay Pipe, Extra Strength, Standard Strength and Perforated	Table 505.1
C 913—08	Specification for Precast Concrete Water and Wastewater Structures	504.2
C 1173—08	Specification for Flexible Transition Couplings for Underground Piping Systems	505.3.1, 505.5.1, 505.7, 505.10.1, 505.12, 505.13
C 1277—08	Specification for Shielding Coupling Joining Hubless Cast-iron Pipe and Fittings	505.6.3
C 1440—08	Specification for Thermoplastic Elastomeric (TPE) Gasket Materials for Drain, Waste and Vent (DWV), Sewer, Sanitary and Storm Plumbing Systems	505.13
C 1460—08	Specification for Shielded Transition Couplings for Use with Dissimilar DWV Pipe and Fittings Above Ground	505.13
C 1461—08	Specification for Mechanical Couplings Using Thermoplastic Elastomeric (TPE) Gaskets for Joining Drain, Waste and Vent (DWV) Sewer, Sanitary and Storm Plumbing Systems for Above and Below Ground Use	505.13
D 1869—95(2005)e1	Specification for Rubber Rings for Asbestos-cement Pipe	505.4, 505.13
D 2235—04	Specification for Solvent Cement for Acrylonitrile-butadiene-styrene (ABS) Plastic Pipe and Fittings	505.3.2, 505.5.2
D 2564—04e01	Specification for Solvent Cements for Poly Vinyl Chloride (PVC) Plastic Piping Systems	505.10.2, 505.11.2
D 2657—07	Standard Practice for Heat-fusion Joining of Polyolefin Pipe and Fittings	505.9.1
D 2661—08	Specification for Acrylonitrile-butadiene-styrene (ABS) Schedule 40 Plastic Drain, Waste, and Vent Pipe and Fittings	Table 505.1, 505.3.2, 505.5.2
D 2665—09	Specification for Poly Vinyl Chloride (PVC) Plastic Drain, Waste, and Vent Pipe and Fittings	Table 505.1
D 2729—03	Specification for Poly Vinyl Chloride (PVC) Sewer Pipe and Fittings	Table 505.1.1
D 2751—05	Specification for Acrylonitrile-butadiene-styrene (ABS) Sewer Pipe and Fittings	Table 505.1
D 2855—96(2002)	Standard Practice for Making Solvent-cemented Joints with Poly Vinyl Chloride (PVC) Pipe and Fittings	505.10.2, 505.11.2
D 2949—01a(2008)	Specification for 3.25-inch Outside Diameter Poly Vinyl Chloride (PVC) Plastic Drain, Waste, and Vent Pipe and Fittings	Table 505.1
D 3034—08	Specification for Type PSM Poly Vinyl Chloride (PVC) Sewer Pipe and Fittings	Table 505.1
D 3212—07	Specification for Joints for Drain and Sewer Plastic Pipes Using Flexible Elastomeric Seals	505.3.1, 505.5.1, 505.10.1
D 4021—92	Specification for Glass-fiber Reinforced Polyester Underground Petroleum Storage Tanks	504.4
F 405—05	Specification for Corrugated Polyethylene (PE) Pipe and Fittings	Table 505.1.1
F 477—08	Specification for Elastomeric Seals (Gaskets) for Joining Plastic Pipe	505.13
F 628—08	Specification for Acrylonitrile-butadiene-styrene (ABS) Schedule 40 Plastic Drain, Waste, and Vent Pipe with a Cellular Core	Table 505.1, 505.3.2, 505.5.2
F 656—08	Specification for Primers for Use in Solvent Cement Joints of Poly Vinyl Chloride (PVC) Plastic Pipe and Fittings	505.10.2, 505.11.2

ASTM—continued

F 891—04	Specification for Coextruded Poly Vinyl Chloride (PVC) Plastic Pipe with a Cellular Core	Table 505.1
F 1488—03	Specification for Coextruded Composite Pipe	Table 505.1, Table 505.1.1
F 1499—01(2008)	Specification for Coextruded Composite Drain Waste and Vent Pipe (DWV)	Table 505.1

CSA

Canadian Standards Assocation
5060 Spectrum Way
Mississauga, Ontario Canada L4W 5N6

Standard reference number	Title	Referenced in code section number
A257.1M—92	Circular Concrete Culvert, Storm Drain, Sewer Pipe and Fittings	Table 505.1
A257.2M—92	Reinforced Circular Concrete Culvert, Storm Drain, Sewer Pipe and Fittings	Table 505.1
A257.3M—92	Joints for Circular Concrete Sewer and Culvert Pipe, Manhole Sections and Fittings Using Rubber Gaskets	505.7, 505.13
B137.3—05	Rigid Poly Vinyl Chloride (PVC) Pipe for Pressure Applications	505.10.2, 505.11.2
B181.1—06	Acrylonitrile-butadiene-styrene (ABS) Drain, Waste, and Vent Pipe and Pipe Fittings	505.3.2, 505.5.2
B181.2—06	(PVC) Polyvinylchloride and Chlorinated Polyvinylchloride (CPVC) Drain, Waste, and Vent Pipe and Pipe Fittings	505.10.2, 505.11.2
B182.1—06	Plastic Drain and Sewer Pipe and Pipe Fittings	505.10.2, 505.11.2
B182.2—06	(PVC) Polyvinylchloride Sewer Pipe and Fittings PSM Type	Table 505.1
B182.4—06	Profile PVC Sewer Pipe and Fittings	Table 505.1
B602—05	Mechanical Couplings for Drain, Waste, and Vent Pipe and Sewer Pipe	505.3.1, 505.5.1, 505.6.3, 505.7, 505.10.1, 505.12, 505.13

CISPI

Cast Iron Soil Pipe Institute
5959 Shallowford Road, Suite 419
Chattanooga, TN 37421

Standard reference number	Title	Referenced in code section number
301—04a	Specification for Hubless Cast Iron Soil Pipe and Fittings for Sanitary and Storm Drain, Waste and Vent Piping Applications	Table 505.1
310—04	Specification for Coupling for Use in Connection with Hubless Cast Iron Soil Pipe and Fittings for Sanitary and Storm Drain, Waste and Vent Piping Applications	505.6.3

ICC

International Code Council
500 New Jersey Avenue, NW, 6th Floor
Washington, DC 20001

Standard reference number	Title	Referenced in code section number
IBC—12	International Building Code®	201.3
IPC—12	International Plumbing Code®	201.3, 505.14

NSF

National Sanitation Foundation
789 N. Dixboro Road
P. O. Box 130140
Ann Arbor, MI 48113-0140

Standard reference number	Title	Referenced in code section number
40—2000	Residential Wastewater Treatment Systems	1102.1
41—1999	Nonliquid Saturated Treatment Systems (Composing Toilets)	1301.2

UL

Underwriters Laboratories Inc.
333 Pfingsten Road
Northbrook, IL 60062-2096

Standard reference number	Title	Referenced in code section number
70—2001	Septic Tanks, Bituminous-coated Metal 504.3	

APPENDIX A
SYSTEM LAYOUT ILLUSTRATIONS

The provisions contained in this appendix are not mandatory unless specifically referenced in the adopting ordinance.

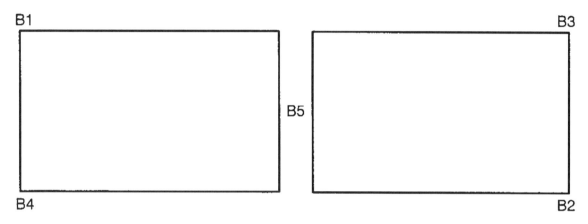

FIGURE A-1 (SECTION 403.1.1)
EXAMPLE OF SOIL-BORING LOCATIONS FOR TWO CONTIGUOUS ABSORPTION AREAS

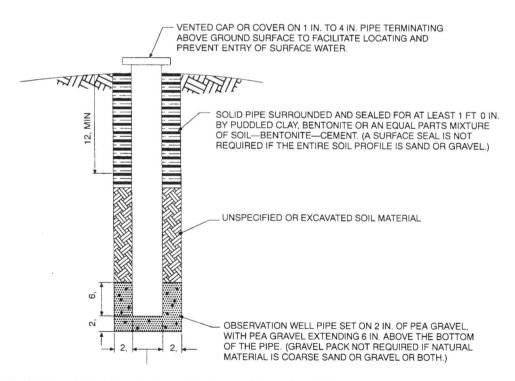

Note: Bore hold shall be 4 inches to 8 inches larger than the outside diameter of observation well pipe size.

For SI: 1 inch = 25.4 mm, 1 foot = 304.8 mm.

FIGURE A-2 (SECTION 405.2.4)
MONITORING WELL DESIGN

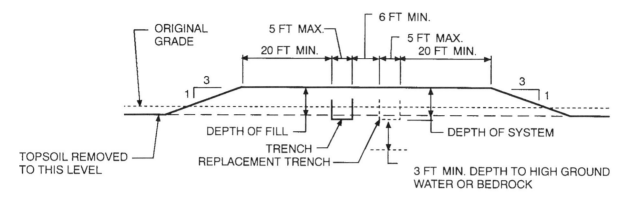

For SI: 1 inch = 25.4 mm, 1 foot = 304.8 mm.

FIGURE A-3 (SECTION 406.6.7)
DESIGN OF FILLED AREA SYSTEM

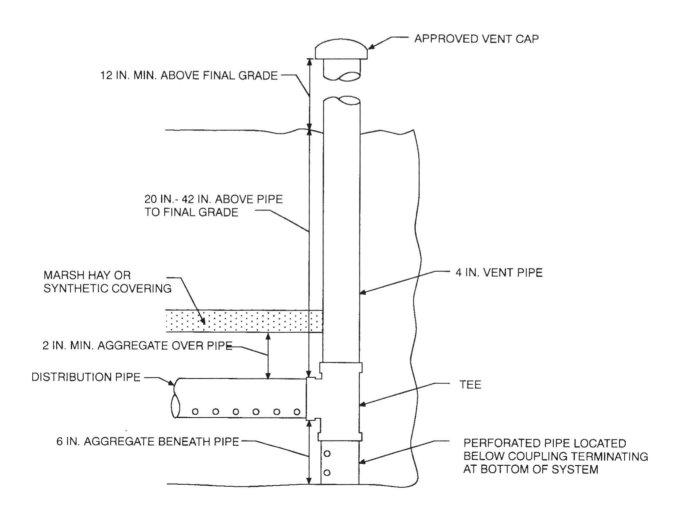

For SI: 1 inch = 25.4 mm.

FIGURE A-4 (SECTION 605.7)
OBSERVATION PIPE

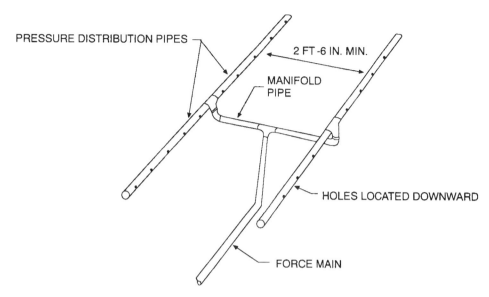

For SI: 1 inch = 25.4 mm, 1 foot = 304.8 mm.

FIGURE A-5 (SECTION 703.1)
PRESSURE DISTRIBUTION SYSTEM DESIGN

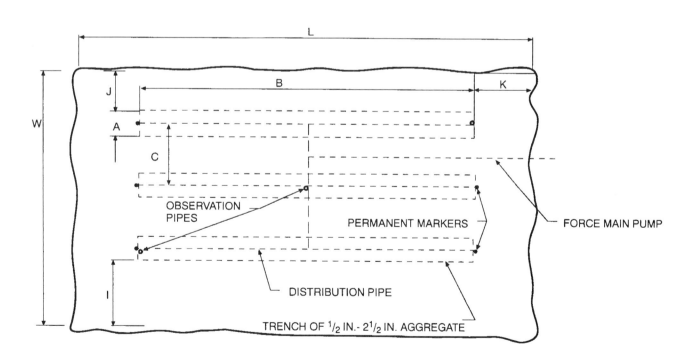

For SI: 1 inch = 25.4 mm.

FIGURE A-6 (SECTION 903.1)
MOUND USING THREE TRENCHES FOR ABSORPTION AREA

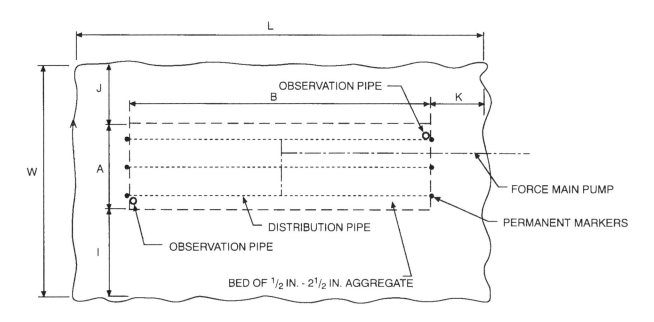

For SI: 1 inch = 25.4 mm.

FIGURE A-7 (SECTION 903.1)
PLAN VIEW OF MOUND USING A BED FOR THE ABSORPTION AREA

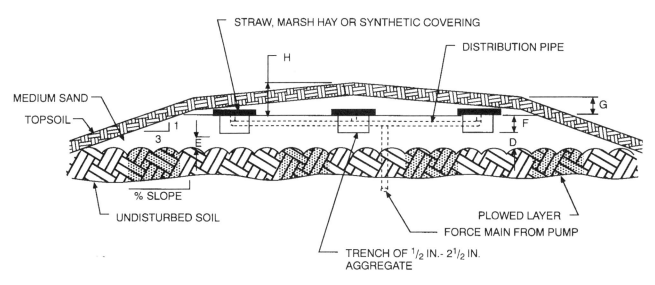

For SI: 1 inch = 25.4 mm.

FIGURE A-8 (SECTION 903.1)
CROSS SECTION OF A MOUND SYSTEM USING THREE TRENCHES FOR THE ABSORPTION AREA

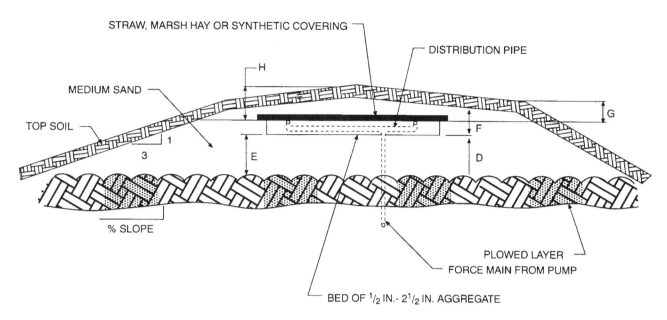

For SI: 1 inch = 25.4 mm.

FIGURE A-9 (SECTION 903.1)
CROSS SECTION OF A MOUND SYSTEM USING A BED FOR THE ABSORPTION AREA

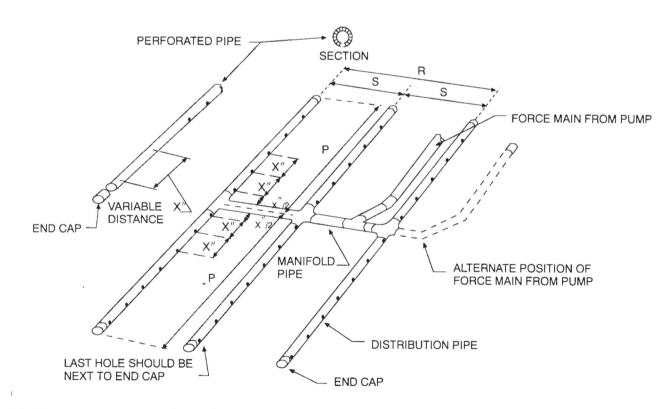

Note: Holes located on bottom are equally spaced.
For SI: 1 inch = 25.4 mm.

FIGURE A-10 (SECTION 903.1)
DISTRIBUTION PIPE LAYOUT

APPENDIX B

TABLES FOR PRESSURE DISTRIBUTION SYSTEMS

The provisions contained in this appendix are not mandatory unless specifically referenced in the adopting ordinance.

TABLE B-1
REQUIRED DISTRIBUTION PIPE DIAMETERS FOR VARIOUS HOLE DIAMETERS,
HOLE SPACINGS AND DISTRIBUTION PIPE LENGTHS (SCHEDULE 40 PLASTIC PIPE)

DISTRIBUTION PIPE LENGTH (feet)	DISTRIBUTION PIPE DIAMETER (inch)																													
	Hole diameter (inch) ¼						Hole diameter (inch) 5/16						Hole diameter (inch) 3/8						Hole diameter (inch) 7/16						Hole diameter (inch) ½					
	Hole spacing (feet)						Hole spacing (feet)						Hole spacing (feet)						Hole spacing (feet)						Hole spacing (feet)					
	2	3	4	5	6	7	2	3	4	5	6	7	2	3	4	5	6	7	2	3	4	5	6	7	2	3	4	5	6	7
10	1	1	1	1	1	1	1	1	1	1	1	1	1	1	1	1	1	1	1	1	1	1	1	1	1¼	1	1	1	1	1
15	1	1	1	1	1	1	1	1	1	1	1	1	1¼	1	1	1	1	1	1¼	1¼	1	1	1	1	1¼	1¼	1¼	1	1	1
20	1	1	1	1	1	1	1¼	1	1	1	1	1	1¼	1¼	1	1	1	1	1¼	1¼	1¼	1	1	1	2	1½	1¼	1¼	1¼	1
25	1¼	1	1	1	1	1	1¼	1¼	1	1	1	1	1½	1¼	1¼	1¼	1	1	2	1½	1¼	1¼	1¼	1¼	2	2	1½	1¼	1¼	1¼
30	1¼	1¼	1	1	1	1	1½	1¼	1¼	1¼	1	1	2	1½	1½	1¼	1¼	1¼	2	2	1½	1¼	1¼	1¼	3	2	2	1½	1½	1¼
35	1½	1¼	1¼	1	1	1	2	1½	1¼	1¼	1¼	1	2	2	1½	1¼	1¼	1¼	3	2	2½	1½	1½	1¼	3	3	2	2	1½	1½
40	1½	1¼	1¼	1¼	1	1	2	1½	1½	1¼	1¼	1¼	3	2	1½	1½	1¼	1¼	3	2	2	2	1½	1½	3	3	2	2	2	1½
45	2	1½	1¼	1¼	1	1	2	2	1½	1¼	1¼	1¼	3	2	2	1½	1½	1½	3	3	2	2	2	1½	3	3	3	2	2	2
50	2	1½	1¼	1¼	1¼	1¼	3	2	2	1½	1½	1¼	3	3	2	2	2	1½	3	3	2	2	2	1½	3	3	3	3	2	2

For SI:1 inch = 25.4 mm, 1 foot = 304.8 mm.

TABLE B-2ᵃ
DISTRIBUTION PIPE DISCHARGE RATE

DISTRIBUTION PIPE OR MANIFOLD LENGTH (feet)

HOLE OR DISTRIBUTION PIPE SPACING (feet)

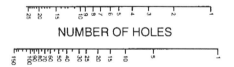

NUMBER OF HOLES

DISTRIBUTION PIPE DISCHARGE RATE (gallons per minute at 2½ feet head)

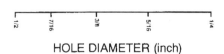

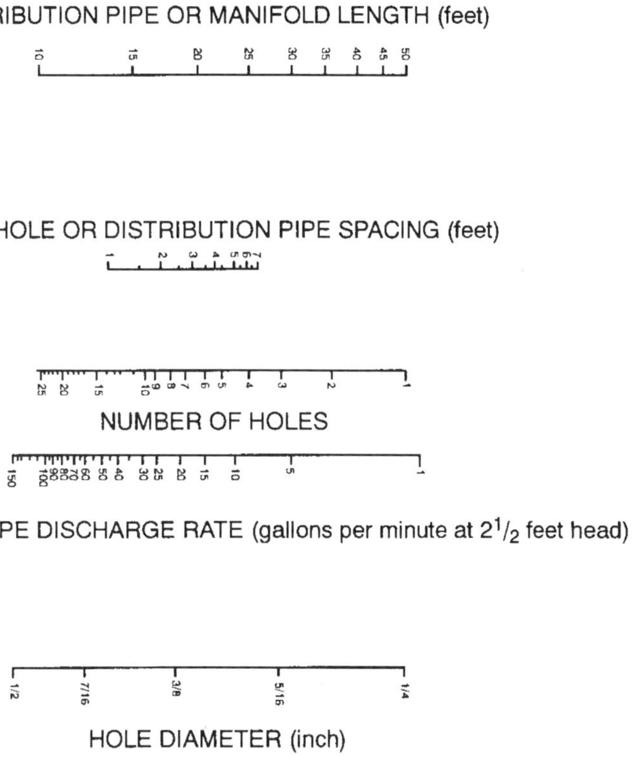

HOLE DIAMETER (inch)

For SI: 1 inch = 25.4 mm, 1 foot = 304.8 mm, 1 gallon per minute = 3.785 L/m.

a. This table, a nomogram, determines the distribution pipe or manifold length, hole or distribution pipe spacing, number of holes, distribution discharge rate and hole diameter of pressure distribution systems by the placement of a straightedge between two known points.

TABLE B-3
RECOMMENDED MANIFOLD DIAMETERS FOR VARIOUS MANIFOLD LENGTHS, NUMBER OF DISTRIBUTION PIPES AND DISTRIBUTION PIPE DISCHARGE RATES (SCHEDULE 40 PLASTIC PIPE)

FLOW PER PIPE (gpm)	MANIFOLD LENGTH (feet)																										FLOW PER PIPE (gpm)
	5		10				15					20					25					30					
	Number of distribution pipes with central manifold																										
	4	6	4	6	8	10	4	6	8	10	12	6	8	10	12	14	6	8	10	12	14	6	8	10	12	14	
	Manifold diameter (inch)																										
5	1	1¼	1¼	1¼	1½	2	1¼	1½	2	2	2	1¼	1½	2	2	3	2	2	3	3	3	2	2	3	3	3	10
10	1¼	1½	1½	2	2	3	2	2	3	3	3	2	3	3	3	3	3	3	3	3	3	3	3	3	4	4	20
15	1½	2	3	3	3	3	2	2	2	2	4	3	3	3	3	4	3	3	4	4	4	3	3	4	4	4	30
20	2	3	3	3	3	3	3	3	3	3	4	3	3	4	4	4	3	4	4	4	4	3	4	4	4	4	40
25	2	3	3	3	3	4	3	3	3	4	4	3	3	4	4	4	4	4	4	4	4	4	4	4	6	6	50
	Number of distribution pipes with end manifold																										
	2	3	2	3	4	5	2	3	4	5	6	3	4	5	6	7	3	4	5	6	7	3	4	5	6	7	

FLOW PER PIPE (gpm)	MANIFOLD LENGTH (feet)																													FLOW PER PIPE (gpm)
	35						40							45								50									
	Number of distribution pipes with central manifold																													
	6	8	10	12	14	16	6	8	10	12	14	16	18	6	8	10	12	14	16	18	20	6	8	10	12	14	16	18	20	22	
	Manifold diameter (inch)																													
5	2	2	3	3	3	3	2	3	3	3	3	3	3	2	3	3	3	3	3	3	3	2	3	3	3	3	3	3	4	4	10
10	3	3	3	3	3	3	3	3	3	4	4	4	4	3	3	3	4	4	4	4	4	3	3	3	4	4	4	4	4	4	20
15	3	3	4	4	4	4	3	4	4	4	4	4	6	3	4	4	4	4	6	6	6	3	4	4	4	4	6	6	6	6	30
20	3	4	4	4	6	6	3	4	4	6	6	6	6	4	4	4	6	6	6	6	6	4	4	6	6	6	6	6	6	6	40
25	4	4	4	6	6	6	4	4	4	6	6	6	6	4	4	6	6	6	6	6	6	4	4	6	6	6	6	6	6	6	50
	Number of distribution pipes with end manifold																													
	3	4	5	6	7	8	3	4	5	6	7	8	9	3	4	5	6	7	8	9	10	3	4	5	6	7	8	9	10	11	

For SI: 1 inch = 25.4 mm, 1 foot = 304.8 mm, 1 gallon per minute = 3.785 L/m.

TABLE B-4[a]
PUMP DOSING RATE

DISTRIBUTION PIPE DISCHARGE RATE (gallons per minute)

NUMBER OF DISTRIBUTION PIPES

DOSING RATE (gallons per minute)

For SI: 1 inch = 25.4 mm, 1 gallon per minute = 3.785 L/m.

a. This table, a nomogram, determines the distribution pipe or manifold length, hole or distribution pipe spacing, number of holes, distribution discharge rate and hole diameter of pressure distribution systems by the placement of a straightedge between two known points.

TABLE B-5
FRICTION LOSS[a] IN SCHEDULE 40 PLASTIC PIPE (C = 150)

FLOW (gpm)	PIPE DIAMETER (inch)								
	1	1¼	1½	2	3	4	6	8	10
1	0.07	—	—	—	—	—	—	—	—
2	0.28	0.07	—	—	—	—	—	—	—
3	0.60	0.16	0.07	—	—	—	—	—	—
4	1.01	0.25	0.12	—	—	—	—	—	—
5	1.52	0.39	0.18	—	—	—	—	—	—
6	2.14	0.55	0.25	0.07	—	—	—	—	—
7	2.89	0.79	0.36	0.10	—	—	—	—	—
8	3.63	0.97	0.46	0.14	—	—	—	—	—
9	4.57	1.21	0.58	0.17	—	—	—	—	—
10	5.50	1.46	0.70	0.21	—	—	—	—	—
11	—	1.77	0.84	0.25	—	—	—	—	—
12	—	2.09	1.01	0.30	—	—	—	—	—
13	—	2.42	1.17	0.35	—	—	—	—	—
14	—	2.74	1.33	0.39	—	—	—	—	—
15	—	3.06	1.45	0.44	0.07	—	—	—	—
16	—	3.49	1.65	0.50	0.08	—	—	—	—
17	—	3.93	1.86	0.56	0.09	—	—	—	—
18	—	4.37	2.07	0.62	0.10	—	—	—	—
19	—	4.81	2.28	0.68	0.11	—	—	—	—
20	—	5.23	2.46	0.74	0.12	—	—	—	—
25	—	—	3.75	1.10	0.16	—	—	—	—
30	—	—	5.22	1.54	0.23	—	—	—	—
35	—	—	—	2.05	0.30	0.07	—	—	—
40	—	—	—	2.62	0.39	0.09	—	—	—
45	—	—	—	3.27	0.48	0.12	—	—	—
50	—	—	—	3.98	0.58	0.16	—	—	—
60	—	—	—	—	0.81	0.21	—	—	—
70	—	—	—	—	1.08	0.28	—	—	—
80	—	—	—	—	1.38	0.37	—	—	—
90	—	—	—	—	1.73	0.46	—	—	—
100	—	—	—	—	2.09	0.55	0.07	—	—
125	—	—	—	—	—	0.85	0.12	—	—
150	—	—	—	—	—	1.17	0.16	—	—
175	—	—	—	—	—	1.56	0.21	—	—

(continued)

TABLE B-5—continued
FRICTION LOSS[a] IN SCHEDULE 40 PLASTIC PIPE (C = 150)

FLOW (gpm)	PIPE DIAMETER (inch)								
	1	1¼	1½	2	3	4	6	8	10
200	—	—	—	—	—	—	0.28	0.07	—
250						—	0.41	0.11	—
300						—	0.58	0.16	—
350						—	0.78	0.20	0.07
400						—	0.99	0.26	0.09
450	Velocities in this area					—	1.22	0.32	0.11
500	become too great for the					—	—	0.38	0.14
600	various flow rates and pipe diameter					—	—	0.54	0.18
700						—	—	0.72	0.24
800						—	—	—	0.32
900						—	—	—	0.38
1,000						—	—	—	0.46

For SI: 1 inch = 25.4 mm, 1 foot = 304.8 mm, 1 gallon per minute = 3.785 L/m.

a. Friction loss expressed in units of feet per 100 feet.

TABLE B-6[a]
MINIMUM DOSE VOLUME BASED ON PIPE SIZE, LENGTH AND NUMBER

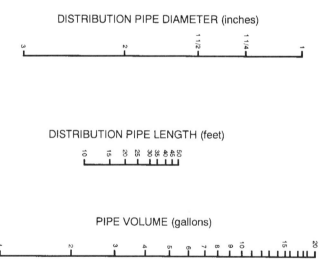

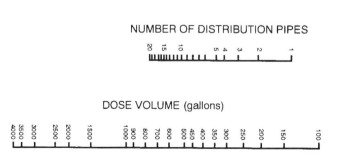

a. This table, a nomogram, determines the distribution pipe or manifold length, hole or distribution pipe spacing, number of holes, distribution discharge rate and hole diameter of pressure distribution systems by the placement of a straightedge between two known points.

INDEX

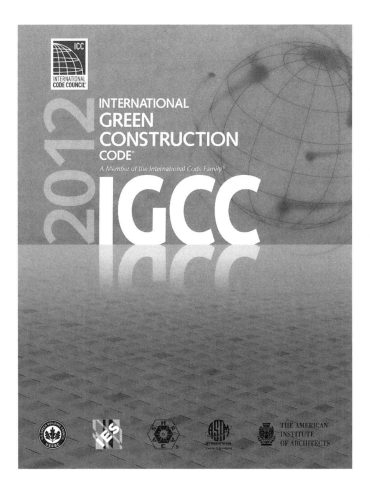